Springer Praxis Books

# Astronautical Engineering

The *Springer Praxis Astronautical Engineering* program covers the very latest applications and systems used in rocket and spacecraft propulsion, spacecraft design, engineering and technology, enabling technologies for current and future space missions both manned and unmanned, including planetary rovers.

Key topics include:

- human missions to the Moon and Mars
- the space debris and radiation environments
- the analysis and design of interplanetary missions, orbital motion, deep space probes and the technologies required for space missions beyond the Solar System into interstellar space and the problems of spacecraft communications

The books are well illustrated with line diagrams and photographs throughout, with targeted use of colour for scientific interpretation and understanding. They feature extensive references and bibliographies, glossaries and appendices.

The books are written for a readership of aerospace and astronautical engineers, space scientists and researchers, spacecraft designers, managers and mission planners in space agencies, space policy makers and postgraduate students in university departments and research institutes in related fields.

Farid Gamgami

# Space Propulsion and Spaceship Design

## A System Perspective

Springer

Published in association with
Praxis Publishing
Chichester, UK

PRAXIS

Farid Gamgami
Innovation Academy for Microsatellites
of CAS (IAMC)
Shanghai, China

Springer Praxis Books
ISSN 2365-9599 ISSN 2365-9602 (electronic)
Astronautical Engineering
ISBN 978-3-031-71335-4 ISBN 978-3-031-71336-1 (eBook)
https://doi.org/10.1007/978-3-031-71336-1



Cover illustration: Depiction of the reference mission C-One. The front cover depicts the interplanetary flight of the solar electric stage and payload to Ceres. The back cover depicts an escape manoeuvre performed by the third liquid booster stage. Credit: Jessica-Christie da Silva Vieira

This Springer imprint is published by the registered company Springer Nature Switzerland AG
The registered company address is: Gewerbestrasse 11, 6330 Cham, Switzerland

*To attain knowledge, add things every day.*
*To attain wisdom, remove things every day.*

*Lao Tzu—6th cent. BCE*

# Foreword

The journey of space exploration, from the early days of competition to the present era of renewed ambition, has always been underpinned by advancements in space propulsion. Dr. Farid Gamgami's book, *Space Propulsion and Spaceship Design—A System Perspective*, is a timely and comprehensive guide that brings clarity and depth to this critical field. Having worked in the field of space technologies and their applications for over 35 years, and as Deputy Director of UNOOSA, I have had the privilege of observing the global collaboration and technological innovations that are shaping our ventures into space. This book is a testament to the progress we have made and the potential that lies ahead. Dr. Gamgami expertly navigates the intricate details of propulsion technologies and integrates them within the broader system of spacecraft design, offering a unique dual perspective that is both technical and holistic.

The importance of space propulsion cannot be overstated. It is the driving force that enables us to escape Earth's gravity, traverse the vast distances of our solar system and envision human settlements on other celestial bodies. This text serves as an essential resource for the engineers, scientists and policymakers who are charting the course for future space missions. By providing a thorough understanding of propulsion systems and their integration into spacecraft design, Dr. Gamgami equips us with the knowledge to overcome the challenges of space travel and to seize its opportunities.

Moreover, the book's emphasis on collaboration reflects a core principle that resonates deeply with the mission of UNOOSA. The exploration of space is not the endeavour of any single nation but a collective journey that requires the ingenuity and cooperation of the global community. As we aim to establish a sustained human presence in space, the collaborative approach advocated by Dr. Gamgami is essential.

Dr. Gamgami's extensive experience and profound insights are evident throughout this work. His background in both aerospace engineering and theoretical astrophysics, coupled with his leadership in international space projects, provides a rich foundation for this book. It is a privilege to present this foreword for a publication that is poised to inspire and guide the next generation of space explorers.

Vienna, Austria
July 2024

Driss El Hadani
Deputy Director/Senior Adviser
United Nations Office for Outer Space Affairs (UNOOSA)

# Preface

The era of the space age would not have been possible without the achievements in space propulsion. In a mere three decades, engineers mastered the required technology to evolve from inefficient engines with limited thrust to highly efficient rocket engines delivering almost 7000 kN. This immense thrust propels launch systems with masses up to 5000 tonnes into space. It goes without saying that space propulsion is indeed an enabling technology. After the great feat of landing humans on the Moon and bringing them safely back to Earth more than half a century ago, mankind is once again targeting space, but this time, with the intent to stay and settle. Once more, propulsion will be the key enabling technology. The motivation for writing this book becomes obvious: to contribute to this new era by equipping future space engineers with the necessary knowledge base. The perspective adopted is unique in a way that it approaches the topic from two sides, the technology and physics of space propulsion as well as the system in which it is embedded. This approach culminates in a comprehensive system study in which the principal propulsion technologies need to be integrated in order to achieve the mission goals.

The early days of space travel were characterised by competition, followed by a phase of cooperation, which resulted in the (almost) International Space Station. We are now back in a phase of intense competition. For humanity to survive and thrive on Earth and in space, it is essential that we avoid the mistakes of past centuries, namely the struggle for resources, and enter a phase of genuine collaboration. The exploration of space, beyond simple probes, is an epochal endeavour and requires humanity's full creative ingenuity and coordinated effort. In other words, it is time for humanity to rise above national prestige and leadership claims, and to collaborate in order to become a spacefaring human nation. The Ceres case study presented in this book is a clear example of how large-scale missions can only be achieved together. And significant is always large when it comes to space.

Kamen, Germany
June 2024

Dr. Farid Gamgami

# Acknowledgements

This textbook would not have been possible without the kind support of many friends and colleagues. I am deeply grateful to Prof. Dr. Ramis Örlü for supplying me with an endless stream of publications, technical and editorial advice, as well as spiritual support 24/7. My deep appreciation to Dr. Rolf Janovsky, my former chief and mentor, for his invaluable and insightful feedback on the entire manuscript. I would also like to give special acknowledgement to Mathias Rohrbeck and Dr. Davide Amato. Mathias performed a thorough reviewed of the taxonomy chapter and the C-One reference mission. Davide commented and corrected the spaceflight chapter and kindly contributed the interplanetary transfer analysis to Ceres. The discussions with you two were, as always, enlightening and a real pleasure—the former padawans became true masters. I feel truly privileged to have high-class experts among my former colleagues and friends who reviewed specific chapters and sections. My sincere thanks to Dr. Birk Wollenhaupt, Stephen Goodburn, Etienne Dumont and Markus Peukert for reviewing the propulsion technology chapters; Navid Fatemi Ph.D. for his valuable comments on solar power technology; Dr. Christoph Kirchberger for reviewing the thermodynamics and chemical energy sections; Nabil Souhair Ph.D. for commenting the rocket equation chapter; and Dr. Solmaz Adeli for her insightful comments on the geology of the solar system. I would like to thank Mrs. Jéssica-Christie da Silva Vieira for the artistic design of the book cover.

Finally, I would like to express my gratitude to Springer Nature and my lector Dr. Ramon Khanna. His calm aura and supportive attitude were reassuring and motivating.

# Contents

# About the Author

**Dr. Farid Gamgami** holds a Master's degree in aerospace engineering from the RWTH Aachen, with a focus on space propulsion and space transportation. Additionally, he obtained a Ph.D. in Theoretical Astrophysics from the University of Heidelberg, which complemented his scientific training with in-depth studies of physical phenomena not commonly found in engineering. During his tenure at DLR and subsequent employment at OHB System AG, he developed a comprehensive understanding of spacecraft design by leading multidisciplinary teams on projects involving space transportation and deep space missions. Dr. Gamgami is currently vice director and head of international collaboration and education programme at the Key Laboratory for Satellite Digital Technology in Shanghai, which belongs to the Innovation Academy for Microsatellites (IAMCAS), an institute of the Chinese Academy of Sciences (CAS). He is the first non-Chinese to assume such a leading position in the field of space engineering in this prestigious scientific association.

# Acronyms

| | |
|---|---|
| AC | Alternating Current |
| AIT | Assembly, Integration and Test |
| AM0 | Airmass Zero |
| amu | Atomic Mass Unit |
| ATV | Autonomous Transfer Vehicle |
| AU | Astronomical Unit |
| BEG | Best Engineering Guess |
| CDR | Critical Design Review |
| CFD | Computational Fluid Dynamics |
| CFRP | Carbon Fibre Reinforced Plastic |
| CLM | Ceres Landing Module |
| CLPM | Ceres Landing Propulsion Module |
| CME | Coronal Mass Ejection |
| CNES | Centre National D'Etudes Spatiales—National Centre for Space Studies |
| CoM | Centre of Mass |
| ConOps | Concept of Operation |
| COPV | Composite Overwrapped Pressure Vessels |
| COTS | Commercial Off-the-Shelf |
| CPM | Ceres Payload Module |
| CPPS | Chemical Propulsion System |
| CSM | Command and Service Module |
| DC | Direct Current |
| DCSS | Delta Cryogenic Second Stage |
| DDP | Design Departure Point |
| DHS | Data Handling System |
| DLR | Deutsches Zentrum für Luft- und Raumfahrt—German Aerospace Centre |
| DPS | Descent Propulsion System |
| DRACO | Demonstration Rocket for Agile Cislunar Operations |
| EFH | Extra Full-Hard (Grade) |

| | |
|---|---|
| EM | Electromagnetic |
| EMA | Electromechanical Actuators |
| EMED | External Momentum Exchange Drive |
| EMP | Electromagnetic Propulsion |
| EOM | End of Manoeuvre |
| EP | Electric Propulsion |
| EPPS | Electric Propulsion System |
| EPR | Electrical Pressure Regulator |
| EPS | Energy and Power Subsystem |
| ESA | European Space Agency |
| ESC-A | Etage Supérieur Cryotechnique de Type A (upper stage) |
| ESM | European Service Module |
| ETA | Electric Thruster Assembly |
| ETP | Electrothermal Propulsion |
| EUS | Exploration Upper Stage |
| FCC | Federal Communications Commission (US) |
| FCV | Flow Control Valve |
| FR | Filter |
| FSA | Flexible Solar Array |
| GCR | Galactic Cosmic Radiation |
| GG | Gas Generator |
| GIT | Gridded Ion Thruster |
| GNC | Guidance Navigation and Control |
| HALEU | High-Assay Low-Enriched Uranium |
| HAN | Hydroxyl Ammonium Nitrate |
| HERMeS | Hall Effect Rocket With Magnetic Shielding |
| HET | Hall Effect Thruster |
| HEU | Highly Enriched Uranium |
| HiPEP | High-Power Electric Propulsion |
| HiRISE | High-Resolution Imaging Science Experiment |
| HPGP | High-Performance Green Propulsion |
| HPIV | High-Pressure Isolation Valve |
| HVB | High-Voltage Bus |
| ICBM | Intercontinental Ballistic Missile |
| ICP | Inductively Coupled Plasma |
| ICPS | Interim Cryogenic Propulsion Stage |
| IME | Internal Momentum Exchange |
| IMED | Internal Momentum Exchange Drive |
| IS | Interstage |
| ISRU | In Situ Recourse Utilisation |
| ISS | International Space Station |
| JAXA | Japan Aerospace Exploration Agency |
| KDF | Knockdown Factor |
| LAD | Liquid Acquisition Device |
| LAE | Liquid Apogee Engine |

| | |
|---|---|
| LCD | Liquid Crystal Device |
| LCH4 | Liquid Methane |
| LH2 | Liquid Hydrogen |
| LIDAR | Light Detection and Ranging |
| LM | Lunar Module |
| LOX | Liquid Oxygen |
| LPIV | Low-Pressure Isolation Valve |
| LPT | Low-Pressure Transducer |
| LV | Latch Valve |
| LVB | Low-Voltage Bus |
| ly | Light Years |
| MARSIS | Mars Advanced Radar for Subsurface and Ionosphere Sounding |
| MECO | Main Engine Cut-Off |
| MEOP | Maximum Expected Operating Pressure |
| MHD | Magentohydrodynamics |
| MHLLV | Medium Heavy Lift Launch Vehicle |
| MIR | Russian Space Station, 'Peace' |
| MJ | Multi-Junction |
| MLI | Multi-layer Insulation |
| MPD | Magnetoplasmadynamic |
| MPDT | Magnetoplasmadynamic Thruster |
| MPR | Mechanical Pressure Regulator |
| NA | Not Applicable |
| NASA | National Aeronautics and Space Administration |
| NC | Nominally Closed |
| NCPV | Nominally Closed Pyrovalve |
| NEA | Near-Earth Asteroid |
| NEO | Near-Earth Objects |
| NEP | Nuclear Electric Propulsion |
| NERVA | Nuclear Engine for Rocket Vehicle Application |
| NETV | Nuclear Electric Transfer Vehicle |
| NEXIS | Nuclear Electric Xenon Ion System |
| NO | Nominally Open |
| NOFBX™ | Nitrous Oxide Fuel Blend |
| NOPV | Nomninally Open Pyrovalve |
| NTP | Nuclear Thermal Propulsion |
| OME | Orbital Main Engine |
| OMS | Orbital Maneuvering System |
| OTS | Off-the-Shelf |
| PCA | Pressurant Control Assembly |
| PDE | Propulsion Drive Electronic |
| PET | Positive Expulsion Tanks |
| PFCV | Proportional Flow Control Valve |
| PGM | Platinum Group |
| PIA | Propellant Isolation Assembly |

| | |
|---|---|
| PIF | Powered Interplanetary Flight |
| PLA | Payload Launch Adapter |
| PM | Payload Module |
| PMAD | Power Management and Distribution |
| PMD | Propellant Management Devices |
| PMS | Propellant Management System |
| ppb | Parts per Billion |
| PPE | Power and Propulsion Element |
| PPS® | Propulseur À Plasma Stationnaire (trade name by Safran for HET) |
| PPU | Power Processing Unit |
| PRU | Power Regulation Unit |
| PSA | Propellant Supply Assembly |
| PT | Pressure Transducer |
| PV | Pyro Valve |
| PVA | Photovoltaic Array |
| PVT | Pressure-Volume-Temperature Analysis |
| PWM | Pulse Width Modulation |
| RACS | Roll and Attitude Control System |
| RAMS | Reliability, Availability, Maintainability and Safety |
| RCD | Reflectivity Control Device |
| RCS | Reaction Control System |
| RCT | Reaction Control Thruster |
| REE | Rare Earth Elements |
| RF | Radio Frequency |
| RFG | Radio Frequency Generator |
| RIT | Radio Frequency Ion Thruster |
| RIU | Remote Interface Unit |
| rpm | Revolutions per Minute |
| RTG | Radioisotope Thermoelectric Generator |
| SADM | Solar Array Drive Mechanism |
| SCC | Staged Combustion Cycle |
| SECO | Second Engine Cut-Off |
| SEP | Solar Electric Propulsion |
| SETV | Solar Electric Transfer Vehicle |
| SHLLV | Supper Heavy Lift Launch Vehicle |
| SI | International System of Units |
| SL | Solenoid Valve |
| SLIM | Smart Lander for Investigating Moon |
| SLS | Space Launch System |
| SMA | Semi-major Axis |
| SOI | Sphere of Influence |
| SPF | Solar Point Failure |
| SPG | Solar Power Generator |
| SRM | Solid Rocket Motor |
| SSA | Space Situational Awareness |

| | |
|---|---|
| SSM | Space Shuttle Main Engine |
| STS | Space Transportation System |
| STT | Surface Tension Tanks |
| TBC | To Be Confirmed |
| TBD | To Be Determined |
| TCP | Thermochemical Propulsion |
| TCS | Thermal Control System |
| TCV | Thrust Control Valves |
| TDB | Temps Dynamique Barycentrique/Barycentric Dynamical Time |
| TECO | Third Engine Cut-Off |
| TID | Total Ionisation Dose |
| TJ | Triple-Junction |
| TOPAZ | Thermionic Experiment with Conversion in Active Zone |
| TRL | Technology Readiness Level |
| TVC | Thrust Vector Control |
| U-235 | Uranium 235 |
| UPS | Unified Propulsion System |
| wt | Percentage by weight |
| XFS | Xenon Feed System |
| XTA | Xenon Tank Assembly |

# Chapter 1
# Introduction

A wealth of literature has been published on space propulsion, including excellent textbooks that have served generations of engineers well. So why a new one? The reason is that these traditional textbooks and monographs are predominantly focusing on the technology-side of the matter. This is perfectly legitimate. Students who wish to become aerospace engineers and professionals who wish to take a deep dive in space propulsion need a sound technological understanding of the subject. However, a sound system understanding is equally important and should be taught at an early stage in the studies. As a matter of fact, the propulsion system interacts strongly with several subsystems, like mission analysis, operations, mechanical design, thermal control, attitude control and the energy and power subsystem. It can therefore be reasonably concluded that the propulsion system plays a pivotal role in the overall architecture, for missions requiring high velocity changes. This deep integration demands that the responsible propulsion engineer has a solid grasp of the system needs and the rest of the engineering team have a good understanding of the basics of the propulsion technology. Such a mutual understanding guarantees a successful project and mitigates misunderstandings and should be established already in the curriculum.

This is the objective of this book, namely to introduce both the technology and physics of space propulsion as well as its relationship to other disciplines in the process of spaceship design–hence, *system perspective*. Part of this system perspective is the system design process, in such a way that the mission objectives are achieved reliably, cost-effectively and on time. To this end, mission and system requirements are defined which are 'broken-down' to subsystem requirements. The recipients of these subsystem requirements are the engineering experts; specialised propulsion engineers, specialised thermal engineers, specialised structural engineers etc. The *requirements breakdown* and the *design development* are collaborative tasks of the a engineering team–a mutual understanding is inevitable to design a good system.

 
F. Gamgami, *Space Propulsion and Spaceship Design*, Springer Praxis Books,
https://doi.org/10.1007/978-3-031-71336-1_1

The intention of this textbook is to bridge the two worlds: The aspiring aerospace engineer will find here the propulsion know-how to form a sound understanding of the basic principles of the different propulsion technologies as well as the–generally missing–link to the system, which is to say the origin of propulsion requirements and the link to other relevant disciplines. For this reason emphasis was given to explain when a specific propulsion type is generally required and what this means for the system.

## 1.1 Scope, Context and Addressee

Following the 'system perspective', it is the ambition of this textbook to draw a thread from the physical working principle of propulsion systems to its impact on system level. Naturally, given the breadth of the subject a selection of technologies is inevitable. The here selected focus is on *space transportation* which comprises transportation from ground to space, in-space and from space to ground of airless bodies, like the Moon. Solid-fuelled and hybrid motors have been omitted due to their limited number of use cases and in order to constrain the scope of the textbook. Similarly, detailed technological aspects have been omitted, including propellant injectors and the theory of combustion instability. Although intellectually appealing, these aspects are beyond the chosen system perspective and in the realm of expert knowledge.

Against the backdrop of a thriving space ecosystem fuelled by visionary entrepreneurs and bold engineers, this textbook places the introduction to space propulsion systems in the context of the burgeoning commercial space exploration industry. Flanked by ambitious national plans, this industry is experiencing considerable growth and attracting engineers and physicists from a variety of backgrounds. Some agency heads speak of even a space race for the resources out there. And there is indeed something to it: solar trade routes will be established along water resources, which can be used to generate rocket propellants.[1] Hence, space propulsion will play a decisive role in this endeavour, determining the speed of progress. For instance, cryogenic propellant management, especially the long-term storage of liquid hydrogen, is a key technology for chemical as well as nuclear propulsion systems.

As anticipated above the textbook shall serve aspiring propulsion engineers as well as systems engineers. It is therefore an ideal textbook in the curriculum of advanced undergraduate as well as postgraduate students of aerospace engineering. Furthermore, the textbook will provide professionals with the knowledge and skills required to transition into the field of space propulsion and spacecraft design.

[1] It is to be hoped that humanity will learn from the past and embark on a joint venture into space to avoid a conflict over resources.

### 1.1.1 Reference Mission to Ceres, C-One

In order to consolidate the understanding of the system perspective, it is beneficial to apply it to a case-study. The here selected case is a large scale exploration journey to Ceres, dubbed 'C-One'.

#### Mission Objectives and Requirements

Ceres is a dwarf planet in the asteroid belt that hosts considerable amounts of water in different forms. Its resources and strategic location make it an ideal outpost for exploratory voyages into the depths of the outer solar system. A first mission has been sent there: Dawn, a space probe with a launch mass of 1,217 kg. It was equipped with remote sensing instruments to analyse the composition of Ceres and map its water resources. There will be many more missions like Dawn to the asteroid belt and in particular to Ceres because of the promising measurement results. The objectives will primarily be of scientific nature–this is the *scouting phase*. Later, the mission objectives will gradually evolve towards institutional and commercial exploitation. As part of this evolution, missions will be carried out to land robotic laboratories to test processes and machines for mining and exploiting resources in this specific environment, with increasing capability and complexity. This will be the start of the *exploitation phase* and this is where C-One will be placed.

Hence a considerably high mass, comprising the actual payload and service provisions, like power and communication, will be required on the surface. As a working hypothesis a mass of 30 metric tonnes (t) shall be assumed. Furthermore a *transfer time* of not more than 5 years (threshold), preferably only 3 years (target).

The execution of such a challenging mission will necessitate the utilisation of a range of different propulsion systems, thereby serving as a fine example for the design and utilisation of different space propulsion technologies. This reference mission is in fact a unique asset of this textbook.

### 1.1.2 Structure of the Book

The book follows a particular structure, progressing from general concepts to specific technical solutions and their use cases.

Chapter 2 provides an introduction into the structure of the *solar system* and the principles of *spaceflight*. Basic knowledge of the scale of the solar system is inevitable to understand its vastness in terms of distances and travel times. Spaceflight is the discipline that defines the performance requirements for propulsion systems, specifically the famous '$\Delta v$ demand', which measures the energy required to change an orbit. We will restrict ourselves here to the two-dimensional case in order to facilitate the understanding of what is otherwise a very mathematical subject.

Just as a submarine must be able to withstand the harsh conditions of pressure and salinity of the deep sea, a spaceship is also greatly affected by the environmental *deep space conditions*. Chapter 3 presents the dominant physical effects that need to be taken into account when venturing into the depths of the solar system.

Chapter 4 sets the aforementioned context of commercial space exploration. The focus will be on resources that will literally fuel an *inner solar ecosystem*. The colonisation[2] of the solar system will take place along water resources and the first outposts in space will be filling stations. It is therefore of vital importance to identify the location of resources needed for this space-based revolution, with a particular focus on those that can be used as propulsion propellants.

Chapter 5 is a space propulsion introductory chapter, that aims to provide a synopsis of propulsion technologies and their basic working principle. The centrepiece of this chapter is a *taxonomy* designed to categorise the extensive range of propulsion technologies. Principle propulsion system performance parameters will be introduced here that are highly relevant for the design of systems in which propulsion dominates.

A textbook on space engineering is not complete without the *rocket equation*. Chapter 6 is fully dedicated to this famous equation and its implications. Concepts like staging and gravity losses are introduced here. Furthermore, the chapter presents the different variants of this equation and their properties in a way that is seldom done in classical textbooks.

Chapter 7 discusses the *acceleration principles* by which propellant is ejected out of the spaceship to generate thrust. Two distinct types will be presented. The first is based on the laws of gas dynamics and is employed in all thermal engines. The second is based on the laws of electrodynamics and is employed in electric engines.

A discussion on propulsion systems would be inadequate without covering the topic of *energy sources and power conversion*. This is subject of Chap. 8. As before, the chapter is subdivided in two distinct parts: chemical energy and solar-power based energy.

Chapter 9 is the technological centrepiece in this textbook. It presents elements of the *propellant management system* which comprises propellant storage and supply. Complete propulsion systems are presented in this chapter following again a categorisation scheme to facilitate understanding of the technology and its application.

The aforementioned *reference mission* to Ceres, C-One, is the subject of Chap. 10 We will see how different types of propulsion technologies need to be combined in a spaceship design to enable travel and landing on Ceres. The trade-off process will be explained by the discussion of architecture relevant aspects. Furthermore, important system engineering concepts are introduced in this chapter.

Chapter 11 finally presents *nuclear-based propulsion systems* for thermal and electric engines that have been already tested and can be readily available in the

[2] This time of an uninhabited area, i.e. without land theft, enslavement and exploitation of people.

next two decades. They promise to give in-space transportation a significant push enabling mission shorter flight times and more frequent departure windows.

The book concludes with several appendices that provide an in-depth dive into selected topics, such as the design of impulsive interplanetary transfers from Earth to Ceres and Mars to Ceres, as well as a derivation of the relativistic form of the Tsiolkovsky equation–hoping that these insights may be employed by readers for future missions.

## 1.2 Space Vehicle Terminology

The field of space engineering is notorious for its incoherent terminology with many synonyms. This can be very confusing for aspiring engineers or novices who enter the field from other domains. A brief overview shall be provided of some space systems in which propulsion, generally, plays a significant role.

**Rocket**: A rocket is a slender body that in the military context is meant to destroy another object and in a civil context to launch a scientific or commercial payload into space. Sounding rockets, for instance, are used to conduct experiments in micro-gravity. A rocket is essentially a flying propulsion system.

**Launcher**: A launcher is a vehicle meant to reach orbital velocity and to place a payload in a stable orbit. Like a rocket, it is a slender body that lifts off vertically and its engines are called rocket motors. The term launcher is preferably used by managers and officials, whereas engineers prefer rocket, because it is rocket science.

**Space Transportation System**: The term Space Transportation System is an umbrella term for all vehicles that deliver payload to space and come back. This comprises classical launchers as well as winged bodies. The space shuttle, for instance, took off vertically and landed horizontally. The holy grail of space transportation is a single stage to orbit, preferably with horizontal take-off and landing. The modern view expands the term and incorporates in-space transportation.

**Satellite**: A satellite in an engineering context is any artificial object that orbits Earth with the purpose to provide a service to the ground, like imaging or broadcasting.[3] A space station is not considered a satellite, nor would a refuelling station be called a satellite even though both orbit Earth. This is because the provided service is an in-space service. The extend and significance of the propulsion system of a satellite is not stringent as for launcher. Some satellites do not even possess a propulsion system, like LEO satellites, while in others it determines the design, like for GEO satellites. New regulations aim to reduce space debris by imposing special regulations: a satellite shall be able to conduct a minimum number of collision avoidance manoeuvres during its lifespan and in addition be capable to de-orbit itself after mission

[3] It shall be noted that for an astronomer, every object that orbits another object is a satellite, i.e. the moon Titan that orbits Saturn is also a satellite.

completion. The first requirement, collision avoidance during nominal operation, demands a propulsion system, while de-orbiting could be in principle also conducted with a drag sail.

**Space Probe**: A space probe is usually sent to 'probe' other celestial bodies. The arguably most famous space probes are Voyager I and II, that have past the heliopause and are the first human made objects to enter interstellar space at an astonishing distance of 155.7 AU and 129.6 AU respectively. The vehicles are equipped with a range of remote sensing instruments and are designed for scientific purposes. Space probes that are meant to enter orbit around a planet have a large propulsion system (including propellant), which commonly comprises up to 60% of the probes' launch mass.

**Spacecraft**: The term spacecraft is an umbrella term for all uncrewed space object. It comprises satellites, space probes, space tugs (logistic vehicle) and so forth.

**Spaceship**: A spaceship could be crewed or uncrewed but in both cases its main purpose is a logistic one, to transport a payload. A crewed capsule is considered a spaceship, as are robotic interplanetary missions that carry a considerable payload. The term is not explicitly defined, yet its common understanding is clear: it denotes a large space vessel capable of travelling through the solar system with large freight loads. Such vehicles do not exist, yet. The title of the book uses this term deliberately to point out that this will change in the next decade or two. Humanity is on the brink of large-scale space exploration, and the good news is that the technology already exists, as will be demonstrated by the C-One reference mission.

Similarly, the terminology for the engine itself is often ambiguous. Large engines that propel rockets and launchers are referred to as 'rocket engine' or 'rocket motor'. They typically have a thrust range between 30–7000 kN. The term 'engine' refers to all devices that eject mass to generate thrust but small engines below 20 N are usually called 'Thruster'. All engines in electric propulsion systems are referred to as thrusters, except when they are not, as illustrated by the 'Hall Effect *Rocket* with Magnetic Shielding' (HERMeS), boasting a modest nominal thrust of 0.59 N.

Terminology is not static, but is constantly evolving. For example, spacecrafts that form a communications constellation around the Moon, as currently envisioned by several space agencies, will not be called 'space probes' because they are not providing a service to humans on Earth, but to a future lunar base. The propulsion system, however, is agnostic to the mission or naming of the vehicle. A rocket motor of the upper stage of a launcher can be readily employed in a spaceship that commutes between Earth and Moon. We will therefore use the entire spectrum of the available terminology, depending on which one is more common in the current context, however, with a preference for *spaceship*.

Disclaimer

To make the content more tangible, real space products were used to illustrate performances and specifications. Publicly available data has been used, primarily from company websites and data sheets. No official confirmation or endorsement has been given by the companies concerned, nor have any financial benefits been received; therefore, no guarantee can be given as to the accuracy and actuality of the information. It is understood that a formal request to the companies is necessary to obtain accurate specifications.

# Part I
# Space, Time and Heavenly Object

# Chapter 2
# The Vast Solar System and Principles of Spaceflight

**Abstract** This chapter presents the structure and dynamics of the solar system and the principles of spaceflight. The link between the two is natural, since the main objective of spaceflight is to visit objects in the solar system, mission requirements are derived from the dynamics of the solar system. The spaceflight section covers the main elements needed to understand the propulsion specific requirements of a spaceship which are mainly due to planetary and interplanetary orbital manoeuvres.

## 2.1 The Scale of the Solar System

Astronomers still debate whether our solar system is the norm or the exception. For instance, we know from observations that stars are seldom single children but are born together with siblings and form a multi-star system. Binaries are the most prevalent form of stellar systems and the Sirius system is a prominent example, where two stars orbit each other with a period of 50.1 years, Sirius A and B.[1] Human writing is known to ancient civilisations. The Alpha Centauri star system is actually a *triple* stars system consisting of a *binary* (Alpha Centauri A & B) in the centre and a third (Proxima Centauri) in a safe distance.[2] A peculiar example of a multiple star system is TYC 7037-89-1. It is a *sextuple* star system consisting of three binaries, all orbiting an immaterial point in space called *barycentre* [2]. The barycentre is the centre of mass around which two or more bodies orbit due to their mutual gravitational attraction. The resulting highly inhomogeneous gravitational field of these multiple star systems makes the formation of planets unlikely, though not impossible. Only from a great

[1] Sirius A is the brightest star in the sky and therefore played a role in several human cultures. The ancient Egyptians, for instance, associated Sirius A with the goddess Isis and its rising marked the annual flooding of the Nile. Sirius B, not visible to the naked eye, was first deduced in 1844 by the German mathematician and astronomer Friedrich Bessel (1846†) based on the apparent motion (wobble) of Sirius A against the fix stars. Observational confirmation was given by the US astronomer Alvan Graham Clark in 1862.

[2] The stellar system belongs to the constellation Centaurus, which is named after the Greek mythological creature that is half man half horse. About half of the visible stars carry actually Arabic names that have been latinised. The famous red star Aldebaran for instance is the Latin version of the Arabic *ad-dabaran*, which literally means ‘the follower’ [1].

 
F. Gamgami, *Space Propulsion and Spaceship Design*, Springer Praxis Books,
https://doi.org/10.1007/978-3-031-71336-1_2

distance would a planet perceive the six stars as a single gravitational object located in the barycentre. This distance is, however, beyond the *habitable zone* as we know it.

In contrast, the solar system appears to be remarkably well ordered, with a dominant star situated at its centre and a number of planets and asteroids positioned around it. The structure of our solar system is depicted in Fig. 2.1.

The *terrestrial* planets are located in the inner solar system while the *gas giants* are in the outer part—the two groups are divided by an asteroid belt. In the outer region, there is a further conglomerate of smaller objects, the *Kuiper belt*, named after the Dutch astronomer Gerard Kuiper.[3] The most prominent representative is the dwarf planet Pluto. The gravitational reach of our Sun extends, however, even further and the most distant group of objects in our solar system form the *Oort cloud*, named after another Dutch astronomer Jan Oort.[4] These objects, remnants of the formation

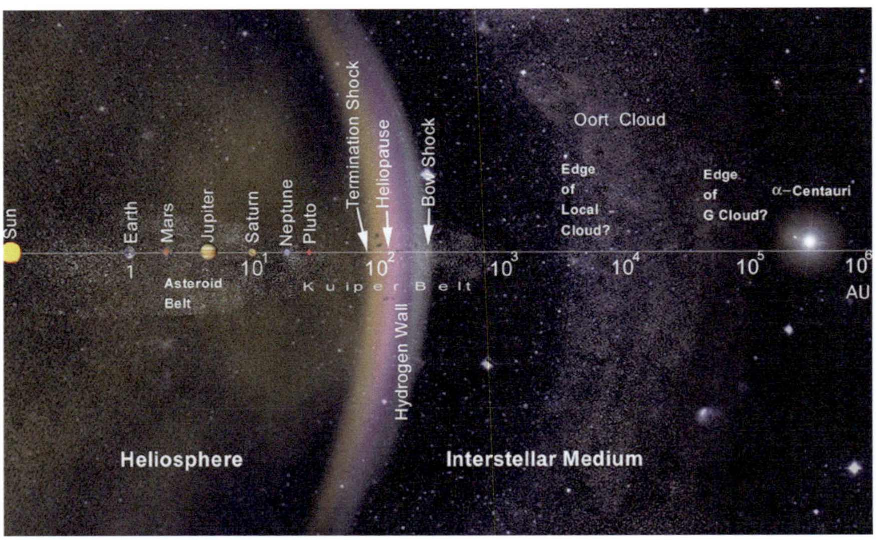


**Fig. 2.1** Composition of the solar system, distance in logarithmic scale, Mercury and Venus not depicted. 1 AU (astronomical unit) is the mean distance from Earth to Sun $149.6 \times 10^6$ km Credit: JPL/NASA

[3] Gerard Kuiper (1976†) made important contributions to the understanding of the solar system in the second half of the 20th century. He was the first to suspect the existence of what is nowadays known as the Kuiper belt by accurate interpretation of the outer planets orbital motion. He further discovered Uranus's moon Miranda and Neptune's moon Nereid and advocated for the advancement of infrared astronomy.

[4] Jan Oort (1992†) contributed to many fields in astronomy noteworthy the rotation curve of the Milky Way and the dynamics of galaxies. He is best known for his speculation about a sphere of ancient remnants dating back from the birth of our solar system, the Oort cloud. Furthermore, he strongly advocated for the development of radio astronomy as a mean of understanding the structure of the universe.

**Table 2.1** Planetary orbital properties and prominent space probes [4]

| Object | Mean sol. distance (AU) | Mean orb. period (yrs) | Max light time (min:sec) | No. of visitors | Prominent visitor | Launch date |
|---|---|---|---|---|---|---|
| Mercury | 0.39 | 0.24 | 11:34 | 2 | Mariner 10 | 1973 |
| Venus | 0.72 | 0.62 | 14:19 | 46 | Venera 8 | 1972 |
| Mars | 1.52 | 1.88 | 20:58 | 50 | Mariner 9 | 1971 |
| Vesta | 2.36 | 4.60 | 27:57 | 1 | Dawn | 2007 |
| Ceres | 2.77 | 3.63 | 31:22 | 1 | Dawn | 2007 |
| Jupiter | 5.20 | 11.86 | 51:34 | 10 | Juno | 2011 |
| Saturn | 9.58 | 29.46 | 87:60 | 5 | Cassini–Huygens | 1997 |
| Uranus | 19.22 | 84.01 | 168:10 | 1 | Voyager 2 | 1977 |
| Neptune | 30.05 | 164.79 | 258:15 | 1 | Voyager 2 | 1977 |
| Pluto–Charon | 39.48 | 248.09 | 336:40 | 1 | New Horizons | 2006 |

process of our solar system, are so distant from our Sun, at about 50,000 AU, that a slight gravitational perturbation of a passing star system could either strip away an object or cause it to fall into the solar system. In the latter case the object becomes a long-period comet. The 'Great Comet of 1882' is a formidable example of a long period comet with an estimated period between 669 and 952 yrs. It was so close to the Sun that it was said to be visible in daylight when the Sun was shadowed by the thumb of a hand [3].

Against the backdrop of this vastness, the inner solar system appears to be right in our front yard–or, depending on the perspective, in our backyard. Table 2.1 shows the mean distances and orbital periods of the planets and dwarf planets in our solar system. Prominent visiting space probes and total visitors (spacecraft and planetary lander) are listed too.[5]

The object with the most visits is the Moon with a total of 150 mission. The maximum light time listed above refers to the time a signal needs from Earth to the object (or vice-versa) when both celestial objects are in conjunction, i.e. in one line with the Sun in middle of them as seen by an external observer. The values show clearly that robotic space exploration requires a high degree of mission planning and spacecraft autonomy.

## 2.2 Principles of Spaceflight

The objective of this section is to provide an introduction to spaceflight in the context of space propulsion. This means that we limit ourselves to an energetic view and therefore to elements of spaceflight that are most relevant for the design of a space

[5] Note that the figure of visitors includes also flybys of spacecrafts with another primary destination, like the US-European mission Cassini–Huygens that performed a flyby at Jupiter on its way to Saturn. For instance, mankind has sent only three orbiters to Jupiter: Galileo (1995), Juno (2016) and JUICE (2031). The gas giant had in total 10 visitors, not all of them stayed.

propulsion system: the required velocity demand for orbit changes and the flight time. Further, to limit the discussion we restricted ourselves to the two dimensional, i.e. planar, case.

This approach is typically used in the early stages of mission design to understand basic orbital geometries and required energies. At a later stage in the design phase, detailed mission planning is required. This necessitates the use of the full range of spaceflight-related data, including the determination of the launch window based on the synodic period of the target orbit and the optimisation of the transfer while respecting communication windows with the ground and planetary flybys. A comprehensive introduction that complements the current propulsion-centric view can be found in [5].

In order to compute the energy demand for orbital changes, it is necessary to mathematically describe the orbits attainable by a spacecraft first.

## 2.2.1 Equation of Motion in a 2 Body Problem

In a pure two-body system the equation of motion of a body with mass $m_2$ is determined by the gravitational force acting on it. The equation can be derived with help of Newton's second law and his famous formulation of gravitational force[6] [7]:

$$m_2 \ddot{\boldsymbol{r}}_2 = -\frac{G m_1 m_2}{r^2} \frac{\boldsymbol{r}}{r}, \tag{2.1}$$

$$\ddot{\boldsymbol{r}}_2 + \frac{G m_1}{r^3} \boldsymbol{r} = 0, \tag{2.2}$$

$m_1$ mass of object 1,
$m_2$ mass of object 2,
$\mathbf{r}_2$ distance of object 2 to the barycentre of the two body system in vector form,
$\mathbf{r}$ relative distance of body 2 to body 1 in vector form,
$G$ Gravitational constant $6.67430 \cdot 10^{-11}\,\mathrm{Nm^2/kg^2}$.

Figure 2.2 illustrates the position vectors of the involved bodies. If $m_2$, is considerably smaller than that of the primary body, $m_1 \gg m_2$, e.g. Earth (1) and Spacecraft (2),

[6] Isaac Newton (1727$^\dagger$) was a brilliant English physicist and mathematician. He is well known for his formulation of the laws of motion and gravity as well as his contributions in optics. He developed independent from the German mathematician Leibniz infinitesimal calculus. As a child of his time, he believed in alchemy and devoted much of his time to these studies [8].

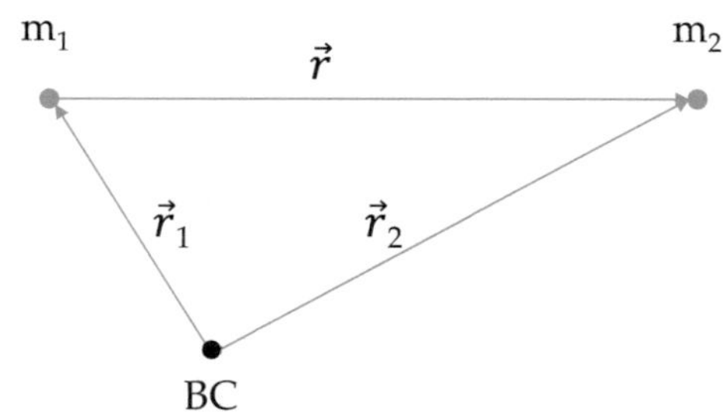


**Fig. 2.2** Geometry of the two body problem

then the motion of $m_2$ will be determined by the primary mass while $m_1$ is not notably[7] affected by $m_2$. In that case, the *barycentre* is practically identical to the centre of mass of $m_1$.

This is not the case if the bodies are comparable in mass. In the Earth-Moon system, for example, the barycentre is almost outside the Earth, at a distance of 4,670 km from the Earth's centre of mass [4]. As a result, the Earth's centre of mass rotates around the barycentre with the same period as the Moon. This causes forces on the Earth, which must be taken into account when discussing tidal effects. Another example in our solar system is the Pluto-Charon system. Both planets have similar masses and their barycentre lies outside of Pluto, causing both objects to orbit around this intangible point in space, bound by the invisible force of gravity. When human-made devices (spaceship, space station etc.) orbit planetary objects, we can safely assume that $\boldsymbol{r_1} = \boldsymbol{0}$ and $\boldsymbol{r_2} = \boldsymbol{r}$. The slight change in the equation leads to:

$$\ddot{\boldsymbol{r}} + \frac{\mu}{r^3}\boldsymbol{r} = 0, \tag{2.3}$$

$\mu$ Gravitational parameter defined as $\mu := G(m_1 + m_2)$ with $G \simeq G(m_1)$, if $m_1 \gg m_2$.

This is a nonlinear differential equation of second order and fortunately, it has a solution in closed form [6], which formulated in polar coordinates is[8]:

$$r = \frac{p}{1 + e\cos\nu}, \tag{2.4}$$

[7] Similarly to a truck hit by a fly.

[8] Isaac Newton was the first to formulate the differential equation for the two body problem and to solve it in his 'Philosophiæ Naturalis Principia Mathematica' (The Mathematical Principles of Natural Philosophy) in 1687.

$p$ semi latus rectum,
$e$ eccentricity,
$\nu$ true anomaly, measures location from pericentre.

This formula describes conic sections. There are four types of conic sections:

- circle,
- ellipse,
- parabola,
- hyperbola.

These geometric curves are called Keplerian[9] orbits, their geometric parameters are depicted in Fig. 2.3. The two focal points are geometric concepts of which only one, F, has a physical meaning: it is the barycentre. The primary object, $m_1$, is located in the focal point F and $m_2$ follows the respective trajectory. The closest approach is

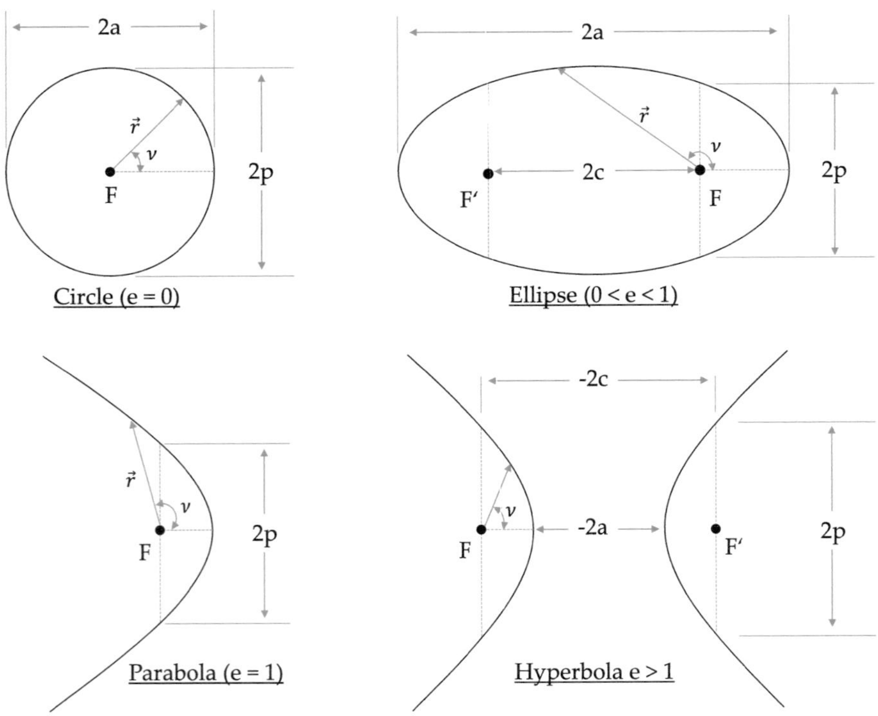


**Fig. 2.3** Main geometric parameters of the conic sections with F' and F being the focal points and F is the barycentre of the Keplerian orbits

[9] Named after the German astronomer and imperial astrologer Johannes Kepler (1630†), who first suggested that the planets move along ellipses around the Sun, with the Sun at one of their focal points, i.e. the barycentre.

referred to as *pericentre*, $r_p$, while the farthest distance is called *apocentre*, $r_a$. The *line of apsides* connects the two opposite locations.

The decisive distinguishing parameter of the different orbit types is the *eccentricity*. The circular and parabolic orbits stand out in that respect. As a consequence, in a two dimensional case at a distance $r$ from the primary object, $m_1$, there exist exactly one circular and one parabolic orbit, whereas there are infinitely many elliptical and hyperbolic orbits, as depicted in Fig. 2.4. However, orbits with eccentricities of exactly 0 or 1 do not occur in nature: firstly, because the gravitational field of the primary body would have to be perfectly symmetrical[10] and secondly the universe is made of more than two bodies such that ubiquitous perturbations eventually lead to deviations from these special cases. The Kepler orbits are therefore idealised mathematical concepts, that grant us valuable insight in orbit dynamics. The parabola for instance is an interesting object as it forms a boundary between bound and unbound orbits. An object revolves in a *bound orbit* if its energy is not sufficient to escape the gravitational field of the primary object. In contrast, objects with sufficiently large energy can follow an unbound trajectory and do not return after one passage.

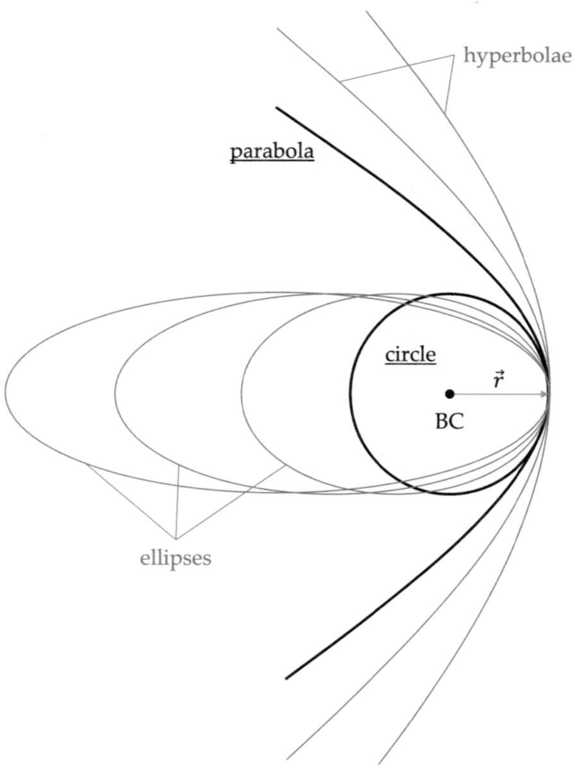


**Fig. 2.4** 2D Manifold of Keplerian orbits at distance **r**

[10] Even the rotation of a gas giant causes a non-symmetric gravitational field due to its ellipsoidal shape.

**Table 2.2** Geometric parameters of Keplerian orbits [7]

| | Circle | Ellipse | Parabola | Hyperbola |
|---|---|---|---|---|
| e | 0 | $0 < e < 1$ | 1 | $> 1$ |
| a | $> 0$ | $> 0$ | $\infty$ | $< 0$ |
| r | $a$ | $p/(1 + e\cos(\nu))$ | $p/(1 + e\cos(\nu))$ | $p/(1 + e\cos(\nu))$ |
| $r_\mathrm{p}$ | $a$ | $a \cdot (1 - e)$ | $p/2$ | $a \cdot (1 - e)$ |
| $r_\mathrm{a}$ | $a$ | $a \cdot (1 + e)$ | $\infty$ | $\infty$ |

The first group consists of ellipses of all forms (a circle is a special ellipse) and the second group of hyperbolae in all forms. The demarcation is formed by a parabola characteristic for each planetary body at a specific distance $r$ in the two dimensional case, very much like a fingerprint.[11] We will quantify the required orbital energy to leave a bound orbit in the following section. Table 2.2 summarises the geometric parameter of the four Keplerian orbits.

In the two dimensional case, the shape of the orbit of mass $m_2$ is fully described by two parameters: the semi-major axis $a$, and the eccentricity, $e$. The position of $m_2$ on the orbit at a specific time is given by the true anomaly, $\nu$. The three parameters can be summarised in a state vector in two dimensions:

$$\begin{pmatrix} a \\ e \\ \nu \end{pmatrix} . \tag{2.5}$$

The Earth's eccentricity ranges between 0.0034 and 0.058 and is currently 0.017. The orbital plane of Earth is called *ecliptic* and the orientations of other planes are measured relative to the ecliptic. Most planetary planes have only a small inclination, which is related to the solar system formation history (protoplanetary accretion disk), see Appendix A. Of all the planets, Mercury has the largest deviation from the ecliptic, 7 degrees. Asteroids and dwarf planets, however, have even larger inclinations. For instance, the dwarf planet Eris, located in the Kuiper Belt, has an inclination of 44.04 deg.

As anticipated above, the real shape of our Earth is not symmetrical and so is not its gravitational field. Figure 2.5 shows Earth's *geoid* (scales are amplified) as measured by the gravimetric satellite GOCE developed by the European Space Agency (ESA) launched in 2009 and decommissioned 2013.

To accurately predict a body's orbit, it is necessary to expand the gravitational model of the primary body. This is usually done in form of perturbation theory as indicated below, see Sect. 2.2.3. As a consequence, the geometry of a Kepler orbit is not invariant in time but changes. These and other perturbations mean that real orbital elements, $a$ and $e$ are not constant but oscillate around mean values [9]. Considering further effects like rest atmosphere and solar pressure leads to non-periodic effects,

[11] In three dimensions this is equivalent to a sphere with radius $r$.

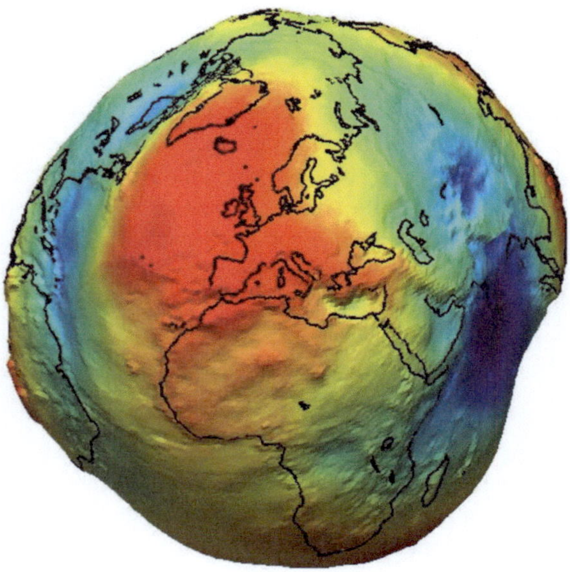

**Fig. 2.5** Earth's geoid–not at scale–as a measure for the gravity field Credit: European Space Agency–ESA/HPF/DLR

called *secular perturbation*, causing the an orbital element to continuously increase or decrease [10]. The resulting orbit is not closed but at any instant a Keplerian orbit can assigned, referred to as *osculating orbit*. In the (hypothetical) sudden absence of any perturbation, the mass $m_2$ would follow this Keplerian orbit.

## 2.2.2 *Vis-Viva Integral*

In physics, an energy analysis is always informative, as it is for the two-body problem in celestial mechanics. To derive the energy equation we need to multiply Eq. 2.3 with the body's velocity ($\dot{\boldsymbol{r}}$) and simplify the result further:

$$\ddot{\boldsymbol{r}} + \frac{\mu}{r^3}\boldsymbol{r} = 0, \tag{2.6}$$

$$\dot{\boldsymbol{r}} \cdot \ddot{\boldsymbol{r}} + \frac{\mu}{r^3}\dot{\boldsymbol{r}} \cdot \boldsymbol{r} = 0, \tag{2.7}$$

$$\frac{d}{dt}\left(\frac{1}{2}\dot{\boldsymbol{r}} \cdot \dot{\boldsymbol{r}}\right) - \frac{d}{dt}\left(\frac{\mu}{r}\right) = 0, \tag{2.8}$$

$$\frac{d}{dt}\left(\frac{1}{2}\dot{\boldsymbol{r}} \cdot \dot{\boldsymbol{r}} - \frac{\mu}{r}\right) = 0, \tag{2.9}$$

$$\frac{d}{dt}\left(\frac{1}{2}v^2 - \frac{\mu}{r}\right) = 0. \tag{2.10}$$

In the last step, we take into account that the vector product of the velocity is the absolute velocity squared: $\dot{\boldsymbol{r}} \cdot \dot{\boldsymbol{r}} = v^2$. The last equation demands that the time derivative of the bracket stays zero for all times. That is only possible, if what is inside the bracket is time-invariant, in simple terms a constant. A close look to the two terms in the bracket reveals that the first term is the specific kinetic energy of $m_2$ and the second term its potential energy in the gravitational field of $m_1$. Hence, the equation demands that the sum of these two energy forms stays constant, which is the definition of the energy conservation equation in a force-free closed system.[12] Force-free means that no other force except of gravity is acting on $m_2$. We now can formulate the famous energy equation of the two body problem, also known as *vis-viva integral*[13]:

$$\frac{1}{2}v^2 - \frac{\mu}{r} = C. \tag{2.11}$$

The specific orbital energy, denoted as $C$, is negative for bound orbits (e.g. ellipses), positive for unbound orbits (i.e. hyperbolas), and zero for parabolic orbits. Therefore, by setting $C$ equal to zero, we can calculate the minimum velocity needed to escape Earth's gravitational pull starting from rest.

$$\frac{1}{2}v_{\text{esc}}^2 - \frac{\mu}{r} = C = 0, \tag{2.12}$$

$$v_{\text{esc}} = \sqrt{2\frac{\mu}{r}}. \tag{2.13}$$

Evaluating the equation[14] for Earth, $\mu_\oplus = 3.99 \cdot 10^{14}\,\text{Nm}^2/\text{kg}$ and $R_\oplus = 6378\,\text{km}$ yields $v_{\text{esc},\oplus} = 11.18\,\text{km/s}$. The parabolic orbit can be regarded as a demarcation line indicating an orbital energy ladder that depends on the geometry of the orbit.

The specific energy $C$ is a function of the semi major axis, which makes perfectly sense as this parameter determines the size of the orbit. Table 2.3 shows the geometric and dynamic relations of the four orbit types discussed. Comparing the velocity on a circle with the same radius as the pericentre of a parabola, $r_\text{p} = r_\text{circ}$, reveals a useful relation:

$$v_{\text{pbl}}(r_\text{p}) = v_{\text{esc}} = \sqrt{2} \cdot v_{\text{circ}}. \tag{2.14}$$

This relation says, that if a spacecraft is orbiting a planet or star in a circle, it takes its current orbital velocity times $\sqrt{2}$ to achieve escape velocity. Our Earth for instance

---

[12] To be precise, a force acting perpendicular to the velocity vector would also not change its energy state but it would have an effect on the angular momentum vector of the orbit.

[13] The term vis-viva originates from the brilliant German mathematician and physicist Gottfried Leibnitz 1716†. He was an equally brilliant contemporary of Isaac Newton 1727† but also a fierce competitor of him.

[14] All computations are performed using the *International System of Units* (SI). The result is then transformed in units typically used in the respective discipline, here km and km/s. This is the case throughout the book.

**Table 2.3** Specific energy and velocity of Keplerian orbits, NA = not applicable

| | C (J/kg) | $v^2$ ($km^2/s^2$) | $v_\infty$ (km/s) |
|---|---|---|---|
| Circle | $-\frac{\mu}{2r}$ | $\frac{\mu}{r}$ | NA |
| Ellipse | $-\frac{\mu}{2a}$ | $\mu\left(\frac{2}{r}-\frac{1}{a}\right)$ | NA |
| Parabola | 0 | $2\frac{\mu}{r}$ | 0 |
| Hyperbola | $\frac{\mu}{2a}$ | $\mu\left(\frac{2}{r}+\frac{1}{a}\right)$ | $\sqrt{\frac{\mu}{a}}$ |

orbits the Sun in an almost circular orbit of currently[15] ($e = 0.017$) and a mean velocity of 29.8 km/s.

To escape the solar system starting from an Earth-like orbit requires a velocity of 42.14 km/s.

The specific energy on a hyperbola trajectory is called *excess energy* since it is more energy than required to reach infinity. For historical reasons the denomination $C3$ has been propagated [12]. It is defined as twice the orbital energy, $C3 = 2 \cdot C$, which comes handy because the hyperbolic velocity in infinity can be then written as:

$$v_\infty = \sqrt{C3}. \tag{2.15}$$

The orbital period, $T$, of a test mass in a circular orbit with radius $r$ can be easily calculated according to the kinematic relation: $T = U/v$, with $U = 2\pi r$ the circumference and $v$ the orbital velocity, which is constant on a circle. Inserting the term for the velocity above yields: $T = 2\pi r/\sqrt{\mu/r}$. It can be shown that this formula is also valid for ellipses, if r is replaced with the semi-major axis $a$ [6]. To avoid the square root, it is often written in squared form:

$$T^2 = 4\pi^2\frac{a^3}{\mu}. \tag{2.16}$$

This is Kepler's famous third law. He formulated it in relative form, by comparing the movement of two planets:

> The squares of the sidereal periods (of revolution) of the planets are directly proportional to the cubes of their mean distances from the Sun [13].

[15] There is a long term variability in Earth's eccentricity (0.0034–0.058), obliquity and precession named after the Serbian scientist Milutin Milankovitch [11].

### 2.2.3 *Orbital Transfer*

The core task of interplanetary spaceflight is to design a mission that flies to a destination with minimum energy demand for given constraints. Starting from Earth, this encompasses in general several orbit changes. The constraint are for instance departure and fFarrival time. Mission analysts distinguish between two types of transfer[16]:

(a) impulsive transfer,
(b) continuous transfer.

The first is a good approximation for chemical propulsion systems while the latter is a representative for low thrust propulsion systems, realised with electric propulsion systems. But also non-conventional systems like solar sails belong to the second category. We will discuss in the following the two types from an orbit mechanics point of view.

#### Impulsive Orbital Change—The Hohmann Transfer

The above established vis-viva integral is very helpful to compute in-plane impulsive orbital changes. Theses changes are related to the spacecraft's velocity and thus linear momentum. The achieved velocity change is referred to as $\Delta v$ (speak 'delta v'). The term impulsive change is a mathematical concept in which the duration of the action is infinitely short and the magnitude infinitely high. In orbit mechanics an impulsive transfer corresponds to a manoeuvre with an infinitely high acceleration. This is of course an idealisation since no propulsion system is capable of this. In reality a sufficiently high acceleration is sufficient to apply this approach without introducing a significant error. However, *very short* and *very high* are relative. Chemical propulsion systems are in general only operated for a fraction of the orbital period, which makes the thrust duration sufficiently short. As a rule of thumb, manoeuvre durations that are short compared to the orbital period ($\leq 10\%$) justify the impulsive approximation. We will see below and in Sect. 10.4.1 how to correct for the introduced error. For the moment we will gratefully accept this practical approximation.

The German civil engineer Walter Hohmann[17] has proven that an energetically optimised transfer between two orbits that are both circular and not inclined to each other, i.e. co-planar, can be achieved by two such impulses [16].

---

[16] A third class has gained popularity in recent years, called *Weak Stability Boundary* transfer (WSB). Suggested in 1987 by the JPL scientist Edward A. Belbruno, it is a low-energy transfer that makes use of the gravitational force of a third body in a two-body system while using minimal thrust [14].

[17] Walter Hohmann (1945†) studied at the Technical University of Munich and earned his degree in civil engineering in 1904, he then earned a doctorate from the Technical University of Aachen in 1919. His pioneering work on planetary landing 'Landing on Other Celestial Objects' served as a guideline in the Apollo program. He died in March 1945 in the German town Essen during an Allied air raid [15].

The intermediate trajectory is half of an ellipse and consequentially called transfer ellipse, the manoeuvre is called *Hohmann transfer*. The distance between the initial and final orbit is twice as large as the semi-major axis of the transfer ellipse. An example is provided in Fig. 2.6, which depicts a transfer from Low Earth Orbit (LEO) to the famous Geostationary Orbit (GEO), where telecommunication satellites are preferably placed to constantly broadcast over the same region. The important feature of this type of transfer is the fact that the transfer ellipse *touches* both circular orbits: the initial orbit and the target orbits. That means that its pericentre coincides with $r_1$ and its apocentre with $r_2$:

$$r_{\mathrm{p}} = r_1,$$

$$r_{\mathrm{a}} = r_2.$$

The figure also reveals that for both manoeuvres the velocity vectors are aligned, meaning that they point into the same direction and we do not need to apply vector calculus. In fact, the required velocity impulsive can be calculated with help of the vis-viva equation. We first determine the velocities of the transfer ellipse at the pericentre and apocentre respectively. We make use of the relations in Tables 2.2 and 2.3:

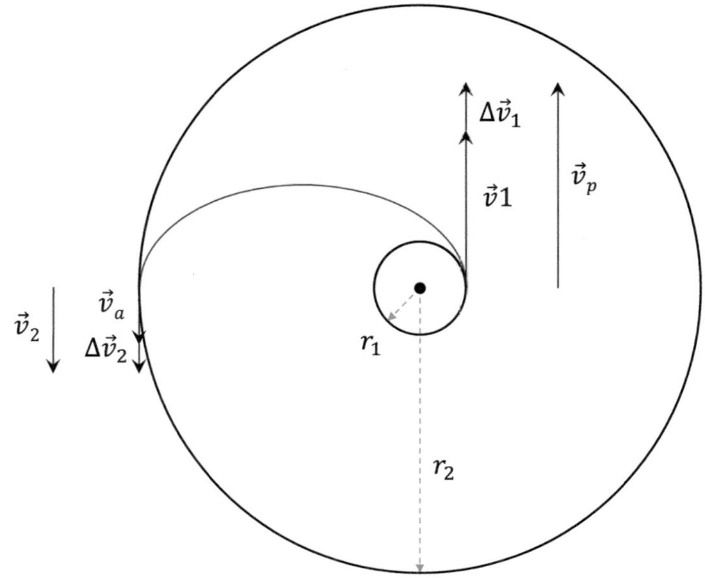


**Fig. 2.6** Hohmann transfer from LEO to GEO–to scale

**Table 2.4** Hohmann transfer from LEO (400 km) to GEO (35,786 km)

| LEO | | Transfer ellipse | | GEO | |
|---|---|---|---|---|---|
| $r_1$ | 6,778 km | $a_t$ | 24,471 km | $r_2$ | 42,164 km |
| $v_1$ | 7,670 m/s | $v_p$ | 10,067 m/s | $v_2$ | 3,075 m/s |
| $\Delta v_1$ | 2,398 m/s | $v_a$ | 1,618 m/s | $\Delta v_2$ | 1,457 m/s |

$$v_p^2 = \sqrt{\frac{2\mu}{r_p} - \frac{\mu}{a_t}}, \tag{2.17}$$

$$= \frac{\mu}{r_p}\sqrt{2 - \frac{r_p}{a_t}}, \tag{2.18}$$

$$= v_1^2 \cdot (1+e), \tag{2.19}$$

$$v_a^2 = \sqrt{\frac{2\mu}{r_a} - \frac{\mu}{a_t}}, \tag{2.20}$$

$$= \frac{\mu}{r_a}\sqrt{2 - \frac{r_a}{a_t}}, \tag{2.21}$$

$$= v_2^2 \cdot (1-e). \tag{2.22}$$

Next we calculate the velocity demand to transfer from the initial circular orbit to the transfer ellipse, position 1, and then from the transfer ellipse to the target circular orbit:

$$\Delta v_1 = v_p - v_1, \tag{2.23}$$

$$= v_1 \cdot (\sqrt{1+e} - 1), \tag{2.24}$$

$$\Delta v_2 = v_2 - v_a, \tag{2.25}$$

$$= v_1 \cdot (\sqrt{1-e} - 1). \tag{2.26}$$

The results are listed in Table 2.4. The orbital velocity $v_1$ must be increased by $\Delta v_1$ to increase the orbital energy and thus the semi major axis. An ellipse with more energy (larger $a_t$) would intersect the target orbit and not touch it. The new velocity at $r_1$ is then equal to the velocity of the ellipse at its pericentre: $v_p = v_1 + \Delta v_1$. The mass $m_2$ reaches the target orbit when it arrives at the apocentre of the transfer ellipse. The velocity is then lower than what would be needed to stay in the target orbit and a second manoeuvre is necessary: $v_2 = v_a + \Delta v_2$

The total $\Delta v_{tot} = \Delta v_1 + \Delta v_2$ is then 3,855 m/s. Space engineers prefer manageable numbers and therefore like to use the unit km/s, i.e. 3.855 km/s. Note that the energy demand is greater than that needed to escape from Earth from the same orbit. A precise analysis shows that the Hohmann transfer is the least energetic transfer, if the ratio of final to initial radius fulfills the condition [6]:

$$\frac{r_2}{r_1} < 11.94. \tag{2.27}$$

Measured from Earth ($r_1 = 1$ AU) this extends well beyond Saturn (9.6 AU) but still below Uranus (19.2 AU). In addition, the Hohmann transfer comes with the disadvantage of a long transfer time and is therefore not applied for interplanetary flights beyond Mars. We will see further below, that there is a method to gain energy along the way to far celestial destinations, the gravity assist manoeuvre.

## Escape Manoeuvre

We stated above that launch providers prefer to provide the escape capacity in terms of excess orbital energy $C3$. We will derive the velocity impulse, $\Delta v_{\mathrm{esc}}$, required to escape the Earth bound circular orbit and transfer to a hyperbolic trajectory with a specified C3. It is instructive to compare the hyperbolic escape with a parabolic escape manoeuvre, as depicted in Fig. 2.7. The parabolic case is straightforward, the spacecraft's velocity needs to be increased to the velocity of a parabolic velocity in its pericentre, $v_{\mathrm{pbl,pc}}$. The required velocity impulse can be calculated by a vector subtraction:

$$\Delta \mathbf{v}_{\mathrm{esc,pbl}} = \mathbf{v}_{\mathrm{pbl,pc}} - \mathbf{v}_{\mathrm{circ}}. \tag{2.28}$$

There are cases, however, in which only $C3$ of a hyperbola is known and a transfer from LEO is required. $C3$ gives us immediately $v_\infty$ but as depicted in Fig. 2.7 we need $v_{\mathrm{esc}}$ to calculate the required $\Delta v$.

The figure also reveals that the final velocity vector $v_\infty$ is not aligned with escape manoeuvre $\Delta v$, meaning that we need the angle between the two vectors for a vector addition. To avoid this cumbersome path, it is more rewarding to take the energy equation as a starting point. The specific kinetic energy, $C_{\mathrm{kin,esc}}$, must equal the energy required to reach infinity on a parabola plus the kinetic energy of the hyperbola:

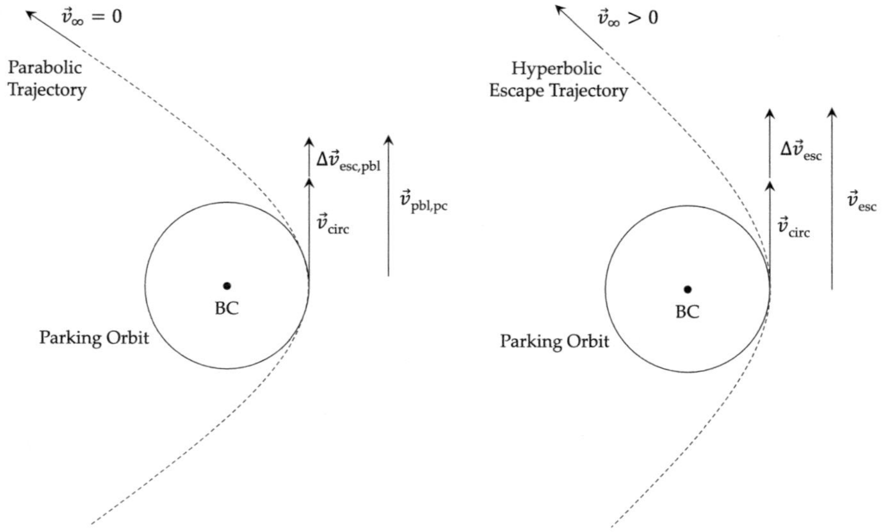


**Fig. 2.7** Orbital transfer: circular→ parabola (left), circular → hyperbola (right)

**Table 2.5** Mission profile Delta-II-rocket for NASA's Dawn mission, final declination angle is 28.5°. *Source* ULA

| Event | Time line | Altitude (km) | Velocity (Km/s) |
|---|---|---|---|
| Lift-Off | | 0 | 0 |
| SRM Jettison (6 of 9) | 01 m 20.5 s | 26 | 1.19 |
| SRM Jettison (9 of 9) | 02 m 39.5 s | 73 | 3.29 |
| MECO | 04 m 22.5 s | 126 | 6.48 |
| Second-Stage Ignition | 04 m 36.5 s | 133 | 6.49 |
| Fairing Jettison | 04 m 41.0 s | 135 | 6.50 |
| SECO-1 | 08 m 58.3 s | 189 | 7.79 |
| SECO-2 | 54 m 14.5 s | 184 | 9.01 |
| Third.Stage Ignition | 55 m 44.7 s | 213 | 8.98 |
| TECO† | 57 m 10.7 s | 275 | |
| Spacecraft Separation | 01 h 2 m 00.8 s | 1021 | |

† altitude represents perigee

$$C_{\text{kin,esc}} = C_{\text{kin,pbl}} + 0.5 \cdot C3, \tag{2.29}$$

$$v_{\text{esc}}^2 = v_{\text{pbl,pc}}^2 + v_{\infty}^2, \tag{2.30}$$

$$v_{\text{esc}}^2 = 2v_{\text{circ}}^2 + v_{\infty}^2, \tag{2.31}$$

$$v_{\text{esc}} = \sqrt{2v_{\text{circ}}^2 + v_{\infty}^2}. \tag{2.32}$$

With $v_{\text{esc}}$, the velocity impulse we are looking for can now be calculated by a simple vector subtraction:

$$\Delta v_{\text{esc}} = v_{\text{esc}} - v_{\text{circ}}. \tag{2.33}$$

NASA's Dawn mission is a good study case. The required $C3$ was 11.4 km$^2$/s$^2$. The used launcher, a Delta II in its most performant nine booster configuration, was still not capable to reach $C3$. An extra solid fuelled kick-stage called Star 48 was required. Table 2.5 lists the main event of the ascent trajectory till spacecraft separation.

Note that the altitude drops between SECO-1 (Second Engine Cut-Off—1) and SECO-2 while the velocity increases. It means that the flight path angle $\gamma$ is negative and the vehicle gains velocity by falling back to Earth during this short ballistic phase. This a typical outcome of trajectory optimisation, if the target orbit is highly energetic, i.e. GEO or escape hyperbola.

### Continuous Orbital Transfer

We mentioned above that impulsive transfers are an appropriate approximation for chemical propulsion systems that are only operated for a relatively short period. This is not the case for all propulsion systems. We will see in Chap. 5 that spacecraft propelled by electric propulsion systems or solar sails experience a small yet continuous acceleration due to a low thrust magnitude rendering the impulsive transfer approximation incorrect. Equation 2.3 is formulated in polar coordinate form, which is not suitable when considering forces acting by and on the spaceship. A *transformation* of the barycentric reference frame into a flight-path-fixed reference frame is required, which will change the state-vector:

$$\begin{pmatrix} a \\ e \\ \nu \end{pmatrix} \Rightarrow \begin{pmatrix} v \\ \gamma \\ r \end{pmatrix}. \tag{2.34}$$

The three variables are the tangential velocity, $v$, the *flight path angle*, $\gamma$, and the radius, $r$. As a result of the coordination transformation, we obtain a new set of differential equations [17], which are then:

$$\dot{v} = -\frac{\mu}{r^2} \sin\gamma, \tag{2.35}$$

$$\dot{\gamma} = \frac{1}{v}\left(\frac{v^2}{r} - \frac{\mu}{r^2}\right) \cos\gamma, \tag{2.36}$$

$$\dot{r} = v \sin\gamma. \tag{2.37}$$

Figure 2.8 depicts the new variables for an elliptic orbit. The velocity vector is tangential to the orbit. To construct the local horizon, we need to draw a circle around the barycentre with radius **r** (location of spacecraft). The tangent to the circle at the intersection between circle and orbit is called the *local horizon* and the angle between the velocity vector and the horizon corresponds to the flight path angle. It is important

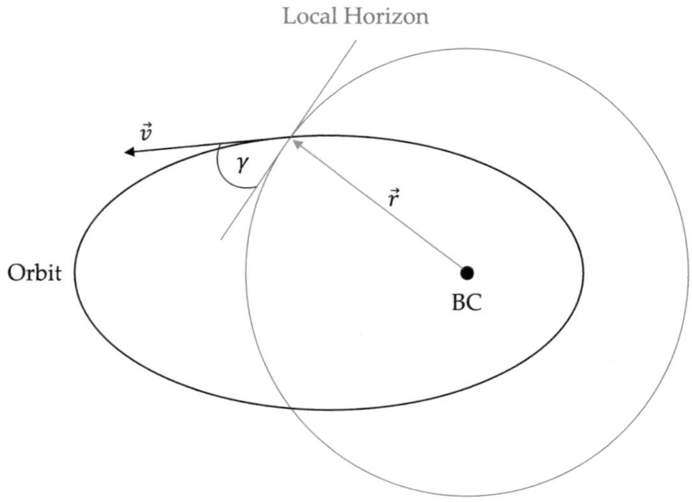


**Fig. 2.8** Definition of local horizon and $\gamma$

to note that the coordinate transformation did not change the physics and that the two forms are fully equivalent. The advantage of the flight-path-fixed reference frame is that the orientation of the spaceship's thrust vector stays constant. Extending the set of equations by the thrust vector and external forces [17], yields then:

$$\dot{v} = \underbrace{-\frac{\mu}{r^2}\sin\gamma}_{\text{ballistic}} + \underbrace{\frac{T_\mathrm{T}}{m_2}}_{\text{thrust}} + \underbrace{f_{\mathrm{p,T}}}_{\text{perturbation}}, \tag{2.38}$$

$$\dot{\gamma} = \frac{1}{v}\left(\frac{v^2}{r} - \frac{\mu}{r^2}\right)\cos\gamma - \frac{T_\mathrm{B}}{v m_2} + f_{\mathrm{p,B}}, \tag{2.39}$$

$$\dot{r} = v\sin\gamma, \tag{2.40}$$

$T_\mathrm{T}$ thrust component acting in velocity direction of $m_2$, i.e. tangential to the trajectory,
$T_\mathrm{B}$ thrust component acting perpendicular to the velocity direction of $m_2$,
$f_{\mathrm{p,T}}$ perturbation force acting in velocity direction of $m_2$, i.e. tangential to the trajectory,
$f_{\mathrm{p,B}}$ perturbation force acting perpendicular to the velocity direction of $m_2$.

We have grouped the right hand side in three terms. The first is called ballistic term and refers to the unperturbed movement of $m_2$ in the gravity field of $m_1$. The second term describes the impact of the spacecraft's thrust vector. As anticipated before, thrust acting perpendicular to the velocity vector, $T_\mathrm{B}$, does not change the orbital velocity and therefore neither the orbital energy. We will consider throughout this book only $T_\mathrm{T}$ and drop in the following the subscription 'T'. The third term summarises all external forces acting on the spacecraft, like atmospheric drag, solar pressure and non-homogeneous gravity contributions.

### 2.2.4 *Multi-body System*

We have considered above a two-body system. We went even one step further and assumed that the second body (e.g. spaceship) is so small that it does not affect the dynamics of the other body (e.g. planet) which we dubbed primary body. We will keep this assumption but extend the picture and allow an additional gravitating body to enter the scene.

#### Sphere of Influence–SOI

The gravitational field of each body extends to infinity and since the solar system consists of several gravitating natural bodies, it is obvious that the resulting gravitational field at a point in space is a superposition of all single fields. The Moon's orbit for instance is dominated by Earth's gravity and not that of the Sun, although the latter's mass is five orders of magnitude larger than the Earth's, $3.33 \times 10^5$ times. The intuitive answer to this is obvious: the Moon is close enough to the Earth, to be dominated by its gravity field. Astronomers call the region in which a body dominates its surrounding *sphere of influence* (SOI), while the term region of influence (ROI) is also popular. The Moon in the scenario above does have a sphere of influence, too. The trajectory of a spacecraft that approaches the Moon is at a certain point dominated by its gravity field.

The immediate question that arises is, how can we calculate the sphere of influence? Before dwelling into equations, it is instructive to spend some thoughts on the matter. The sphere of influence of a planet orbiting the Sun must depend on both an intrinsic and extrinsic parameter. The intrinsic parameter is obviously the planet's mass, $m_\mathrm{p}$, as it determines the strength of its gravitational field. The extrinsic parameter is the Sun's mass $m_\odot$. Both gravitational fields extend to infinity and thus overlap. Therefore, there is a transition region in which both forces are similar in strength. Adopting a formal approach by introducing a test mass, $m_\mathrm{t}$, that is much smaller than the gravitating bodies $m_\mathrm{p}$ and $m_\odot$, the equations of motion in both SOI's can be expressed as:

$$\begin{array}{ll} \text{SOI Sun} & \text{SOI Planet} \\ m_\mathrm{t}\ddot{\mathbf{R}} = \mathbf{F}_\odot + \mathbf{F}_\mathrm{p}, & m_\mathrm{t}\ddot{\mathbf{r}} = \mathbf{F}_\mathrm{p} + \mathbf{F}_\odot, \\ \ddot{\mathbf{R}} = \mathbf{A}_\odot + \mathbf{P}_\mathrm{p}, & \ddot{\mathbf{r}} = \mathbf{a}_\mathrm{p} + \mathbf{p}_\odot, \end{array}$$

(a) If $P_\mathrm{p}/A_\odot$ is *large* then the orbit of the test mass (e.g. spacecraft) around the Sun is significantly perturbed by the planet.
(b) If $p_\odot/a_\mathrm{p}$ is *large* then the orbit of the test mass (e.g. spacecraft) around the planet is significantly perturbed by the Sun.

The ratio of the accelerations within its sphere of influence can be used to derive a criterion for the transition from one SOI to the other. The criterion says that if the Sun's relative perturbation is smaller than the planet's relative perturbation than the test mass is within the planet's SOI:

$$\frac{p_\odot}{a_\mathrm{p}} < \frac{P_\mathrm{p}}{A_\odot}. \tag{2.41}$$

$m_t$ small test mass with no recognisable gravitational effect on other masses,
$\mathbf{R}$ distance of $m_t$ to the Sun,
$\mathbf{F}_{\odot}$ gravitational force of the Sun on $m_t$ measured in the SOI of the Sun,
$\mathbf{F}_p$ gravitational force of the planet on $m_t$ measured in the SOI of the Sun,
$\mathbf{r}$ distance of $m_t$ to the planet,
$\mathbf{A}_{\odot}$ dominant acceleration due to Sun,
$\mathbf{P}_p$ perturbative acceleration due to Earth,
$\mathbf{a}_p$ dominant acceleration due to Earth,
$\mathbf{p}_{\odot}$ perturbative acceleration due to Sun.

A quantitative analysis [6] reveals that the mathematical criterion we are looking for is:

$$r_{SOI} = R\left(\frac{m_p}{m_{\odot}}\right)^{\frac{2}{5}} . \tag{2.42}$$

Table 2.6 lists the SOIs for selected planetary objects in our solar systems. The Moon's sphere of influence normalised to its distance to Earth, is very large due to its size. Ceres has a very large SOI in absolute figures but a very low relative SOI, i.e. normalised to its distance to the Sun. Earth reigns about a sphere with an approximate radius of 924,416 km, which is 6% of its distance to the Sun. Neptune has the largest SOI, which extends 6 AU into space. Finally, the term sphere is actually misleading as it suggests a region that is symmetric. In reality, the shape is more oval than circular, and its structure becomes even more complex when considered in three dimensions. The edge is then defined where the ratios of perturbing accelerations are equal.[18]

**Table 2.6** Sphere of influence (SOI) for selected bodies in the solar system to primary body, Earth for Moon, Sun for the other

| Object | Symbol | Mass (kg) | Distance* (km) | $r_{SOI}$ (km) | $r_{SOI}$/Distance (%) |
|---|---|---|---|---|---|
| Mercury | ☿ | 3.30E+23 | 5.8E+07 | 1.1.E+05 | 0.19 |
| Venus | ♀ | 4.87E+24 | 1.1E+08 | 6.2.E+05 | 0.57 |
| Earth | ♁/⊕ | 5.97E+24 | 1.5E+08 | 9.2.E+05 | 0.62 |
| Moon | ♂ | 7.30E+22 | 3.8E+05 | 6.6.E+04 | 17.18 |
| Mars | ☾ | 6.42E+23 | 2.3E+08 | 5.8.E+05 | 0.25 |
| Ceres | | 9.10E+20 | 4.1E+08 | 7.6.E+04 | 0.02 |
| Jupiter | ♃ | 1.90E+27 | 7.8E+08 | 4.8.E+07 | 6.19 |
| Saturn | ♄ | 5.68E+26 | 1.4E+09 | 5.5.E+07 | 3.82 |
| Uranus | ⛢ | 8.68E+25 | 2.9E+09 | 5.2.E+07 | 1.80 |
| Neptun | ♅ | 1.02E+26 | 4.5E+09 | 8.7.E+07 | 1.92 |

[18] This perspective renders the designation 'region of influence' actually more appropriate but changing historic names in science is a futile endeavour.

The concept of SOI is important for interplanetary spaceflight. In the initial analysis, the trajectory of the spaceship is divided into several segments, each corresponding to a specific sphere of influence. This approach is known as *patched-conic* method. Whenever the spaceship crosses this imaginary border, the reference frame needs to change. This will result in new (relative) orbital parameters. An example helps to understand the relations: a spaceship in the heliocentric reference frame might be on an elliptic orbit, but as soon as it enters the SOI of a planet, the reference frame needs to be changed. From the planet's *point of view* the spaceship is coming from infinity and has an excess velocity $v_\infty > 0$, which is the condition for a hyperbolic trajectory. Contrary, a spaceship that leaves Earth on a parabolic trajectory with an asymptote that points in direction of Earth's orbital velocity will end in an orbit around the Sun that is almost identical to Earth's orbit.

### Gravity Assist Manoeuvre–Swing-By

In interplanetary spaceflight there is a phenomenon called *gravity assist* or *swing-by*, that helps to gain velocity at the expense of the primary body's orbital energy.[19] The same phenomenon is responsible for kicking-out stars from a stellar cluster and asteroids out of their orbital planes. Astronomers call this scattering. The arguably first to study this phenomenon was the Ukrainian engineer and mathematician Yuri Kondratyuk,[20] and he did it in the context of interplanetary spaceflight. The principle is equivalent to the momentum exchange a ball experiences that is thrown against the side of a moving train, only that the gravitational interaction in a gravity assist is intangible. Figure 2.9 shows the kinematics of the gravity assist manoeuvre.

From the planet's reference frame the spacecraft enters and leaves its sphere of influence without a change in the spacecraft's velocity magnitude. The same manoeuvre in the Sun's reference frame differs because the planet's own orbital velocity $\mathbf{v}_\mathrm{p}$ needs to be considered. The resulting exit velocity, $|\mathbf{v}_2|$, is therefore larger in magnitude than the entry velocity $|\mathbf{v}_1|$. The loss of planetary momentum and thus velocity is not measurable, but the spaceship's gain in velocity is significant. The effect of the gravity assist manoeuvre is stronger if:

(a) the mass of the planet is large,
(b) the approach is close, i.e. $r_\mathrm{p}$ close to gravity centre [7].

The technique of gravity assist manoeuvre was used for a number of deep space missions. Table 2.7 shows a selected number of spacecraft and the achieved velocity gain. Note that en route to the outer solar system, Jupiter and beyond, swing-by

---

[19] A further denotation for gravity assist is 'flyby'. The term flyby, however, is not necessarily indicative of a gravity assist manoeuvre. It can simply refer to a flyby without changing of orbital parameters.

[20] Yuri Kondratyuk (1942†), born as Aleksandr Shargei, studied at the Polytechnic School in St. Petersburg, Russia. He is another visionary space pioneer who laid out fundamental technical ideas for interplanetary spaceflight. Among them the idea to use the gravitation of a primary body to enhance the spaceship's velocity [18].

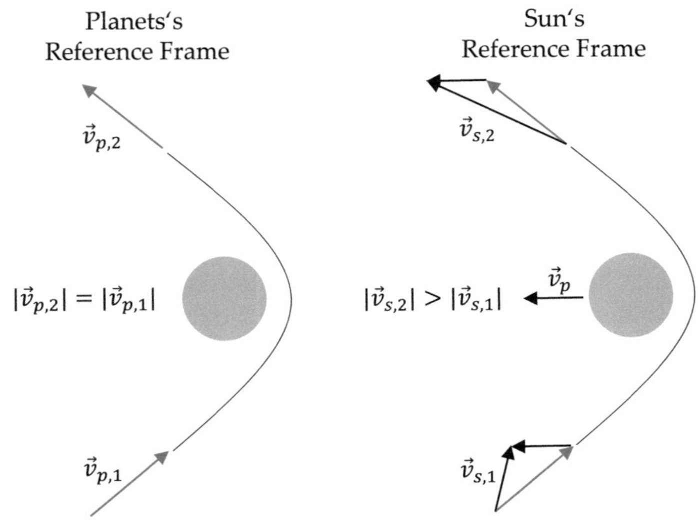


**Fig. 2.9** Gravity assist in a planetocentric and a heliocentric reference frame

**Table 2.7** Velocity gain of selected spacecraft utilising swing-by manoeuvre, velocity gain in km/s [19]

| Spacecraft | Type | Venus | Earth | Mars | Jupiter | Saturn | Uranus |
|---|---|---|---|---|---|---|---|
| Pioneer 10 | flyby | | | | 15.3 | | |
| Voyager I | flyby | | | | 16.0 | | |
| Voyager II | flyby | | | | 10 | 5 | 2 |
| Cassini-Huygens | Saturn | 6.8* | 4.1 | | 2.1 | | |
| New Horizons | flyby[†] | | | | 3.8 | | |
| Dawn | Vesta & Ceres | | | 1.81 | | | |

* performed two consecutive swing-bys at Venus: 3.7 km/s (I), 3.1 km/s(II)

[†] destination was Pluto, SC did not enter orbit, which currently is energetically impossible

manoeuvres at Mars are not favoured by mission analysts. This is due to the planet's small mass.

Finally, it shall be noted that swing-by manoeuvres are not only used to achieve velocity gains but also to deflect the spacecraft's trajectory. A famous mission in which the trajectory was deflected was Ulysses a joint mission by NASA and ESA. Here the spacecraft was sent first to Jupiter to perform an inclination change of 80.2° to the ecliptic. A manoeuvre that is otherwise not feasible by a propulsion system alone due to its high energy demand (Fig. 2.10).

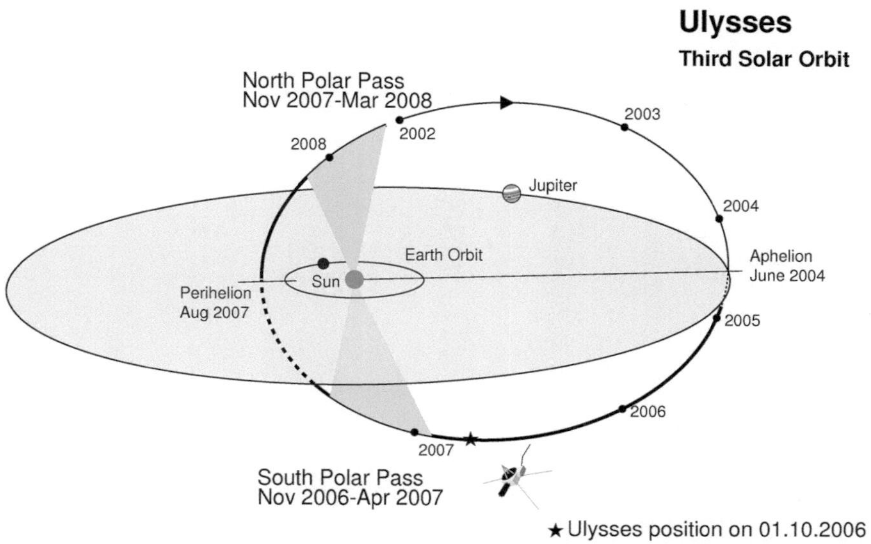


**Fig. 2.10** Third solar orbit of Ulysses spacecraft. Credit: European Space Agency—ESA

### *2.2.5 Impulsive Transfer to Ceres*

We have learned above that the classical Hohmann transfer is between two co-planar and circular orbits. Reality, though, is slightly more complex than this: the orbits are not circular and also not co-planar, but elliptic and inclined to each other. Ceres' inclination to the ecliptic is 10.6°. The latter has a significant impact on the frequency of energy-efficient transfer options, expressed as the frequency of start windows within a given time period, e.g. a decade. Appendix B provides a detailed two step analysis for an impulsive transfer from Earth to Ceres starting with a co-planar and circular case and then increasing the complexity towards a realistic 3 dimensional case using the patched-conic method. The analysis, however, does not consider gravity assist manoeuvre at Mars or Earth in order to keep the example comprehensible.

A realistic trajectory analysis should, however, consider swing-bys since they help to reduce the required $\Delta v$ significantly as we have seen in Sect. 2.2.4. The NASA online tool 'Trajectory Browser'[21] provides this capability. Table 2.8 shows the results for a rendezvous missions to Ceres in the time span of 2030 to 2040, optimised for minimum $\Delta v$.

Some missions require a swing-by at Mars, others suggest to leave Earth and come back for a swing-by. The total required velocity demand is between 8.96 km/s and 9.84 km/s. The transfer time is surprisingly small with 3.02–3.85 yrs. In order to pick a baseline mission, several considerations need to be made. First, the baseline mission must envelope at least one further mission in terms of performance requirements.

[21] https://trajbrowser.arc.nasa.gov/traj_browser.php.

**Table 2.8** Orbital transfer from LEO (200 km) to Ceres (rendezvous), all $\Delta v$ in km/s, = Earth departure, EMC = Mars swing-by, EEC = Earth swing-by, Ceres arrival

| ID | Departure date | Duration yrs | Energy $C_3$ | Escape $\Delta v_{esc}$ | Injection $\Delta v_{inj}$ | Total $\Delta v_{tot}$ | Route |
|---|---|---|---|---|---|---|---|
| 1. | Jan-03-2030 | 3.55 | 33.8 | 4.66 | 4.49 | 9.16 | EMC |
| 2. | Jul-03-2032 | 3.42 | 25.7 | 4.34 | 5.18 | 9.52 | EEC |
| 3. | Jan-17-2035 | 3.02 | 24.9 | 4.30 | 5.52 | 9.82 | EEC |
| 4. | May-27-2036 | 3.85 | 34.7 | 4.70 | 4.26 | 8.96 | EMC |
| 5. | Jul-17-2037 | 3.15 | 25.9 | 4.35 | 5.49 | 9.84 | EEC |

The reason is obvious: missing one launch opportunity, shall not render the entire programme in vain due to incompatibility with subsequent launch windows. This is a kind of mission redundancy that needs to be formulated in a mission requirement. Second, the escape manoeuvre poses requirements on the thrust as will be discussed in Sect. 10.4.1 and tends to drive the entire mass of the spaceship. A not too high $\Delta v_{esc}$ shall be therefore adopted. Given these minimum considerations, mission ID-2 (Jul-03-2032) is a reasonable choice. The time launch window repeats 3 years later (Jan-17-2035) and than again 2 years later (Jul-17-2037). The required velocity demand to perform the escape manoeuvre in mission ID-2 (4.34 km/s) is of the same magnitude as required for the back-up missions, i.e. mission ID-3 (4.30 km/s) and ID-5 (4.35 km/s). The final figure should numerically envelope ID-5 and can be considered as *growth potential* for ID-2.

## References


1. Adams, D. (2018). Whose stars? Our heritage of Arabian astronomy. https://www.planetary.org/articles/whose-stars-arabian-astronomy
2. Powell, B. P., Kostov, V. B., Rappaport, S. A., Borkovits, T., Zasche, P., Tokovinin, A., Kruse, E., Latham, D. W., Montet, B. T., Jensen, E. L. N., Jayaraman, R., Collins, K. A., Mašek, M., Hellier, C., Evans, P., Tan, T.-G., Schlieder, J. E., Torres, G., Smale, A. P., âŁ¦ Villasenor, J. (2021). TIC 168789840: A sextuply eclipsing sextuple star system. *The Astronomical Journal, 161*(4), 162. https://doi.org/10.3847/1538-3881/abddb5
3. Orchiston, W., & Drummond, J. (2020). Observations of the great September comet of 1882 (C/1882 R1) from New Zealand. *Journal of Astronomical History and Heritage, 23*, 628–658. https://doi.org/10.3724/SP.J.1440-2807.2020.03.10
4. Faure, G., & Mensing, M. (2007). *Introduction to planetary science: the geological perspective* (p. 526). https://doi.org/10.1007/978-1-4020-5544-7
5. Vallado, D. A. (2007). *Fundamentals of astrodynamics and applications* (3rd ed.). Berling: Springer.
6. Curtis, H. D. (2009). *Orbital mechanics for engineering students* (2nd ed.). Amsterdam: Elsevier Ltd.
7. Bate, R. R., Mueller, D. D., & White, J. E. (n.d.). *Fundamentals of astrodynamics*
8. Losure, M. (2017). *Isaac the alchemist: secrets of Isaac newton*. Somerville, MA: Candlewick Press. ISBN: 978-0763670634.

9. Spiridonova, S., & Kirschner, M., (2014, May). Precise mean orbital elements determination for LEO monitoring and maintenance.
10. King-Hele, D. G. (1987). *Satellite orbits in an atmosphere. Theory and applications*. https://api.semanticscholar.org/CorpusID:118625616
11. NASA Science Editorial Team. (2020). Milankovitch (orbital) cycles and their role in earth's climate. https://science.nasa.gov/science-research/earth-science/milankovitch-orbital-cycles-and-their-role-in-earths-climate/
12. Uphoff, C. (2001). The history of the term C3. https://cbboff.org/UCBoulderCourse/documents/history_c3.pdf
13. Britannica, T. (2024, April 15). *Kepler's laws of planetary motion*. Encyclopedia Britannica. https://www.britannica.com/science/Keplers-laws-of-planetary-motion
14. Belbruno, E. (1987, May). Lunar capture orbits, a method of constructing earth moon trajectories and the lunar GAS mission. In *Proceedings of AIAA/DGLR/JSASS International Electric Propulsion Conference*. AIAA Paper No. 87-1054
15. New Mexico Museum of Space History. (2024). https://www.nmspacemuseum.org/inductee/walter-hohmann/
16. Hohmann, W. (1925). Die Erreichbarkeit der Himmelskörper: Untersuchungen über das Raumfahrtproblem. R. Oldenbourg.
17. Tewari, A. (2007). *Atmospheric and space flight dynamics—modeling and simulation with MATLAB and Simulink*. Boston: Birkhäuser. ISBN: 978-0-8176-4437-6
18. New Mexico Museum of Space History. (2024). https://www.nmspacemuseum.org/inductee/yuri-vasilievich-kondratyuk/
19. Franc, T. (2011). The gravitational assist. In *WDS'11 Proceedings of Contributed Papers, Part III* (pp. 55–60). ISBN 978-80-7378-186-6 MATFYZPRESS

# Chapter 3
# Deep Space Conditions

**Abstract** We refer to the environment of interplanetary space as deep space, in contrast to the space in the vicinity of a planet, such as Earth, Venus, or Jupiter. This chapter highlights the dominant factors to consider when designing a mission that traverses interplanetary space, with emphasis on solar flux and harmful radiation. Solar flux is particularly important as it defines the habitable zone and the frost line, both of which are crucial in the search for water. In technical terms, solar flux is a decisive factor in the development of solar-electric propulsion systems. Radiation presents a severe challenge for both humans and machines.

## 3.1 Fire and Ice

The Sun is the dominant object in the solar system, in terms of gravitational pull as well as heat source. Being of spectral class G and surface temperature of about 5,800 K our Sun is in its prime. Born 4.6 billion years ago, the Sun has steadily increased its luminosity by about 10% on average each billion years. In about 2.4 billion years, give or take a few million years, its evolution will start to accelerate, due to the depletion of its prime energy source hydrogen deep in its centre. The Sun will rapidly develop towards a red giant swallowing the inner planets, Mercury, Venus and potentially Earth, too.[1] This evolution is accompanied by a stark increase in luminosity, $L_{\odot}$, thereby eating up Earth, which will then resemble Venus today.

The Sun is currently emitting $3.828 \times 10^{26}$ W into space of which 342 W/m$^2$ reaches Earth's surface on average, considering mean cloud coverage [1]. This sums up to $4.3 \times 10^{16}$ W for the entire sun-facing side of the Earth, which is 2130 times more than the instant average world power consumption of 2022,[2] $20.4 \times 10^{12}$ W. The potential is therefore immense and the concept of Space Based Solar Power (SBSP) has attracted the attention of several agencies and private companies. This is because the solar flux (Watt per square meter) is four times higher in Earth's vicinity

[1] An interesting theory is that the Earth will spiral away from the Sun's barycentre as the Sun loses mass due to the increased solar wind that is part of the evolution towards the red giant.

[2] https://ourworldindata.org/energy-production-consumption.



F. Gamgami, *Space Propulsion and Spaceship Design*, Springer Praxis Books,
https://doi.org/10.1007/978-3-031-71336-1_3

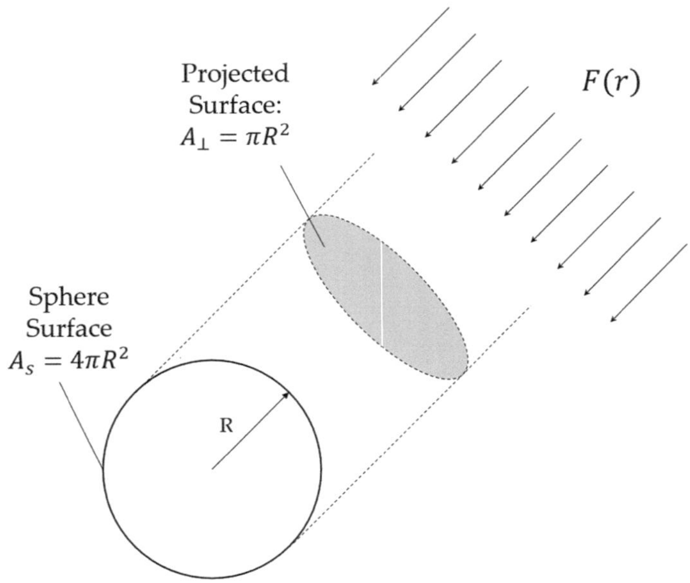


**Fig. 3.1** Black body of a sphere experiencing solar irradiance at distance r from the Sun

(1 AU), about 1350 W/m$^2$, leading to a factor of 8400 of received power to consumed power.[3]

A simple black body model of a sphere shows the heating caused by the Sun as depicted in Fig. 3.1. We assume that the sphere is painted with a perfectly deep black colour. The result is that all incident radiation will be fully absorbed, $Q_\text{a}$, which then heats the body. Since the sphere is placed in the cold universe[4] it will lose energy to its surrounding. The only way to do so in vacuum is again by radiation, $Q_e$. The energy balance yields:

$$\begin{aligned} Q_\text{a} &= Q_\text{e}, \\ \text{absorbed flux} &= \text{emitted flux}. \end{aligned} \tag{3.1}$$

Due to its perfectly deep black colour we may well assume that this happens entirely via thermal radiation according to the Stefan-Boltzmann law:

$$\alpha \cdot A_\perp \cdot F(r) = \epsilon \cdot \sigma \cdot A_\text{s} \cdot T_{eq}^4, \tag{3.2}$$

[3] The famous Russian astronomer, Kardashev (1932–2019) ranked civilizations in a scale from I to III. Type I civilizations are capable of using energy resources of a single planet. Type II civilizations use the full energy of a star. Type III civilizations have access to the energy of an entire galaxy [2].

[4] We ignore the microwave background temperature of 2.726 K caused by photons floating the universe since the big bang.

$\alpha$ absorption coefficient, ($\alpha = 1$) for a black body,
$F(r)$ solar flux at distance $r$, ($W/m^2$),
$\epsilon$ thermal emissivity, ($\epsilon = 1$) for a black body,
$T_{eq}$ equilibrium temperature.

Solving for the temperature yields:

$$T_{eq} = \left(\frac{\alpha}{\epsilon}\frac{F(r)}{4 \cdot \sigma}\right)^{1/4} \tag{3.3}$$

Obviously the solar flux, $F(r)$, dilutes while spreading into space and we need to consider that it is a function of distance, $F(r)$. The conservation of energy helps us to determine the flux dependency on distance $r$ based on the Sun's luminosity $L_\odot$:

$$L_\odot = 4\pi r^2 F(r), \tag{3.4}$$

$$F(r) = \frac{L_\odot}{4\pi r^2}. \tag{3.5}$$

Inserting the mean distance of Earth to Sun yields the well-known *solar constant*, 1350 $W/m^2$. Table 3.1 depicts both, the solar flux and the equilibrium temperature of a *black body* at different distances in the solar system.[5] Note that the Flux drops sharply with distance.

Jupiter for instance receives only 3.7% of Earth's solar flux and Neptune merely 1%. Jupiter is so far the maximum distance of what is currently doable by photovoltaic powered spacecraft. All missions with destinations beyond Jupiter have had nuclear based energy sources, that are distance independent. The Voyager space probes, for instance, had radioisotope thermoelectric generators (RTGs), so did Cassini-Huygens and New Horizon. In contrast, space probes with destinations close to the Sun require a heat shield to protect themselves from the high solar flux, like BepiColombo and the Parker Solar Probe.

### The Solar Frost Line

The decrease in solar flux affects the equilibrium temperature and implies an important concept known as the *frost line*, also called the snow line. Below this limit, water cannot exist when exposed to direct sunlight. This is due to the high-energy ultraviolet (UV) part of the solar spectrum, which is capable to dissociate water molecules into its constituents. The Earth's upper ozone layer protects life from harmful UV radiation. It is therefore the UV flux that determines the location of the frost line, beyond which the temperature is low enough that water and volatile elements

[5] Since the distance of Earth to the Sun changes over one revolution, the solar constant is actually not constant but changes periodically around this value, with a maximum of 1408 $W/m^2$ at perihelion and a minimum of 1314 $W/m^2$ at aphelion.

**Table 3.1** Mean solar flux and black body temperature at distances equivalent to celestial objects

| Solar distance (AU) | Sol. flux ($W/m^2$) | $T_{BB}$ (K) | Equivalent distance |
|---|---|---|---|
| 0.39 | 9012 | 446 | Mercury |
| 0.72 | 2581 | 327 | Venus |
| 1.00 | 1350 | 278 | Earth |
| 1.52 | 581 | 225 | Mars |
| 2.76 | 177 | 167 | Ceres |
| 5.20 | 50 | 122 | Jupiter |
| 9.57 | 15 | 90 | Saturn |
| 19.16 | 4 | 63 | Uranus |
| 30.18 | 1 | 51 | Neptun |

condense into their solid state. In principle, each element has its own frost line, with that of water estimated to be situated at approximately 4 AU–between the asteroid belt and Jupiter's orbit. There the UV flux is low enough to prevent sublimation (phase change from solid to gas form) and photo-dissociation. Despite the lack of a protective atmosphere, favourable conditions, however, allow the presence of water and volatiles within the frost line, which will be detailed in the next chapter.

## 3.2 Corpuscular Radiation

In addition to the stream of photons flowing continuously from our Sun, the Sun also emits high energetic particles (protons, electrons and heavy Ions) that travel through the solar system, forming the *solar wind*. According to the solar orbiter Ulysses, Fig. 2.10, wind speeds are not isotropic but depend on (a) the solar latitude, at high latitudes it reaches up to 800 km/s and is around 400 km/s around the ecliptic [3], and (b) the solar cyclic activity. The stream of these particles is referred to as *corpuscular*[6] *radiation* in distinction to radiation that consists of photons which are mass-less. Following common practice, we will refer to particle-based radiation as radiation only. In other parts of the book, when both effects are discussed in the same context, we will be more precise.

It frequently occurs that the Sun ejects large amounts of high energetic particles in a burst. The phenomenon is known as *coronal mass ejection* (CME). The particles are particularly energetic, up to 100 MeV (proton velocities above 3000 km/s) compared to the bulk energy of 10 keV of the steady particle stream [4]. We speak of a solar storm, if such an ultra high energetic cloud hits Earth. Solar storms can have multiple

[6] Corpuscule comes from latin and stands for small body. The term dates back to a natural philosophy dominant in the 17th century called Corpuscularianism. It believed that all phenomena can be explained by the interaction of particles. Newton used the term to establish a theory for optics.

negative effects. The high energetic particles can enter through the poles into our atmosphere and interfere with the terrestrial electrical grid causing blackouts in wide areas. The electronics in satellites can be destroyed by short circuits on the circuit boards (known as *latch-up*) and software *bit flips* (0 to 1 and vice versa) can cause unintended activation of the propulsion system or shut-down of other units with severe consequences. In addition to the harmful radiation from our Sun, there is also the *galactic cosmic radiation* (GCR). This radiation consists of heavy ions with extremely high energy levels of 100 MeV to 10 GeV exceeding those from CMEs by three orders of magnitude. As the name suggests GCR originates from cataclysmic events in the cosmos such as supernovae, pulsars, black holes, and active galactic nuclei.

In essence, this is a similar radiation that is released in nuclear reactions, which is lethal in high doses. Earth's magnetosphere as well as our atmosphere protects us and to a good extend also cosmonauts in Earth's vicinity from this deadly fate. However, a closer look reveals that this shield comes with a price: the magnetic fields trap these high energetic charged particles in specific radial regions around Earth forming the radiation belt, named after van Allen [5]. Satellites avoid this region and other spacecraft try to traverse this region fast to minimise the total ionisation dose (TID). The biological damages are: DNA damage leading to mutations and cancer as well as damages to the central nervous system [6]. Obviously, humans are not made for such harsh conditions and a major requirement for any human long-term extraterrestrial presence or even settlement will be the quick erection of protective shelter from the harsh radiation environment[7]–besides producing nutrition and propellant. So far only 24 humans have left Earth's protective shield for maximum 12 days and they were in fact lucky not to encounter a CME [7]. An encounter with such a high-energy cloud en route to the Moon or Mars would cause radiation poisoning similar to that by an atomic bomb or nuclear reactor failure, unless the crew is warned in time to seek shelter in special compartments.[8] Space components, too, must be shielded and designed in redundancy.[9]

Finally, it shall be emphasised that the total effect on humans and technology does not only depend on the energy level of the single particle but also on the flux and exposure time. Both factors drive the likelihood of a detrimental effect. An important quantity that makes the impact of radiation comparable is the *fluence*. It is defined as the integral of the flux of particles over a period of time:

---

[7] It is indeed an irony of fate that humans invent sophisticated technologies to venture out into the universe only to find a cave for shelter at the first opportunity.

[8] The key word is 'warning'. This is achieved by a dedicated network of ground-based observation telescopes and space-based satellites that constantly observe the Sun and establish predictive models. Global collaboration is essential to save human lives and space assets. This task belongs to the domain called Space Situational Awareness (SSA).

[9] Hydrogen and hydrogen-based protection is the preferred choice as shielding material. Heavier elements like lead cause secondary order elements, which are less but still harmful. New materials are currently subject of intense research to mitigate this threat and enable long duration crewed space travel [8].

$$\text{Fluence} = \int_0^{\Delta t} \frac{N}{A \cdot \Delta t} \mathrm{d}t. \tag{3.6}$$

$N$ number of particles,
$\Delta t$ time span,
$A$ penetrated surface.

Technical fluence values consider monoenergetic particles. We will discuss this quantity in detail in the context of solar array degradation in Sect. 8.3.

## 3.3 Physiological and Psychological Challenges of Space Travel

The physiological problems a human body faces in space is intense. In addition to the above discussed exposure to extremely high energetic radiation the body is also exposed to a low gravity environment, which during the transfer is actually zero. This causes muscle and bone loss and daily training will be essential to counteract the weightlessness.

Therefore, a legitimate and important question to be answered in the next years will be the acceptable mission duration for crewed interplanetary spaceflight. The first crewed mission to land on the Moon, Apollo 11, lasted in total not more than 8 days including a residence time of 21 h. With the Earth always in sight, the astronauts could take comfort in knowing that home was only a 3-day journey away. Assuming a Hohmann transfer, see Sect. 2.2.3, and optimal planetary alignment, a round-trip to Mars from Earth would require 1.5–2 years. The Earth, though still visible to the naked eye, would be no more than a blueish dot in the dark universe and the Sun would shrink to 62% of its angular diameter as seen from Earth–the small finger would be sufficient to cover it. Contact to Earth will be heavily delayed since the round-trip time for an electromagnetic wave from Earth to Mars varies between 3 to 22 min depending on the planetary constellation. This does not seem like a lot, but a live conversation with family and friends is not practical nor enjoyable. Instead, cosmonauts will most likely send messages and receive messages on a daily basis, before and after work.

The psychological challenges might seem insurmountable and very special training will be needed. Some psychologist suggest to recruit the first crews from the military as all space agencies did in the pioneering years of spaceflight. Nonetheless, given the fact that humans have demonstrated amazing psychological resilience in various occasions in history, the challenges of spaceflight will be surely met with human ingenuity and creativity.[10] The challenge will be to design a mission and system, that is robust enough to avoid or at least limit critical situations from the outset.

Irrespective of the technological developments in the next two decades, crewed interplanetary missions will always be limited by psychological and physiological constraints. These limits will drive the mission architecture, the development of innovative propulsion systems and ultimately the accessible destinations. Robotic missions on the other hand are more flexible and will take the role of scouting and technology demonstration in this harsh environment.

## References


1. NASA Facts. (2024). *The balance of power in the earth-sun system.* https://www.nasa.gov/wp-content/uploads/2015/03/135642main_balance_trifold21.pdf
2. *Astronomical and Astrophysical Transactions (AApTr), 31*(3), 399–402 (2019). ISSN 1055-6796. Photocopying permitted by license only, Cambridge Scientific Publishers
3. Wang, Y.-M., & Sheeley, N. (2008). Sources of the Solar Wind at Ulysses during 1990–2006. *The Astrophysical Journal, 653*, 708. https://doi.org/10.1086/508929
4. Papaioannou, A., Sandberg, I., Anastasiadis, A., Kouloumvakos, A., Georgoulis, M. K., Tziotziou, K., Tsiropoula, G., Jiggens, P., & Hilgers, A. (2016). Solar flares, coronal mass ejections and solar energetic particle event characteristics. *Journal of Space Weather and Space Climate, 6*, A42. https://doi.org/10.1051/swsc/2016035.
5. Kennel, C., & Petschek, H. (1969). Van Allen belt plasma physics.
6. Chancellor, J. C., Scott, G. B., & Sutton, J. P. (2014) Space radiation: the number one risk to astronaut health beyond low earth orbit. *Life (Basel), 4*(3), 491–510. PMID: 25370382; PMCID: PMC4206856. https://doi.org/10.3390/life4030491.
7. Nasa (2012) Space faring: the radiation challenge introduction and module 1: radiation, educator guide
8. Nasa (2012) Space faring: the radiation challenge module 3.


[10] Most of these occasions were extreme situation during war or airplane crashes in the wilderness in which the instinct of survival takes control. Mission control, however, needs rational thinking personal under life-threatening circumstances.

# Part II
# Cost and Reward of Space Exploration

# Chapter 4
# Inner Solar Resources

**Abstract** This chapter examines the structure and composition of the solar system from the perspective of scientific space exploration and commercial exploitation. It postulates that future solar trade routes will run along water resources, much like the early spread of humans across the world along rivers. The chapter covers the inner solar system with a focus on the potential for resource utilisation on the Moon, asteroids, comets, Mars and the dwarf planet Ceres.

## 4.1 Towards a Solar Ecosystem

Terrestrial resources and the engineering knowledge to harvest and use them have enabled the civilization we know today on Earth. The solar system harbours a multitude more resources than Earth and humanity will sooner or later learn how to extract them, too. This chapter is to provide an overview of the existing resources in the solar system and the cost in terms of delta-v to access these objects and areas in the inner solar system.[1] It is reasonable to assume that the first planetary outposts will be established where water can be extracted. Water is essential for humans to survive but also as rocket fuel. The first space stations will be close to where this precious resource is gained and will serve as refuelling stations. From then on, more planetary outposts will be built, primarily for raw material mining.

The Japanese Lunar Industry Vision Council has summarised this succinctly in a vision entitled 'planet 6.0' [1]. The council, composed of industry leaders, legislators, and scientists, intends to foster Japan's participation in cislunar space, which they describe as 'the frontline of a new space ecosystem'. The council has, therefore, coined the term planet 6.0 as the next step of humanity's societal evolution and expansion of its sphere of influence, depicted in Fig. 4.1. Such an alliance with this focus is indeed unique in the world and demonstrates foresight.

[1] Note that there is no universally defined boundary for the extent of the inner solar system. In this textbook, the inner solar system comprises the terrestrial planets and the asteroid belt, beyond which lies the realm of the gas giants.



F. Gamgami, *Space Propulsion and Spaceship Design*, Springer Praxis Books,
https://doi.org/10.1007/978-3-031-71336-1_4

**Fig. 4.1** Human societal and technological evolution according to the Japanese Lunar industry vision council [1]

It is worth delving a bit further into this concept. The vision describes the evolution of humanity through their understanding of the laws of the universe and their ability to apply them to shape the environment through technological progress. As a result, a dependency has been created at each technological evolutionary step, making life hard to imagine without these achievements, whether it be agriculture, industry, the information age, or soon, AI. Space for instance has already become part of our economy and life without a global navigation satellite system (GNSS) and satellite based communication and imagery is indeed hard to imagine. According to the council's vision, the subsequent phase will bring the Moon within humanity's sphere of influence. The exploitation of water ice and Helium 3 will create a sustainable and irreversible Earth-Moon economy,[2] dubbed as *cislunar economy*.

The flow of goods within the inner solar system will obey the principles of economic viability meaning that, generally speaking, it is uneconomical and certainly unsustainable to transport resources from Earth to the destinations of interest, e.g. Moon, Mars and Asteroids. Vice versa, it is not always economical to bring harvested resources back to Earth. Therefore, local resources will be assessed in view of the following two use cases:

(a) support continued local settlement,
(b) enhancement of an in-space economy,

The first is obvious and necessary to establish a sustainable base. The second case is needed to fuel an economic cycle that encompasses other celestial settlements, e.g. asteroid resources for a Mars colony, or lunar resources for Earth. An ecosystem that also encompasses the *outer solar system* is not feasible without a permanent and sustainable human presence within the inner solar system, which should extend as far as the asteroid belt. Furthermore, venturing into the deep spaces of the outer solar system requires advanced propulsion systems, which will be discussed in Chap. 11.

The discipline dealing with resource extraction from an extraterrestrial object is called ISRU, *in-situ resource utilisation*. Processes relevant for propellant extraction will be provided in the subsequent sections.

Fortunately, our solar system contains abundant precious resources. In the following we will identify and map resources of the inner solar system. This is a first and necessary step to understand future interplanetary trade routes. We can group them according to their use:

[2] The physicist Gerald Kulcinski, who is Professor of Nuclear Engineering-Emeritus and the Director of the Fusion Technology Institute at the University of Wisconsin-Madison, was the first to point out the advantage of Helium-3 compared to Deuterium and Tritium for nuclear fusion [3]. The latter generate high-energy neutrons, which contaminate the plant radioactively over time. This is not the case with Helium-3. However, the idea is not free from controversy and has also disadvantages, as pointed out by the German physicists Frank Close, Professor-Emeritus at the University of Oxford [4].

*consumables*: to be used as rocket propellant or drinking, breathing, agriculture,

*civil engineering*: to construct housing, radiation shelters, launch pad, roads and other infrastructure,

*energy production*: to power life support systems, ISRU facilities and other machinery,

*high-tech industry*: to manufacture goods in space.

As pointed out earlier, the first sustainable outposts will be established where water (i.e. consumables) can be directly found or relatively close to it in stable orbits: the Lagrange points in the Earth-Moon or Earth-Sun system, for instance. The other use cases will be developed gradually once this infrastructure is in place. A more generic grouping of resources of interest follows a geological perspective. We may distinguish between the following five categories:

*volatiles*: these are mainly gases, like hydrogen, helium, carbon and oxygen, but also compounds like $CO_2$ and $SO_2$, they can be processed to be used in life support systems, as rocket propellant, for energy production or other industrial processes,

*water*: the most precious resource in the universe for humans. It exists in many forms in our solar system.[3] On Moon, Mars and asteroids it occurs either in form of water ice or in form of hydrated minerals. The first form is more valuable as it is easier to access. Once purified, it can be used directly for drinking or it can be decomposed by electrolysis to produce oxygen for breathing and for agriculture. The obtained hydrogen can be used as rocket propellant fuel in a mono-propellant system or in combination with oxygen for a hydrolox based propulsion system,

*rare earth elements*: REEs such as lanthanum, cerium and neodymium are highly valuable for high-tech applications, e.g. electronics, magnets and lasers. The extraction of REEs is an inevitable step towards a technological high-end space-based industry,

*platinum group metals*: PGMs, such as iridium, osmium and platinum, are the most valuable resource after water. Due to their production cost on Earth, most researches agree that this could be the only extraterrestrial resource worth to bring back to Earth from an *economic viability* point of view [9],

*nuclear fuel*: like Uranium and Plutonium that can be used for power generation and propulsion in a nuclear reactor.

Due to the importance of water and volatiles as consumables for human survival and propellant for rocket engines, special focus will be given to these resources in the following sections.

## 4.2 The Moon

We shall briefly discuss the resources on our Moon, where they can be found and how they fit into the emerging space economy. Emphasis is given to consumables, especially rocket propellant fuels. It is trivial but important to highlight the non-uniform distribution of resources on the Moon. This will require a *massive infrastructure* stretching from the equatorial highlands, across the far side to the lunar south pole.

### 4.2.1 Volatiles

Due to the lack of a lunar atmosphere or a global magnetic field *solar wind* can penetrate lunar soil unhindered, creating so-called *solar wind implanted volatiles*.[4] Investigations of Apollo samples (382 kg were collected in total) from the old highlands and the younger mare have shown, that regolith is very efficient in retaining solar particles as listed in Table 4.1. To obtain most of these elements, it is necessary to heat up the regolith to 900 °C, while a temperature of 700 °C is sufficient to release trapped hydrogen and helium [7].

As pointed out by Crawford [26] all samples have been retrieved from low latitudes which could lead to the conclusion that the retrieved data might be biased. The reason is simple: lower latitudes experience a higher solar wind flux than high latitudes. In consequence the concentration of deposited volatiles should be larger. Crawford points out that an important counteracting effect: the regolith in higher latitudes has a lower temperature, which favours retention of volatiles, thereby balancing the surface distribution of implanted volatiles. More measurements are needed to understand where the equilibrium of these competing effects lies. Furthermore,

**Table 4.1** Concentrations of solar wind implanted volatiles in lunar regolith [8], errors within ± one standard deviation. Assumed bulk density is 1.66 g/cm$^3$ [10]

| Volatile | ppm ($\mu g/g$) | $\bar{g}/m^3$ |
|---|---|---|
| H | 46 ± 16 | 76 |
| C | 124 ± 45 | 206 |
| N | 81 ± 37 | 135 |
| F | 70 ± 47 | 116 |
| Cl | 30 ± 20 | 50 |
| $^4$He | 14 ± 11.3 | 23 |
| $^3$He | 0.0042 ± 0.0034 | 0.007 |

[4] The Moon exhibits several local magnetic anomalies of which the Descartes and the Reiner Gamma anomalies are the strongest. They are characterised by anomalous high-albedo markings and known as lunar swirls [2].

orbital neutron spectroscopy measurements (Sinitsyn, 2014), give reason to believe that the concentration of hydrogen reaches 100 ppm in some highlands areas. All in all, our neighbour is still terra incognita from a geochemical point of view but current measurements give reason for a promising future.

### Helium-3

Of the volatiles in the lunar regolith, $^3$He is perhaps the most valuable resource because of its potential as a fusion fuel.

$$D + {}^3\mathrm{He} \longrightarrow {}^4\mathrm{He} + p + 18.4\,\mathrm{MeV}, \tag{4.1}$$

D deuterium (heavy hydrogen), a stable hydrogen isotope with 1 proton and 1 neutron in its core,
$^3$He helium-3, a stable isotope of helium, with 2 protons but only a single neutron in its core.

Following [5] a regolith thickness of 3 m together with a concentration of 20 ppb of $^3$He for the Mare Tranquillitatis and Oceanus Procellarum could be assumed. This is a total mass of $2 \times 10^8$ kg and corresponds to an energy reservoir of $5 \times 10^{16}$ MJ. Assuming an average terrestrial consumption of 877 kWh/month, this is enough energy to power 6 Million households for 1000 years. The advantage of using $^3$He in nuclear fusion lies in the ability to directly convert plasma energy into electricity. This is in stark contrast to traditional deuterium- tritium fusion $(D - T)$. In the latter case an energetic neutron is generated that in turn heats a medium and electrical energy is generated via the detour of thermal energy, see Sect. 11.3. In contrast, fusion processes in which not more than 1% of the released energy is carried away by neutrons is referred to as *aneutronic*. The above described $D - {}^3$He process belongs to this category [6], which brings a further advantage: high energetic neutrons have detrimental effects on the surrounding structure and their deceleration by a moderator (e.g. water) causes harmful gamma rays. Both effects are avoided by default in $D - {}^3$He fusion.

To generate electrical energy, so called direct conversion technologies are required for aneutronic nuclear fusion [12]:

- direct electrostatic conversion,
- direct electromagnetic conversion,
- high temperature thermal cycles using microwave heating.

It needs to be stressed here that these advantages come with significant implications: $D - {}^3$He fusion requires 4 times higher plasma temperature and plasma confinement than $D - T$ fusion. Furthermore, the power density is 50 times lower which would require a plasma volume that is 50 times larger or a magnetic field that is 2.6 times larger than in case of $D - T$ fusion to achieve the same fusion power [12]. Finally,

**Fig. 4.2** Shackleton crater at the lunar south pole. Credit: European Space Agency—ESA, Jorge Mañes Rubio in collaboration with DITISHOE

it shall be mentioned that to harvest 1 kg of $^3$He a total of 38 metric tonnes of lunar regolith need to be processed [12].

### 4.2.2 Water

The frost line discussed in Sect. 3.1 speaks against the existence of water at the surface of *airless bodies*[5] within the inner solar system. There are two ways in which water can exist in this hostile environment, namely in form of ice in cold regions inside permanently shadowed craters, see Fig. 4.2, and in mineral hydrated form. Both shall be discussed briefly in the following.

Water Ice

The small tilt of airless bodies like the Moon, Ceres and even Mercury in combination with deep craters cause topographic depressions with permanently shadowed and thus cold regions [13]. These regions are referred to as *cold traps*. Arguably the most prominent of these cold traps is the Shackleton crater depicted in Fig. 4.2. Direct evidence of surface exposed water ice on the Moon was found with help of the NASA instrument M(3), Moon Mineralogy Mapper, onboard the Indian lunar

[5] 'Airless body' is a term used by planetologists to describe a celestial body that has no significant atmosphere.

orbiter, Chandrayaan-1. As anticipated, the results show that water ice does indeed exist at the lunar poles. The good news is that these cold traps, contain up to 30 wt% of water ice mixed with regolith. The less good news is that only 3.5% of the cold trap area contains water ice [14].

In absolute figures, a total of up to 6.6 billion tons of water ice could be deposited at the poles [15]. However, there is a catch to the matter hidden in the word *permanently shadowed*. These locations epitomize pitch darkness. The interior of the steep craters, $\simeq 40°$, lies in eternal darkness with an estimated temperature of less than 110 K (−163 °C). These extreme conditions will pose enormous challenges for humans and machines to work there.

#### Hydrated Minerals

Besides this water ice, there is the possibility to extract water from the regolith itself. In certain areas regolith contains significant amount of OH and $H_2O$ bearing minerals. The German-US flying observatory SOFIA (Strategic Observatory for Infrared Astronomy) has discovered water even in sunlit areas such as the Clavius crater. If processed, one cubic meter (1.6 t) of regolith could yield 355 ml of water. These numbers legitimately raise the question of the economic viability of processing hydrated minerals in view of energy and machinery demand. Given high concentration of water ice at the poles, hydrate minerals will not be the first choice of lunar water exploitation.

## 4.3 Asteroids

A significant number of asteroids exist in our solar system, and it is believed that they are predominantly remnants of a cataclysmic event that occurred approximately 3.8 billion years ago. This event is thought to have destroyed the majority of planetesimals and proto-planets at the early stages of solar system formation. Those Asteroids in close proximity to Earth are referred to as Near Earth Asteroids (NEAs). They are of importance for two reasons: firstly, they could pose a threat to the human civilisation in case of a collision with Earth and secondly, they are relatively easy to access and exploit.

Asteroids have been classified into several types and subtypes, each possessing distinct characteristics in terms of composition. We will focus on the three most common types.

*C-type* asteroids contain hydrated minerals and other contained volatile elements, organic carbon and phosphorus. Their mean density is about 1.7 g/cm$^3$ which is close to the regolith density of 1.66 g/cm$^3$ on the Moon. A prominent example is 101955 Bennu,

*S-type* asteroids consist predominantly of silicates and metals, such as iron-silicates or magnesium-silicates. They have a mean density of 3 g/cm$^3$, which is slightly above Earth's average crust density, 2.6 g/cm$^3$. A prominent example is 433 Eros.,

*M-type* asteroids have 10 times more metals than S-type do. In addition, they possess precious elements like cobalt, gold, platinum and rhodium. They are believed to be remaining cores of early failed planet formation. Their mean density is 5.3 g/cm$^3$. Prominent examples are Psyche 16 and 22 Kalliope.

C-type and M-type asteroids are obviously interesting for resource mining. The good news about the two types is their abundance in the solar system, approximately 75% of all objects in the asteroid belt are volatile-rich C-type asteroids and 8% are metal-rich M-type asteroids [16]. The less encouraging aspect of this situation is that the asteroids with the highest concentration of water are primarily situated at the periphery of the asteroid belt, at a distance of 3.5 AU, in proximity to the frost line. Furthermore, they have a low albedo, between 0.03 and 0.10, which means they reflect only 3% to maximum 10% of the already little light they receive from the Sun—dark and cold objects are hard to detect in the universe.

### Evidence for Water

Presence of water on asteroids been identified using a range of techniques, including analyses of asteroid spectra and in-situ measurements by space probes. Two prominent examples of sample-return mission are NASA's recent OSIRIS-REx mission to the near-Earth asteroid Bennu (September 2023) and JAXA's Hayabusa 2 mission[6] to Ryugu (December 2020). Bennu is a B-type asteroid which is a sub-group of C-type. It has a mass of $7 \times 10^{10}$ kg and first conservative remote sensing data suggests a water content of 1% by mass leading to 700,000 tonnes of water, if fully processed.[7] This number[8] could even increase as less conservative models yield a 5–10 times higher water content in addition to water ice deposits at the poles [17]. These results give us confidence that there are more water rich asteroids in Earth's vicinity than

---

[6] OSIRIS-REx spacecraft had a launch mass of 2.1 t of which 1.3 t was propellant, Hayabusa2 had a launch mass of 600 kg of which 110 kg was propellant mass.

[7] This is equivalent to the complete destruction of the asteroid, which raises ethical questions about the responsible use of resources inn our solar system. Given the amount of micrometeoroid creation there is also the question of sustainability that has to be addressed by a future legal framework.

[8] OSIRIS-REx brought back a sample of 250 grams from Bennu and preliminary analysis indeed suggest a higher $H_2O$ content of this water-bearing clay. Further investigation of these samples may answer fundamental questions regarding the origin of life on Earth based on the content of water and organic compounds on these asteroids.

**Table 4.2** Time line Psyche Mission [18]

| Event | Time line |
|---|---|
| Launch | Aug. 2022 |
| Mars gravity assist (500 km) | May 2023 |
| Psyche capture (2.7 AU) | Jan. 2026 |
| End of orbit ops (3.3 AU) | Oct. 2027 |

expected. Some might even house water ice well protected beneath a mantle of dust and clay. Therefore, characterization and mapping of NEAs in the near future will be of utmost importance for asteroid mining and sustainable space exploration.

Metal-Rich Asteroids

M-type asteroids on the other hand are true treasure chambers. The quantity of rare metals, especially platinum, is immense and prompted the famous astrophysicist Neil deGrasse Tyson to make the statement, 'The first trillionaire there will ever be, is the person who exploits the natural resources on asteroids'. The asteroid 16 Psyche has been reported to contain US $700 quintillion worth of gold, enough for every person on earth to receive about US $ 93 billion, assuming no value depreciation of gold [11]. These numbers show in a compelling way the fortune that is out there.[9] It is therefore unsurprising that the first private companies have emerged that are working on the commercial exploitation of these extraterrestrial resources.

NASA has sent 2023 a Spacecraft of the same name as its destination, Psyche, to further investigate this peculiar planetesimal. Table 4.2 shows the mission profile of Psyche. Note that despite a gravity assist manoeuvre at Mars, the journey still takes 5 years and 10 months until Psyche is reached, in August 2029. The distance to the Sun will then be 2.7 AU and at the end of the mission, in November 2031, the distance will have increased to 3.1 AU. Hence, a commercial endeavour will take decades before any return on investment materialises. A public private partnership model including risk- and profit sharing might a viable approach.

## 4.4 Comets

Comets are generally referred to as dirty snow balls. The content of water ice, volatiles and rock differs from comet to comet but this depiction is in general correct. Measurements of comet's bulk density varies from 0.3 to 1.5 g/cm$^3$ with a peak around

[9] Value depreciation will inevitably occur, and depending on the cost of extracting these now less rare minerals, a significantly lower price will result. Nonetheless, the technical devices that require these metals would also become cheaper.

0.6 g/cm$^3$ [19]. This is lower than the density of water ice, 0.9 g/cm$^3$, and suspiciously close to that of alpine glaciers [20]. The rock content of most comets must therefore be very small and the objects are rather made of water ice in which a significant amount of volatiles is trapped.

There are about 4000 known comets in our solar system. The majority of them originates from the cold outer regions well beyond the frost-line, presumably from the Kuiper belt (50 AU). This explains the very long orbital periods, like the famous Halley's comet whose period is 75 years. Others coming from the Oort cloud with periods of thousands of years, Sect. 2.1. They belong to the long-period group. The group of short-period comets, with orbital periods shorter than 200 years are further subdivided into the Halley-type comets with periods longer than 20 years but shorter than 200 years and the Jupiter-family comets with periods shorter than about 20 years [21]. On September 2014, the Rosetta Spacecraft of the European Space Agency visited the comet 67P/Churyumov-Gerasimenko and studied the object from close proximity for 2 years. The comet has a perihelion that is within Mars' orbit and an aphelion slightly above that of Jupiter. One important scientific results was the in-situ measurement of the comet's bulk density. The results deduced from gravitational attraction by the comet yielded a value of $0.532 \pm 0.007$ g/cm$^3$ [22], which is almost half of the water ice density, 0.916 g/cm$^3$, indicating a high amount of volatiles.

Hence, exploiting water ice and volatiles from comets in general promises to be more rewarding, from a pure compositional point of view, than asteroids. There are severe challenges, however. Firstly, they have in general highly elliptical orbits, which are much more difficult to reach by a spacecraft than asteroids. Secondly, a direct consequence of the high eccentricity are the varying environmental conditions on comets. 67P/Churyumov-Gerasimenko, for example, has an aphelion just beyond Jupiter's; at these distances, the temperature drops significantly and the available light from the Sun is extremely low, making resource extraction impossible with current technology. Close to perihelion the comets become active and start outgassing forming the characteristic tails, which endanger machines and operation on the surface. Another challenge is the short and infrequent departure windows (outbound and inbound) accompanied with high elliptical orbits and the associated long turnaround times. These challenges surely limit the number of suitable comets for resource exploitation but the reward might be worth the hustle.

## 4.5 Mars

There are several parameters that could make the planet habitable to humans; a surface gravity of 3.71 m/s$^2$ (38% of Earths), a solar day of $24^h\ 37^m$ and an average midday temperature of 27 °C at the equator in Martian summer. But the most striking argument for settling on Mars are the water resources on the planet. The planet is rich in water ice. In fact, if all water on Mars melted, it would create an ocean of approximately 35 m depth covering the entire planet. All in all, the planet holds abundant resources that will enable *living off the land*, meaning that all resources are available to support a permanent human presence. Unlike the objects discussed previously, Mars is a veritable planet whose gravitational pull allows it to have an atmosphere and the tilt of its rotational axis allows seasons. If the planet had a magnetic field similar to Earth's, its atmosphere could have been thicker, allowing liquid water to exist and potentially enabling the development of life. The search for life or remnants of life on the Red Planet is motivated by the possibility that this was in fact the case at some point of the planet's previous evolution. There is ample evidence, geological and mineralogical, that water flowed on the surface of Mars [23].

The current absence of a strong magnetic field has left the planet unprotected against the solar flux, especially the UV radiation, and resulted in a relatively thin atmosphere with a surface pressure 160 times lower than on Earth. This low pressure causes water ice to sublimate immediately when exposed to the atmosphere except at the poles where temperature is low enough to prevent this mechanism. Water can be therefore found either at the poles or in the soil. The latter is called subsurface ice. We will discuss both occurrences in the following.

#### Subsurface Ice

Large amounts of hydrogen have been discovered in the upper layers of the Martian soil by NASA's Mars Odyssey gamma-ray and neutron spectrometers. The concentration and distribution are depicted in Fig. 4.3. The detection of hydrogen is a marker for water. The measurements show water concentrations above 20% in latitudes beyond $\pm 60$ deg and up to 100% at the poles. Note that the map does not distinguish between hydrated minerals and water ice. Closer to the equator, higher water concentrations are less frequent but irregular. At the first glance, it seems that the Martian soil is indeed dust dry at the equator but images from NASA's Mars Reconnaissance Orbiter have revealed the contrary. The spacecraft was able to detect fresh impact craters and the aftermath of the impact. The high resolution camera (HiRISE) witnessed the emergence of bright white patches of ice inside the crater or around it. The spacecraft also witnessed how the ice sublimated away after a few days. This is a strong prove that subsurface water ice is more wide spread than so far suggested by the data [24].

#### Polar Ice

The amount of water at the Martian poles is very large. The northern cap has a diameter of 1000 km and a mean thickness of 2 km. It contains $1.6 \times 10^6\,\text{km}^3$ of water [25], twice the ice volume of the Greenland ice sheet. The southern cap is smaller in diameter, 350 km, and has a similar mean thickness of 3 km. Both caps undergo seasonal changes of which the water ice is unaffected but impacts the dry ice. $CO_2$, at the poles. In their respective winters, both poles are covered under a layer

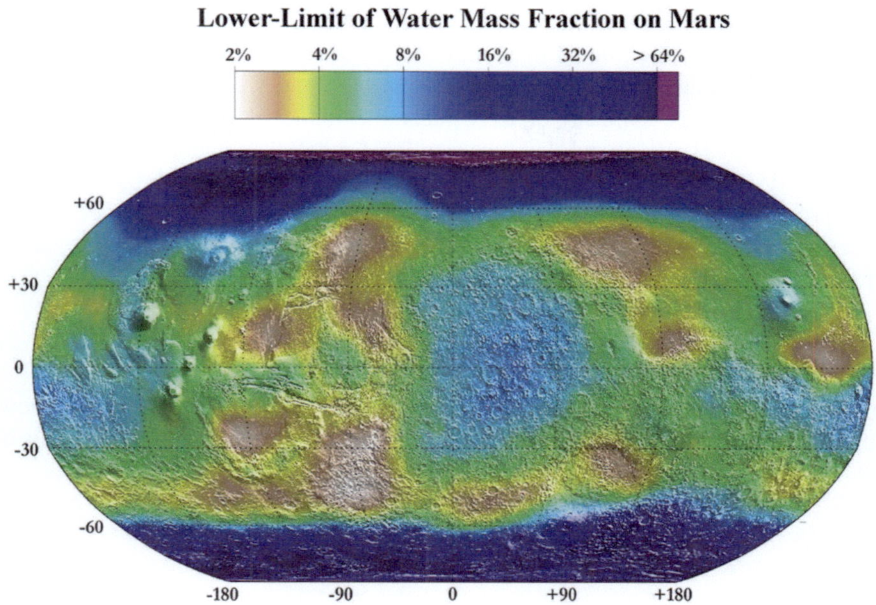


**Fig. 4.3** Distribution of ground water measured by Mars-Odyssey's neutron spectrometer component of the gamma ray spectrometer suite. Credit: NASA/JPL/Los Alamos National Laboratory

of 1.5–2 m of dry ice which completely sublimates again in summer. Measurements have shown that one third of the $CO_2$ atmosphere condenses during this process.

The probably most exciting discovery in recent years has been achieved by an Italian science team using the Mars Advanced Radar for Subsurface and Ionosphere Sounding (MARSIS) aboard Mars Express of ESA. They found a salty subglacial lake approximately 1.5 km beneath the surface close to the northern cap [27]. This is a significant discovery that gives new impetus to the search for extraterrestrial life.

### Resource Utilisation

The elevation of both caps differs, which has meteorological implications and affects the accessibility of their resources. The southern cap sits on a plateau 1000 m above the surface and stretches 3500 m up. Contrary, the ice cap in the north lays in a valley with a base at −5000 m and stretches up to −2000 m [28]. Landing directly on high altitude plateaus poses a particular technical challenge for the landing system. Parachutes require a dense atmosphere and are only capable to decelerate a limited amount of mass per deployed square meter of parachute and total available deceleration path. The already low Martian atmospheric density in combination with the reduced deceleration path to ground limit strongly the mass that can be landed by this method on the southern plateau. Accessibility and amount of available resources favour initial water exploitation in the northern hemisphere close to the cap. A challenge for the operating mining machines and all related operations will be the low temperature and the short sunlight hours.

The private space company SpaceX plans to colonise Mars and extract Oxygen and Methane, $CH_4$, to propel their spacecraft, Starship. The intention is to harvest water by ISRU from the soil and to collect carbon dioxide from the atmosphere. The process by which $CH_4$ is gained is the often discussed Sabatier process. It is a catalytic chemical reaction in which hydrogen gas reacts with carbon dioxide to produce methane and water vapour:

$$4\,H_2 + CO_2 - > 2\,H_2O + CH_4. \tag{4.2}$$

A catalyst is required, for instance nickel, to facilitate the reaction at relatively low temperatures. Furthermore, oxygen and hydrogen could be gained by electrolysis and also used as propellant.

All in all, given these striking evidences of the existence and amount of Martian water ice, it is obvious that the planet will play a pivotal role in the human exploration plans. Mars will become a transshipment point and the first permanent and sustainable extraterrestrial settlement discovery of the solar system.

## 4.6 Ceres

Ceres has been promoted by the International Astronomical Union (IAU) from an asteroid to a dwarf planet. With a diameter of 940 km, it is the biggest object of the main asteroid belt followed by other dwarf planets with sonorous names such as Pallas, Vesta and Hygiea. Encompassing 35% of the total asteroid belt mass, one may consider Ceres as a failed planet, hampered in its evolution by the disturbing gravitational forces of Jupiter.

Ceres exhibits a small axis tilt, called obliquity, of merely 4 deg. Like for the Moon this creates permanently shadowed craters that form cold traps for water ice. The water was either directly deposited by impacting comets or migrated there via the exosphere [29]. Dawn's data support the assumption of subsurface water ice at a shallow depth. First, its bulk density is 2.08 g/cm$^3$. This is a very low value, below that of silicates (2.65 g/cm$^3$ for quartz to 3.37 g/cm$^3$ for olivine) and metals (2.69 g/cm$^3$ for Aluminium to 8.9 g/cm$^3$ for Nickel). Models with reasonable compositions of silicate, metals, water ice and volatiles indicate a 25% water ice content at the cold poles [30]. The surface distribution of water equivalent hydrogen has been measured with Dawn's Gamma Ray and Neutron Detector (GRaND). The results show a concentration of up to 29wt.% (percent per weight) at the poles. Second, in-situ gravitational measurements by Dawn suggest moments of inertia that are only compatible with a differentiated structure.[10] Meaning that heavier elements, metals and silicates, migrated to the centre while lighter elements, water ice and volatiles, remained close to the surface. The abundance of water ice on Ceres makes this object an important outpost and resource source in the outer rim of the inner solar system. Being an airless dwarf planet with an escape velocity of merely 516 m/s, it is energetically cheap to land and leave the object. Resources are therefore easy to access once the expedition is there. Ceres will therefore play a vital role in any exploration architecture, serving as a transshipment hub that bridges the inner and outer solar system.

## References


1. Lunar Industry Vision. (2024). https://www.lunarindustryvision.org/about
2. Hood, L. L. (2021). Lunar magnetic anomalies. In B. Cudnik (Ed.), *Encyclopedia of Lunar Science* (pp. 1–8). Springer International Publishing. https://doi.org/10.1007/978-3-319-05546-6_4-1
3. Taylor, L. A., & Kulcinski, G. L. (1999). Helium-3 on the moon for fusion energy: the Persian Gulf of the 21st century. *Solar System Research, 33*, 338.
4. Frank Close. (2007). Fears over factoids. https://physicsworld.com/a/fears-over-factoids/
5. Fa, Wenzhe, & Jin, Ya-Qiu. (2007). Quantitative estimation of helium-3 spatial distribution in the lunar regolith layer. *Icarus., 190*, 15–23. https://doi.org/10.1016/j.icarus.2007.03.014
6. Chapman, J. J. (2011). Advanced fusion reactors for space propulsion and power systems. *IEEE International Conference on Plasma Science*, 1–1. https://doi.org/10.1109/PLASMA.2011.5992998
7. Haskin, L. A., & Warren, P. (1991). Lunar chemistry. In G. H. Heiken, D. Vaniman, & B. M. French (Eds.), *The Lunar sourcebook: A user's guide to the Moon* (pp. 357–474). Cambridge: Cambridge University Press.
8. Fegley, B., & Swindle, T. D. (1993). Lunar volatiles: implications for lunar resource utilization. In J. Lewis, M. S. Matthews, & M. L. Guerrieri (Eds.), *Resources of near earth space* (pp. 367–426). Tucson: Tucson University Press.
9. Cannon, K. M., Gialich, M., & Acain, J. (2023). Precious and structural metals on asteroids. *Planetary and Space Science, 225*, e105608. https://doi.org/10.1016/j.pss.2022.105608
10. Carrier, D. W., Olhoeft, G. R., & Mendell, W. (1991). Physical properties of the lunar surface. In G. H. Heiken, D. Vaniman, & B. M. French (Eds.), *The Lunar sourcebook: A user's guide to the Moon* (pp. 475–594). Cambridge: Cambridge University Press.
11. Gamillo, E. (2022). Smithsonian Magazine—Smart News. https://www.smithsonianmag.com/smart-news/asteroid-16-psyche-may-be-worth-more-than-planet-earth-at-10-quintillion-in-fine-metals-180979303/
12. Lunar Helium-3 and Fusion Power, NASA Technical Reports Server (NTRS). (2013). https://ntrs.nasa.gov/citations/19890005471 (accessed 29-03-2024)
13. Urey, H. C., & Korff, S. A. (1952). The planets: Their origin and development. *Physics Today, 5*(8), 12–12. https://doi.org/10.1063/1.3067687
14. Li, Shuai, Lucey, Paul, Milliken, Ralph, Hayne, Paul, Fisher, Elizabeth, Williams, Jean-Pierre., Hurley, Dana, & Elphic, Richard. (2018). Direct evidence of surface exposed water ice in the lunar polar regions. *Proceedings of the National Academy of Sciences, 115*, 201802345. https://doi.org/10.1073/pnas.1802345115


[10] A differentiated structure is a primary distinguishing criterion between (dwarf) planets and asteroids, alongside the spherical shape.

15. NASA Goddard Space Flight Center. (2024). Ice on the moon—A summary of clementine and lunar prospector results. https://nssdc.gsfc.nasa.gov/planetary/ice/ice_moon.html
16. MIT—Massachusetts Institute of Technology. (2024). Asteroid mining. https://web.mit.edu/12.000/www/m2016/finalwebsite/solutions/asteroids.html
17. Nuth, J. A., Abreu, N., Ferguson, F. T., Glavin, D. P., Hergenrother, C., Hill, H. G. M., Johnson, N. M., Pajola, M., & Walsh, K. (2020). Volatile-rich asteroids in the inner solar system. *The Planetary Science Journal, 1*(3), 82. https://doi.org/10.3847/PSJ/abc26a
18. Oh, D. Y., Collins, S. M., Goebel, D. M., Hart, B., Lantoine, G., Snyder, S., Whiffen, G. J., Elkins-Tanton, L. T., Lord, P. W., Pirkl, Z., & Rotlisburger, L. (2017). Development of the psyche mission for NASA's discovery program. https://api.semanticscholar.org/CorpusID:31454933
19. Paradowski, M. L. (2022). A new indirect method of determining density of cometary nuclei. *Acta Astronautica, 72*(2), 141–159. https://doi.org/10.32023/0001-5237/72.2.4
20. Donn, B. (1963). The origin and structure of icy cometary nuclei. *Icarus, 2*, 396–402. https://doi.org/10.1016/0019-1035(63)90068-X
21. Lowry, S., Fitzsimmons, A., Lamy, P., & Weissman, P. (2008). Kuiper Belt objects in the planetary region: The jupiter-family comets. The Solar System Beyond Neptune.
22. Jorda, L., Gaskell, R., Capanna, C., Hviid, S., Lamy, P., Ďurech, J., Faury, G., Groussin, O., Gutiérrez, P., Jackman, C., Keihm, S. J., Keller, H. U., Knollenberg, J., Kührt, E., Marche, S., Mottola, S., Palmer, E., Schloerb, F. P., Sierks, H., ... Wenzel, K. P. (2016). The global shape, density and rotation of Comet 67P/Churyumov-Gerasimenko from preperihelion Rosetta/OSIRIS observations. *Icarus, 277*, 257–278. https://doi.org/10.1016/j.icarus.2016.05.002
23. Bibring, J.-P., & Langevin, Y. (2008). Mineralogy of the Martian surface from Mars Express OMEGA observations. In J. Bell III. (Ed.), *The Martian surface-composition, mineralogy, and physical properties* (pp. 153–168). New York, NY, USA: Cambridge University Press.
24. Stuurman, C. M., Osinski, G. R., Holt, J. W., Levy, J. S., Brothers, T. C., Kerrigan, M., & Campbell, B. A. (2016). SHARAD detection and characterization of subsurface water ice deposits in Utopia Planitia, Mars. *Geophysical Research Letters, 43*, 9484–9491. https://doi.org/10.1002/2016GL070138
25. Carr, M. H., & Head III, J. W. (2003). Oceans on mars: An assessment of the observational evidence and possible fate. *Journal of Geophysical Research: Planets, 108*(E5). https://doi.org/10.1029/2002JE001963
26. Crawford, I. (2015). Lunar resources: A review. *Progress in Physical Geography, 39*, 137–167. https://doi.org/10.1177/0309133314567585
27. Orosei, R., Ding, C., Fa, W., Giannopoulos, A., Hérique, A., Kofman, W., Lauro, S. E., Li, C., Pettinelli, E., Su, Y., Xing, S., & Xu, Y. (2020). The global search for liquid water on mars from orbit: Current and future perspectives. *Life, 10*(8). https://doi.org/10.3390/life10080120
28. Fishbaugh, K. E., & Head, J. W. (2001). Comparison of the North and South polar caps of mars: New observations from MOLA data and discussion of some outstanding questions. *Icarus, 154*(1), 145–161. https://doi.org/10.1006/icar.2001.6666
29. Zhang, J. A., & Paige, D. A. (2009). Cold-trapped organic compounds at the poles of the Moon and Mercury: Implications for origins. *Journal of Geophysical Research Letters, 36*(16), eL16203. https://doi.org/10.1029/2009GL038614
30. De Sanctis, M. C., Ammannito, E., Raponi, A., Marchi, S., Mccord, T., McSween, H., Capaccioni, F., Capria, M., Carrozzo, F., Ciarniello, M., Longobardo, A., Tosi, F., Fonte, S., Formisano, M., Frigeri, A., Giardino, M., Palomba, E., Turrini, D., & Russell, C. T. (2015). Ammoniated phyllosilicates with a likely outer Solar system origin on (1) Ceres. *Nature, 528*, 241–244. https://doi.org/10.1038/nature16172
31. Prettyman, T., Yamashita, N., Toplis, M., McSween, H., Schorghofer, N., Marchi, S., Feldman, W., Castillo-Rogez, J., Forni, O., Lawrence, D., Ammannito, E., Ehlmann, B., Sizemore, H., Joy, S., Polanskey, C. A., Rayman, M., Raymond, C., & Russell, C. (2016). Extensive water ice within Ceres' aqueously altered regolith: Evidence from nuclear spectroscopy. *Science, 355*. https://doi.org/10.1126/science.aah6765

# Part III
# Space Propulsion Technology and Architecture

# Chapter 5
# Taxonomy and Fundamentals of Space Propulsion

**Abstract** Space propulsion systems perform a variety of tasks, each with unique requirements reflected in a thrust range of over eight orders of magnitude. These systems employ diverse technical solutions, including chemical combustion, nuclear heating, and electrostatic acceleration. This chapter aims to introduce these propulsion systems, explaining their basic working principles and typical applications. To facilitate understanding and innovation, a categorisation scheme that encompasses all conceivable propulsion systems is presented. This framework offers order and guidance for future developments in space propulsion technology.

## 5.1 Space Propulsion Taxonomy

Despite the extensive literature on space propulsion, there is currently no standardised scheme for categorising and classifying the underlying principles and technologies utilised in this field. Reasons for this lack are potentially manifold—but most likely engineers do not perceive the necessity to discuss commonplace aspects within their field plus the fact that actually *all* currently used propulsion systems are based on *mass ejection* of either chemical or electrical energy sources. Notwithstanding this general reluctance, establishing a taxonomy of propulsion principles and technologies is helpful in navigating the myriad of applications and concepts within the field.[1] Moreover, such a taxonomy may serve as a catalyst for driving innovation. Figure 5.1 shows the proposed categorisation scheme. It follows a matrix structure and shall be read outside-in. The intersections of horizontal and vertical lines within the matrix create nodes, which serve to represent the propulsion categories. The vertical category indicates the *principles of acceleration*, whereas the horizontal category represents *sources of energy*. The principles of acceleration can be further classified into two distinct classes:

[1] Taxonomy plays an important role in fundamental physics since the right scheme may have the potential to shed light into the hidden structure of nature. Even if space engineers are not seeking *the* truth of nature, it is believed that a sound categorisation scheme is helpful to navigate through the sometimes confusing forest of space propulsion systems.

 
F. Gamgami, *Space Propulsion and Spaceship Design*, Springer Praxis Books,
https://doi.org/10.1007/978-3-031-71336-1_5

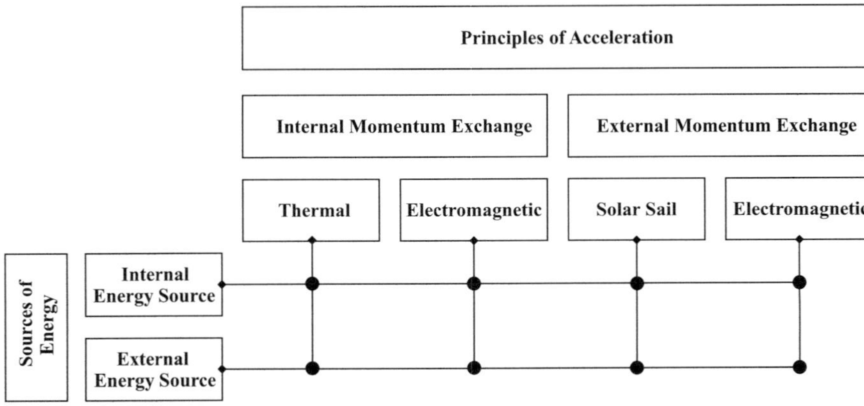


**Fig. 5.1** Space propulsion categorisation in internal and external momentum exchange, acceleration mechanism and location of energy source

(a) internal momentum exchange,
(b) external momentum exchange,

Some combinations (cross sections of the matrix) as we will see below are not yet flight proven and still subject to theoretical and experimental investigation. Others are in principle feasible but not (yet) efficient. The generic propulsion architecture of the two classes and the energy sources with which they can be combined will be explained in the subsequent sections.

The energy source can be either inside the spaceship or outside. The solar flux collected by a solar generators is a prime example for an external energy source. Contrary, the term internal energy source refers to an energy source inside the spaceship. There is, however, a further differentiation to be made, namely as to whether the source is within the propellant and released in a chemical reaction, or released by an external device (e.g. nuclear reactor) and *transferred* to the propellant. In the first case we say the energy is *propellant intrinsic* in the latter it is *propellant extrinsic*. This distinction is important to fully grasp all forms of the famous Tsiolkovsky equation as will be detailed in Sect. 6.2.

The main focus of this book is the first category, *Internal Momentum Exchange Drive*, as it forms the bulk of current space propulsion applications. The second category *External Momentum Exchange Drive* is still in an experimental stage and its application restricted to small system masses. We will present this field in Sect. 5.7.

Internal Momentum Exchange Drive—IMED

The fundamental working principle of propulsion systems based on internal momentum exchange is to *eject* propellant from the spaceship. Consequently, these systems *consume* onboard mass. Hence, mass ejection and mass consumption are two fundamental features of these propulsion systems. According to the law of *momentum conservation*, the momentum of the ejected propellant creates a momentum in the

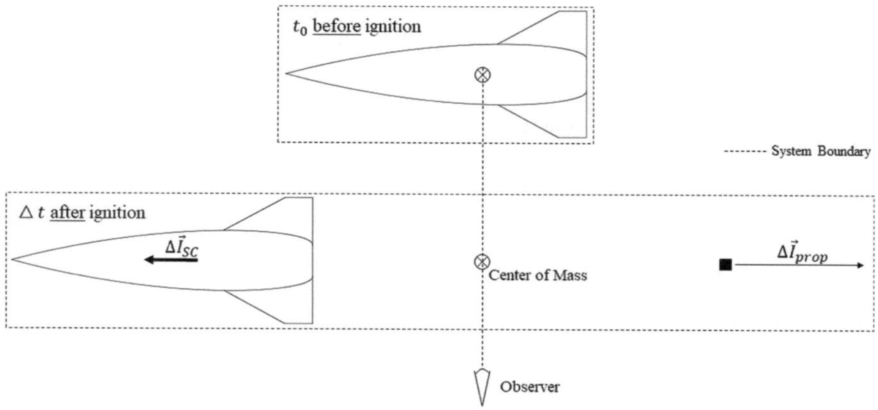


**Fig. 5.2** Spaceship in ejects propellant and gains an equal amount of momentum in the opposite direction in a force-free system, i.e. free-space

opposite direction within the spaceship. From a physics standpoint, this argumentation may suffice; however, some find it lacking as it does not incorporate the concept of force.[2] Nevertheless, what holds true for the fundamental law of momentum conservation should also apply to the concept of force interaction.

Figure 5.2 illustrates the perspective of an observer at rest relative to a spaceship during a time span $\Delta t$ when the propulsion system is instantaneously activated and deactivated. The spaceship shall be in a force-free environment, meaning that no external forces act on it—no gravity, no atmosphere etc.

After accurate measurements and meticulous budgeting of each single momentum contribution the observer will conclude that the centre of mass of the entire system, comprising the spaceship and the ejected propellant, remained unchanged in position. This is fully in line with the law of momentum conservation, which states that the total momentum a system posses does not change, if no interaction with external forces occur. The mass within the system boundaries remains constant as no mass has exited or entered the system. However, due to the spacecraft and ejected mass moving in opposite directions, the system boundaries must consequently expand over time.

The movement of the spaceship is solely caused by *internal redistribution* of momentum, requiring no external forces to act upon it. Hence, the term *internal* momentum *exchange* drive (IMED).

The momentum and eventually the velocity of the spaceship is high, if the momentum of the ejected propellant is high:

$$\Delta I_{\text{prop}} = \Delta m_{\text{prop}} \cdot u_{\text{ex}}, \tag{5.1}$$

[2] It was falsely believed for a long time in the 19th century that the rocket exhaust gas pushes against the atmosphere and that in the vacuum of space a rocket ceases to work.

$\Delta I_{prop}$ momentum of the ejected propellant mass,
$\Delta m_{prop}$ ejected propellant mass in the time span $\Delta t$,
$u_{ex}$ mean velocity of the ejected propellant mass,

Obviously, in order to save onboard mass it is the general objective of propulsion engineers to design systems with high exhaust gas velocities $u_{ex}$—the lower the required propellant mass, the higher the available mass for other necessary systems and payload. Hence, IMEDs with high exhaust gas velocities are called more *efficient* than those with low $u_{ex}$. There are cases, however, where efficiency is not the dominant design objective as we will see further below. The implications of the law of momentum conservation in relation to propulsion technology based on internal momentum exchange are profound: the Tsiolkovsky equation is a direct consequence of this principle.

Movement represents kinetic energy, and since creating movement requires the release of energy, it becomes imperative to consider another fundamental law: the law of energy conservation. In the following section, we will explore the energetic principles underlying thermal and electrostatic acceleration. It is important to emphasise that these acceleration principles remain unaltered regardless of whether the energy source is situated within or external to the spacecraft.

## 5.2 Thermal Acceleration

The core idea of this propulsion principle involves heating up the propellant and then converting the thermal energy of the gas into *directed* kinetic energy, resulting in a high-velocity *jet stream*. Figure 5.3 depicts the process of thermal based acceleration in three steps. In each step the state of the medium (i.e. propellant) is depicted within stationary and non-moving system boundaries. The boundaries are permeable, allowing atoms to freely enter and exit the system in all directions. The first

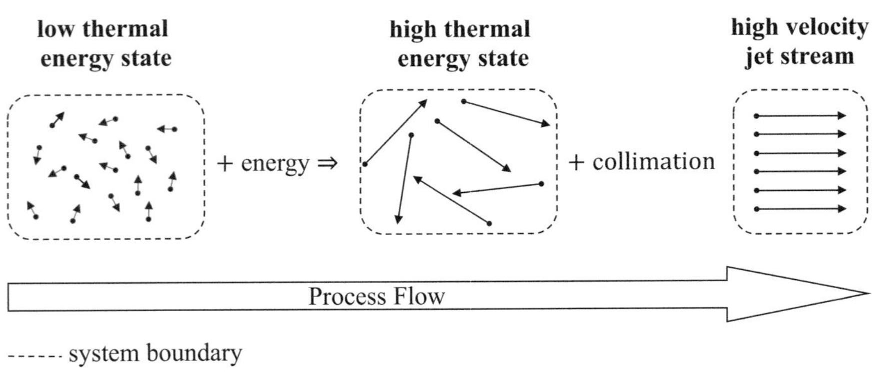


**Fig. 5.3** Change of thermal energy state by temperature increase and formation of jet stream (i.e. collimation)

state is a *low thermal energy state*. It is characterised by small velocity vectors of the molecules, that point randomly in space. Propellant is stored in tanks in this low energetic state—the system boundary is equivalent to the tank inner surface. When energy is introduced into the system as heat, the thermal energy state increases, causing atoms to move faster, yet they remain oriented randomly in space. This is the second state, the *high thermal energy state*. This process does not occur where the propellant is stored. Instead, it takes place in a dedicated chamber to which the propellant must flow or is transported by means of a pump. In the third step, the bulk of the molecules' velocities become *collimated*, meaning that all velocity vectors now point in one direction. This creates a high-momentum jet capable of propelling the spaceship.[3] This jet, or beam as it is referred to in electric propulsion systems, is known as the *exhaust force*. It is directed out of the spaceship, and according to Newton's third law (action equals reaction), an opposing force is generated. This opposing force pushes the spaceship in the opposite direction and is called *thrust*. Since force is defined as the time derivative of the linear momentum, we can immediately derive an expression for it in this simple case:

$$T = \frac{\Delta I_{\mathrm{prop}}}{\Delta t} = \frac{\Delta m_{\mathrm{prop}}}{\Delta t} \cdot u_{\mathrm{ex}} = \dot{m} u_{\mathrm{ex}} \tag{5.2}$$

| | |
|---|---|
| $T$ | thrust magnitude |
| $\dot{m}$ | ejected propellant mass flow |
| $u_{\mathrm{ex}}$ | mean velocity of the ejected propellant mass |

A formal introduction to this propulsion property will be provided in Sect. 5.4.

In conclusion, the art of rocket science is to transform the random velocity distribution of individual molecules into an aligned stream of particles flowing in a single direction with minimal randomness through the process of collimation. This is the third state in the process: formation of a *high velocity jet stream*. In fact, temperature and thus thermal energy are defined by the randomness and magnitude of the molecules' velocity vectors. If the molecule is composed of a single atom, its thermal energy is equivalent to the sum of the translational kinetic energies in all three dimensions. For more complex molecules, additional contributors need to be considered, as will be detailed in Sect. 8.2.1.

The jet depicted to the right of Fig. 5.3 represents an ideal case in that it no longer exhibits any randomness (i.e. perfect alignment). Consequently, the temperature and pressure measured by a co-moving observer (sitting on a molecule) are then zero. This is because the co-moving observer—as the name suggests—has the same velocity as the jet's bulk velocity and the fluid is not in motion from the observer's point of

[3] In physics, collimation refers to the process of aligning particles or radiation into parallel paths. It involves controlling the direction and spread of the emitted or scattered particles or radiation. Collimated beams are important in many applications, such as optics, particle accelerators, and medical imaging.

view.[4] The method by which collimation is achieved will be discussed in detail in Chap. 7.

We continue the energetic analysis and seek to establish a relationship between jet velocity $u_{ex}$ and the fluid's energy state. Equation 5.3 is a simplified form of the fundamental law of energy conservation:

$$h_0 = h + \frac{u^2}{2} = const, \tag{5.3}$$

$h_0$ enthalpy of a fluid, when the fluid is at rest, i.e. $u \equiv 0$,
$h$ enthalpy of a fluid that flows, i.e. $u \neq 0$,
$u$ bulk velocity of the fluid, measured by an observer at rest.

We know from thermodynamics, that the thermal energy state of a fluid is comprehensively described by its *enthalpy* $h$, Eq. 5.3. Enthalpy is a powerful thermodynamic concept. It quantifies the total energy possessed by a thermodynamic system. This energy consists of the sum of the fluid's internal energy $e$ and the work $(pV)$ the fluid has performed against the environmental pressure to create volume for its system. At any instant in the process described above, the *total enthalpy* $(h_0)$ is made up of the *actual* enthalpy $(h)$ and the bulk kinetic energy of the fluid. The total enthalpy $(h_0)$ is constant and can only be changed by supplying energy to the fluid. Enthalpy can, however, be converted into kinetic energy, if the fluid flows. A *co-moving observer* would measure $h$ instead of $h_0$. But the sum of enthalpy $h$ and kinetic energy $u^2/2$ remains constant, as long as the system does not interact with its environment by heat or work transfer (i.e. compression or expansion).

We have seen that collimation reduces the random velocity portion. We have also stated that temperature and pressure are an expression of this random molecular movement. Collimation consequently leads to a pressure and temperature reduction with increasing fluid velocity, Sect. 7.1. Therefore, an accelerated fluid flow is said to expand. Imagine the fluid expands so strongly that $h \ll u^2/2$. The total enthalpy would then be completely converted into the bulk velocity of the jet and Eq. 5.3 yields then:

$$u_{max} = \sqrt{2h_0}. \tag{5.4}$$

For an ideal[5] fluid, enthalpy is solely a function of temperature. If we further demand that the fluid is perfect, the relation becomes linear, with $c_p$ as the proportionality factor:

$$h = c_p T, \tag{5.5}$$

[4] The bulk velocity refers to the measurable macroscopic velocity of the flow that is representative of the fluid's properties. The local velocity in a flow (here jet stream) is in general not known and only partially measurable. This statement is also true for other properties like density and temperature.

[5] The idealisation of reality is a common practice in physics, as it allows for insights to be gained where, in reality, there is a substantial amount of noise (e.g. second-order effects) that obscures the underlying truth. Idealisation is the realm of theorists whereas real effects are the day-to-day struggle of engineers.

$c_p$ heat capacity at constant pressure, in general a function of temperature and pressure $c_p(T, p)$, for an *ideal gas* $c_p(T)$ and constant for a *caloric perfect gas*,
$T$ temperature of the fluid.

Inserting Eq. 5.5 into Eq. 5.4 leads a more tangible expression for the fluid's maximum achievable velocity:

$$u_{max} = \sqrt{2c_p T_0}, \tag{5.6}$$

Considering further that the heat capacity $c_p$ is inversely proportional to the mean molecular mass $\mathcal{M}$, we can write [12]:

$$u_{max} \propto \sqrt{\frac{T_0}{\mathcal{M}}}, \tag{5.7}$$

$\mathcal{M}$ mean molecular mass of the fluid,
$T_0$ temperature of the fluid, when the fluid is at rest, i.e. $u \approx 0$.

Equation 5.7 shows clearly that there are *two* decisive design parameters when it comes to propulsion efficiency of thermal-based acceleration: *temperature* and *molecular mass*. Hence, to achieve a high exhaust velocity (equivalent to an efficient propulsion system) the temperature of the propellant before collimation (i.e. expansion) should be high and the fluid (i.e. propellant) should have a low molecular mass. For instance, a hydrogen atom has the lowest atomic mass of 1 amu (atomic mass unit), while an oxygen atom has an atomic mass of 16 amu. Assuming the same total temperature, the exhaust velocity of atomic oxygen is 4 times lower than of pure hydrogen. In chemical propulsion systems these two design parameters are not entirely independent, though, as the medium is also the energy source. A thorough discussion is provided in Sect. 8.2.2. In case of nuclear thermal propulsion (NTP) systems, the two parameters are independent, which simplifies the system analysis, refer to Chap. 11.

### 5.2.1 Examples of Thermal Propulsion Systems

In principle, any conceivable energy source can be applied to heat up the propellant. Rocketry started simple, utilising solid propellant. Chinese chemists invented gun-powder[6] and propelled primitive but yet effective rockets. In the 20th century engineers started experimenting with liquid fuel due to its higher energy density. Figure 5.1 categorieses the energy source in two groups: *internal* and *external*. Internal means that the energy source is carried inside the spaceship, either stored in the propellant (e.g. chemically) or in another device (e.g. nuclear reactor). In contrast, the energy source could be located outside the spaceship. This is the case for

[6] Presumably in the 9th century AD at the beginning of the Tang Dynasty (618–907 AD).

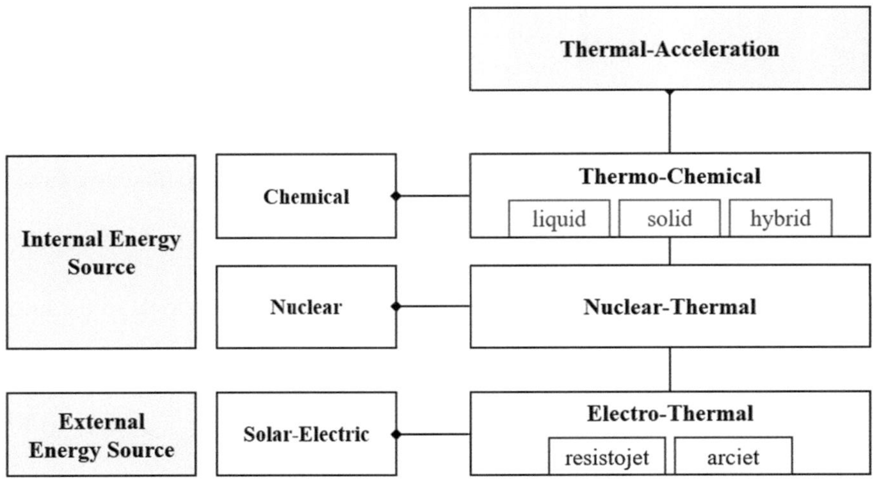


**Fig. 5.4** Categorization of propulsion technologies based on thermal acceleration (i.e. thermal rockets)

electric propulsion systems. In the following overview, we restrict ourselves for the sake of clarity to technologies that are flight proven or have been demonstrated on a representative scale. We will group them according to the energy source:

- chemical energy (combustion)
- nuclear energy (fission)
- electric energy (solar).

In Fig. 5.4 the aforementioned nodes are expanded, revealing technical realisations. The first node represents the combination of an IMED based on thermal acceleration with an *internal* energy source. It has two realisations: chemical and nuclear-thermal propulsion systems. The second node in which an *external* energy source is utilsed has only one demonstrated realization: electrothermal propulsion systems.

There are currently several new space companies working on propulsion systems that fit into the latter combination but are not electrothermal. Some intend to develop *solar-thermal* propulsion system. In this concept, solar radiation is collected and focused to heat-up the propellant for further acceleration (i.e. collimation) [1]. However, their superiority against classical systems has not been proven in terms of a business case for standard space applications.[7]

All propulsion systems of the thermal acceleration category with liquid propellant share the same functionalities of the *thrust-train*. Figure 5.5 depicts a generic functional block diagram of this category. The gray functional blocks in the upper part show the flow of propellant, common to all technologies of this category. Those

[7] The current business case of solar-thermal propulsion systems is centred around space mining of water and volatiles. The idea is to immediately utilise the extracted material in a simply and robust way.

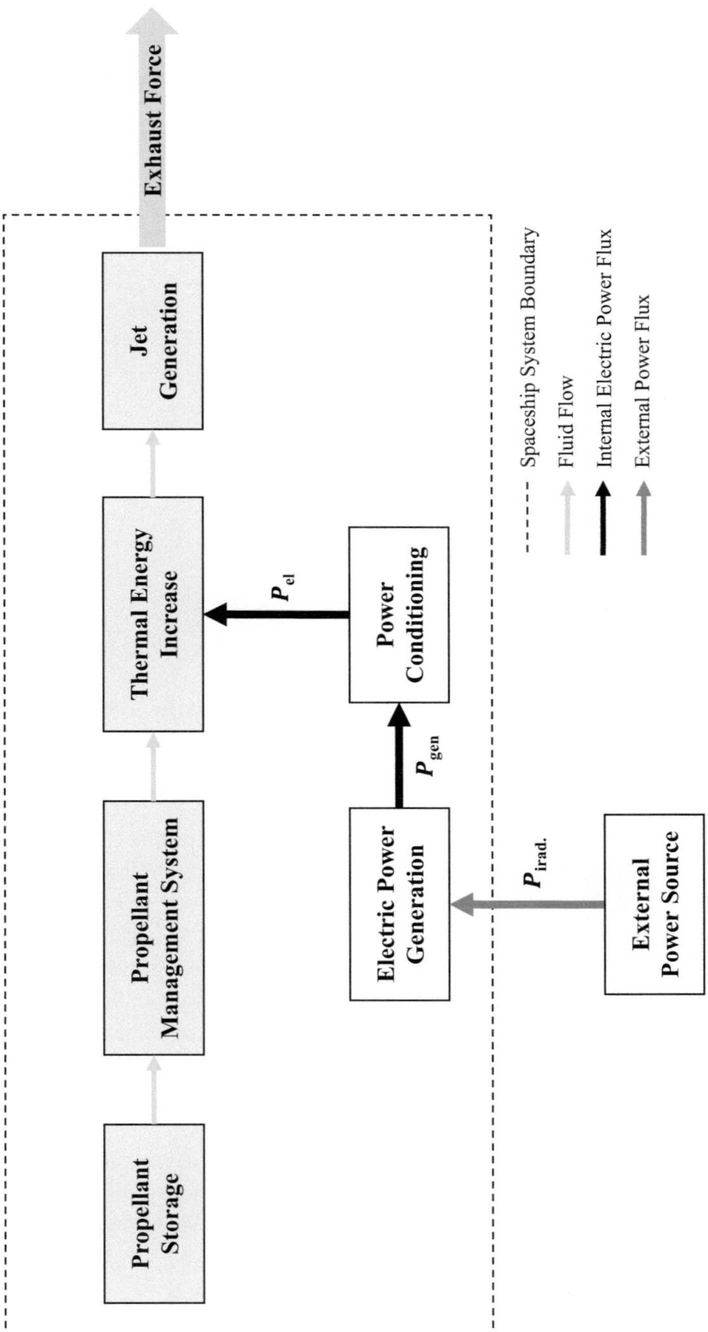


**Fig. 5.5** Generic functional block diagram of an IMED power-train based on thermal acceleration with *liquid* propellant, white blocks required for electrothermal systems

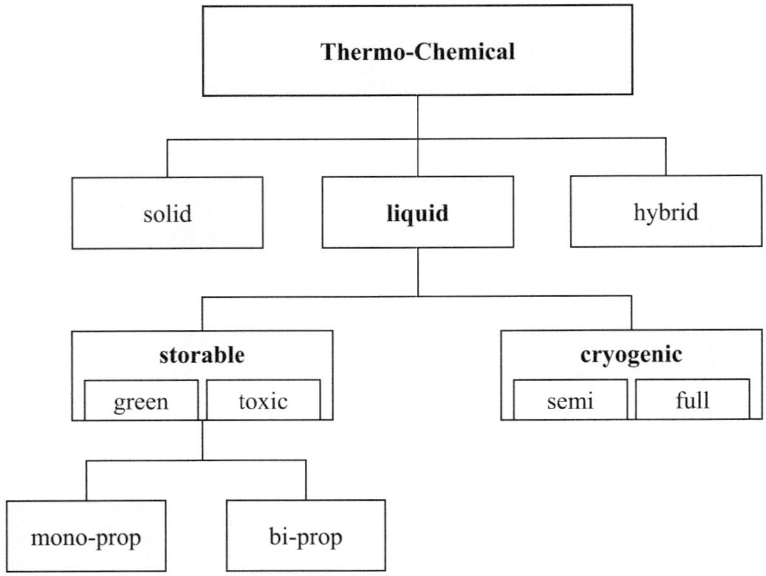


**Fig. 5.6** Classification of thermo-chemical propulsion systems

that need an external power source require in addition the white functional blocks: power generator and conditioner.

The finally generated exhaust contains three fluxes: mass, energy and momentum. The momentum flux integrated over the exhaust is a force and the counter force—according to Newton's third law—is defined as thrust.

Thermo-Chemical Engines

The probably oldest and easiest way of heating up a medium is by simply igniting it thereby releasing chemical bond energy. The technical discipline that studies and optimises this form of energy release is called *combustion*. All rocket engines used in launchers are combustion engines. The reason is the ability of combustion engines to generate very high thrust forces—ideal to overcome Earth's gravitational pull. The propellant can either be liquid, solid or a combination of both, called hybrid.[8] In the context of thermo-chemical engines, the focus of this book is on liquid propulsion systems as they form the vast body of propulsion applications. Figure 5.6 shows the types of liquid propulsion systems in use. They are usually distinguished by the propellant they employ.

The first level makes a distinction between *storable* and *cryogenic* propellant. The criteria here is the storability of the propellant under standard[9] temperature and pressure conditions: 298.15 K (15.15 °C) and 1 bar (100 kPa). Within this category there

[8] The only system currently using hybrid propulsion technology is the suborbital plane SpaceShipTwo, designed to carry space tourists just above the Kármán line, and thus legally into outer space.

[9] This definition comes from thermodynamics and is sometimes referred to as *normal* conditions.

is a distinction between *green* and *toxic* propellant. Almost all storable propellants used in the past and still in use today are toxic to humans. There is a clear trend towards green propellant worldwide.[10] The second layer within this branch makes a distinction on how many propellants are used by the propulsion system. If a single propellant is used, the entire propulsion system is referred to as a *mono-prop*, short for mono-propellant. Energy is therefore not released in a classical combustion process (reaction between fuel and oxidant), but by the *decomposition* of the fuel in the presence of a catalyst, which is often a special alloy. The lack of a second propellant means that these propulsion systems require a minimum of flow control equipment, making them less heavy, comparatively inexpensive and reliable, see Sect. 9.2. The decomposition of hydrazine, $N_2H_4$, yields $H_2 + NH + 3 + N_2$ and reaches a temperature of merely 868 K leading to a moderate exhaust velocity of 2158 m/s in current engines. Hydrazine has been for more than half a century a popular propellant, but need to be replaced due to its toxicity.

A special type of mono propellant is the *blended* mono-prop. This type combines fuel and oxidiser in a single storage tank, hence blended. The idea dates back to the German rocket laboratory in Peenemünde, where Wernher von Braun and other prolific figures in rocket history worked on Aggregat 4.[11] A US company has patented a fuel called NOFBX™ with an impressive exhaust velocity projection of up to 3190 m/s [5]. Other companies, like the Japanese IHI, work on hydroxyl ammonium nitrate based mono-propellant (HAN) and blends of HAN with hydrazine [6]. Mono-props are used in spacecrafts as main or auxiliary propulsion but also in launcher. The Italian medium class launcher, VEGA, is an example that uses 220 N hydrazine thruster for its roll and attitude control system (RACS), which shall be soon replaced by hydrogen peroxide ($H_2O_2$), which is a green alternative.

The counterpart of a mono-prop is a *bi-prop*, short for bi-propellant. For historic reasons, this term is exclusively used for storable propellant combinations used in orbital systems, like satellites. It referrers to the classical fuel and oxidiser combination and is the predominant choice for in-space propulsion systems with significant mass, like GEO satellite (2,000–8,000 kg), and large velocity demands ($> 100$ m/s). These are merely indications, a proper trade-off needs to be established in the gray area to determine whether the added complexity of a bi-prop over a mono-prop is justified. Both systems are subject of Sects. 9.2 and 9.3.

---

[10] In the European Union there are general regulations, restricting the use of toxic substances called REACH, which is short for Registration, Evaluation, Authorisation and Restriction of Chemicals. Space propulsion is directly affected by this regulation, which stirs new developments in this field.

[11] Wernher von Braun (1977†) was a seminal figure in the fields of rocket development and space exploration in the 20th century. He was the technical director of the rocket laboratory in Peenemünde, Germany, where the the A4 (Aggregat) was developed. The A4 (later dubbed V-2, 'vengeance 2' for propaganda purposes by the Nazis) was the first veritable rocket featuring all aspects of modern rocketry from gyroscopic stabilisation to regenerative thrust chamber cooling. It was further the first human made vehicle to reached space ($> 100$ km). Following his capture and relocation to the United States, he and his team underwent rehabilitation and assumed a pivotal role in the advancement of the Redstone rocket and later the Saturn V rocket, which successfully brought the first humans to the Moon. His contributions significantly advanced both, military and space exploration technologies, establishing him as a leading architect of modern rocketry [2].

**Fig. 5.7** Staged combustion engine 'E2' of the US company Launcher. Courtesy: John Kraus

Cryogenic propellants are not storable under normal environmental conditions. The word 'cryo' is Greek and means frost. It implies that the propellant is cooled down to very low temperatures. The need for this can be readily understood when considering the high energetic chemical reaction of hydrogen ($H_2$) and oxygen ($O_2$). Both are gaseous at standard conditions. Because large quantities are needed, and to save storage volume, it is necessary to cool the substances to just below their boiling temperature, which is 20 K (−253 °C) for hydrogen and 90 K (−183 °C) for oxygen.

This is why they are referred to as LH2 and LOX—'L' stands for liquid and describes their aggregate state. Semi-cryogenic propellant combinations are those in which one is cryogenic and the other not. A prominent example is kerolox which stands for the combination of kerosene (fuel), that is storable, and LOX (oxidiser), that is cryogenic. We will discuss in detail the most common propellant combinations in Chap. 8.

Figure 5.7 depicts a ground test of the rocket motor E2 of the US company Launcher, which is a compact oxidizer-rich staged combustion engine, see Sect. 9.5.3, of 100 kN thrust and exhaust velocity of 2844 m/s at sea level. Figure 5.8 compares a range of high thrust liquid rocket engines used for the first stage of launch vehicles. The thrust ranges from almost 8000 kN down to about 800 kN. Unlike first stage engines, upper stage engines operate at significantly lower thrust levels, typically ranging from 200 kN down to 2 kN, Fig. 5.9. This disparity arises from the launcher's operational requirements: launching vertically, i.e. with a flight path angle $\gamma$ of 90 deg, the launcher must rapidly gain speed and potential energy to escape the

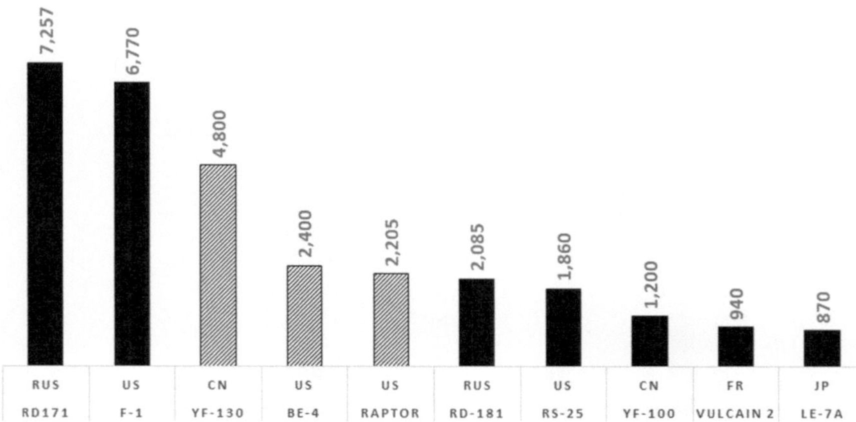


**Fig. 5.8** Sea level thrust (kN) of high thrust first stage rocket engines with liquid. Engines in development are marked by a dashed bar

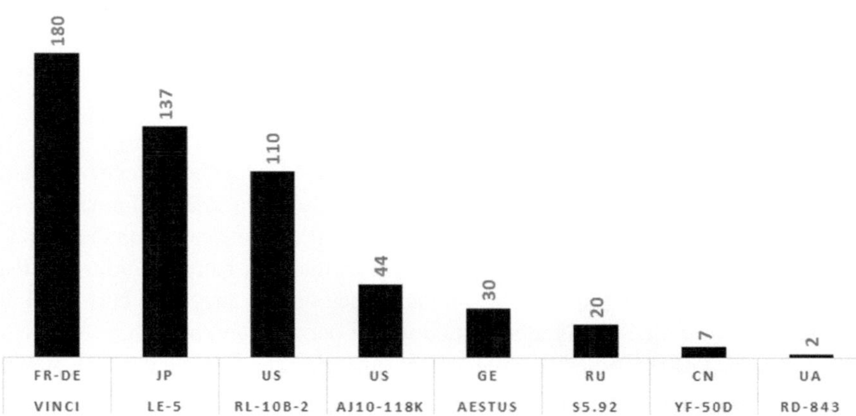


**Fig. 5.9** Vacuum thrust level (kN) of medium to low thrust upper stage engines with liquid propellant. Engines in development are marked by dashed bar

dense atmosphere before initiating the *gravity turn*.[12] In this attitude the launcher is more inclined to the local horizon (small flight path angle) its prime task is to gain velocity. Although the launcher did not yet reach orbital velocity, the required thrust to accelerate the remainder of the rocket is considerably lower, especially since the *dead mass* of the first stage has already been jettisoned.

Some launcher designer rely on *solid booster* to lift the vehicle in the vertical ascent phase. While solid boosters do have lower efficiency due to their lower exhaust velocities, their affordability and reliability make them an optimal choice when high

[12] On the other hand, a launcher cannot accelerate too quickly, as this would cause high thermo-mechanical loads due to the atmosphere, known as aerodynamic pressure $Q_{\max}$ and heating $\dot{Q}_{\max}$.

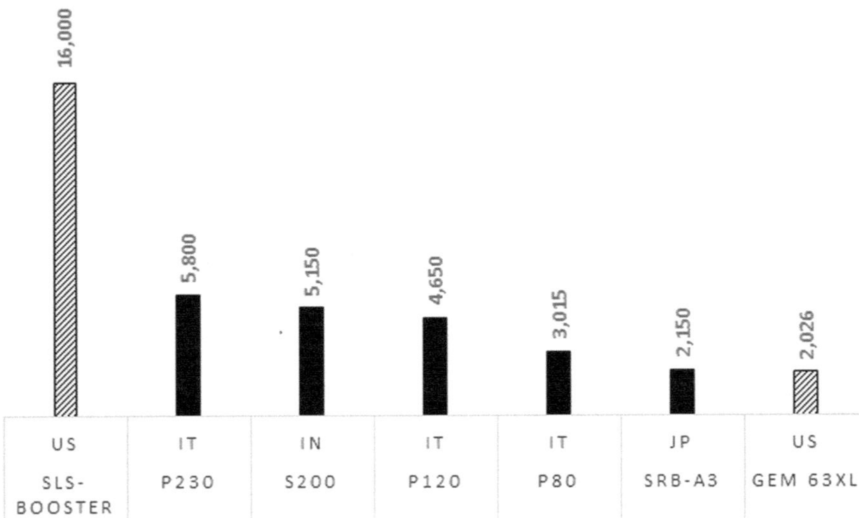


**Fig. 5.10** Sea level thrust (kN) of a range of solid rocket motors used as either as first stage or strap-on booster. Motors in development are marked by dashed bar

thrust levels are required for relatively short durations. A further advantage on system level is the high energy density of solid fuel, which leads to a smaller diameter compared to liquid-based boosters, thereby reducing the atmospheric drag on the vehicle. An important peculiarity of solid booster is the fact that they can neither be *throttled* (i.e. active thrust modulation) nor *re-ignited*. Once ignited they will burn until all propellant is consumed. Designers of solid rocket boosters are, however, capable of modulating the thrust profile (magnitude over time) through specific internal geometries. For instance, a star-shaped internal geometry with many spikes offers a high initial burn surface, resulting in a high initial thrust that declines slowly. Conversely, a simple round geometry has an initially small burn surface, which progresses, leading to a thrust increase towards the end [12]. Still, the thrust profile is pre-programmed and cannot be changed after ignition. A range of solid rocket motors is depicted in Fig. 5.10.

Chemical *in-space* propulsion systems, used in orbital systems such as satellites, space tugs, and space probes, generally do not have high acceleration requirements like launch vehicles. For orbital change manoeuvrers of satellites and cargo modules, thrust levels are considerably lower, typically ranging from 10 N to 1000 N, with re-ignitability being an important requirement.

In space engineering jargon, these smaller engines are referred to as *thruster* to distinguish them from their larger counterparts, which are typically called engines or rocket motors. If these thrusters are used for *attitude control*, the thrust requirement is even smaller and challenging with respect to thrust profile reproducibility and accuracy. These requirements typically lead to thrusters of 1 N down to 0.02 N. Hence,

**Table 5.1** Exhaust velocity in m/s of selected engines in their respective operating environment

| Engine | Propellant | $u_{ex}$ | Environment |
|---|---|---|---|
| RL-10B-2 | Full cryogenic | 4562 | Vacuum |
| RS-25 | Full cryogenic | 3590 | Sea-level |
| Raptor | Full cryogenic | 3237 | Sea-level |
| RD-180 | Semi cryogenic | 3050 | Sea-level |
| GEM 63XL | Solid | 2747 | Sea-level |
| SLS-booster | Solid | 2619 | Sea-level |
| P230 | Solid | 2541 | Sea-level |

liquid chemical propulsion systems span a thrust range from 0.02 N to 8000 kN, which results in a demonstrated *scalability* of a staggering eight orders of magnitude.

Table 5.1 lists selected engines sorted with respect to their exhaust velocity. The RL-10B-2 engine, an evolution of the legendary RL-10 engine family, features the highest exhaust velocity demonstrated in flight. This engine powers the equally legendary Centaur upper stage. Solid propellant based chemical propulsion systems are obviously inferior, if propellant efficiency is required.

More exotic propellant combinations use three constituents, so called *tri-pops*. They are in general composed of fuel, oxidiser and a performance enhancing additive of metallic nature. They are referred to as metallised propellant. Aluminum (Al) and beryllium (Be) additives have been thoroughly studied with partially promising results. For instance an increase in exhaust velocity to 680 m/s for Be-hydrolox is theoretically possible [4]. Beryllium is, however, a toxic substance and necessitates enhanced safety procedures and thus complex operation. The system-level impacts and the extra research effort did not justify yet the adoption of this technology.

Besides combustion, *heat transfer* represents the second method to increase the temperature of the reaction mass. Technologies that utilise this mechanism will be discussed in the following.

Nuclear-Thermal Engines

Fission based *nuclear-thermal* propulsion systems (NTP) belong to the most promising propulsion technologies for interplanetary flights due to their excellent propulsion characteristics: high exhaust velocity, large thrust and moderate propulsion system mass. The principle is fairly simple: the waste heat produced in the fission process of radioactive material (called fuel) is used to heat-up the propellant to very high temperatures. Solutions for the technological challenges to manufacture and operate a nuclear-thermal engine have already been laid down in the late 60 s of the 20 th century by both space powers, the US and the USSR. Figure 5.11 depicts the experimental nuclear-thermal engine NERVA, (Nuclear Engine for Rocket Vehicle Application), designed by the Los Alamos Scientific Laboratory in the Rover/NERVA programme. The demonstrated performance of NERVA is remarkable: it boasts a jet exit velocity of 8333 m/s, more than twice as high as chemical engines at seal level, while offering

**Fig. 5.11** NERVA dimensions and layout. Credit: Atomic Energy Commission/NASA

a noteworthy thrust magnitude of 30 kN [9]. The NERVA project was successful as it led to a flight representative qualification model of the engine, that was close to in-orbit experiments. History took another trajectory when the NERVA project was cancelled in 1973 by the Nixon administration, as the space race was considered over and public enthusiasm for space declined. An alternative trajectory could have provided humanity with a technology that would have made human exploration of the inner solar system possible half a century ago. Shortly after the US decision, the Soviet Union also cancelled its nuclear engine program which had culminated in the demonstrator engine RD0410 of similar characteristics as its US counterpart. Fifty years later this promising technology is experiencing a renaissance, see Chap. 11.

### Electro-Thermal Engines

Another, less spectacular, but easier method to heat up a medium is by electricity. There are two realisations in use: the *Resistojet* and the *Arcjet*. Both will be presented in the following. While the comparison may seem unconventional, a resistojet can be compared to a sophisticated kettle, since both operate on the same physical and technical principle: an electric current flows through a conductive coil and electric energy dissipates by resistive heating. This heat is transferred via convection to the propellant flow on its path to the nozzle where it expands into space. In order to achieve high temperatures, the resistor (i.e. coil) must be heat-resistant while offering adequate conductive properties to allow the electric current to flow. Refractory metals fulfill these two requirements, among them tungsten, tantalum and rhenium.

**Table 5.2** Propulsion characteristic of resitojets and arcjets with hydrazine as propellant. Credit: L3Harris™ (formerly Aerojet Rocketdyne)

| Engine | Type | F (mN) | $P_{dc}$ (W) | $V_{dc}$ (V) | $u_{\mathrm{ex}}$ (m/s) | $\eta$ (%) |
|---|---|---|---|---|---|---|
| MR-510 | Arcjet | 222 | 2,200 | 100 | 6000 | 30 |
| MR-502A | Resistojet | 500 | 885 | 29.5 – 24.5 | 3000 | 85 |

All of which have melting points well above 3,000 K. This limit, however, is reduced by the addition of further elements to create ductile alloys since refractory metals are too brittle in pure form. We have seen that a chamber temperature limit is equivalent to a limit in jet velocity, Eq. 5.6. This is why, even if operated with hydrogen as propellant (lowest molecular mass), jet velocities above 10,000 m/s cannot be achieved [6]. One advantage of Resistojets is their capacity to be combined with a wide range of fluids that can be converted into thrust—including bio-waste in crewed vehicles. Additionally, electrothermal thrusters offer the advantage of operating at low voltages, making them compatible with LEO satellite buses that typically operate at 28 V DC without the need for a dedicated converter.

The second electrothermal propulsion type is the *Arcjet*. As the name suggests, an arc jet generates a powerful electric arc that heats the fuel stream as it passes through it. A temperature of about 20,000 K can be achieved before the fluid expands in the nozzle. Compared to its siblings (chemical and nuclear-thermal engines), an Arcjet is able to create significantly higher temperatures locally. The walls are protected by a cold gas flow, which reduces the average temperature and also the exit velocity. A too high average temperature is also not desired as this leads to molecular dissociation and excessive atomic ionisation of the propellant. Both are forms of energy-losses in a thermal thruster and reduce their efficiency, $\eta_{\mathrm{T}}$, as will be detailed in Sect. 8.2.1. Arcjet designer tune the thruster towards an operating point in which the ionisation rate is just as high as to keep the arc stable but not too high in order to limit the losses [13].

The achievable thrust level of both electrothermal engine types is small compared to chemical and nuclear-thermal engines. Table 5.2 lists two demonstrated and flight proven thruster. Due to their low thrust level but comparably high exhaust velocity, electrothermal engines are generally used for satellite attitude control and in case of GEO satellites for orbital station-keeping.[13] The table also reveals that the efficiency of arcjets is remarkably low compared to the resistojet. This is a general drawback of this propulsion system and not specifically related to the here selected model [8]. The comparatively higher efficiency of resistojets is due to the lower temperature, which causes lower dissociation and ionization losses. The currently used power

[13] Thruster that require an electric power source are almost only an option, if the main payload has a high power demand. It is not just about leveraging synergies; there's a clear logical order to consider. If the payload is not power-hungry and the thrusters are the primary power consumer, these thrusters are unlikely to be selected as the baseline—unless a mission enabling mass margin can be achieved. This is because the power subsystem is typically the most expensive component in the spacecraft bus and the monetary invest must payoff accordingly.

source of electrothermal propulsion systems (as of all electric engines) is based on photovoltaic arrays (PVA), also referred to as solar power generator (SPG). The main advantage of solar power generation is that no consumables are required. However, due to the decrease in solar flux, this entails limitations at mission level, which must be taken into account for missions to the outer solar system, see Sect. 8.3.

## 5.3 Electrostatic Acceleration

The idea to use electrostatic field to accelerate charged particles to high velocities was first formulated by the brilliant Russian scientist Konstantin Tsiolkovsky in the late 19th century.[14] Tsiolkovsky proposed the idea of electric propulsion for space travel in his work 'The Exploration of Cosmic Space by Means of Reaction Devices', published in 1903. Around two decades later, two other space pioneers independently proposed this principle: the American physicist Robert Goddard,[15] who proposed the idea in his treatise 'A Method of Reaching Extreme Altitudes', published 1919 and the scientist Hermann Oberth,[16] who wrote in his book 'Die Rakete zu den Planetenräumen' (1923) The Rocket into Interplanetary Space, about the use of electric propulsion. All three recognised the potential of this acceleration principle for space travel at an early stage. Over the past 70 years, there has been a plethora of technical realisations of electric propulsion systems. To discuss them all would be a book in itself. We limit ourselves to those that have been demonstrated in flight and have the potential to enable deep space travel.

Electromagnetism is often regarded as more complex than thermodynamics, and indeed, an appreciation for mathematics is essential to grasp the beauty of Maxwell's famous equations and to navigate through the equations describing non-idealized Magnetohydrodynamics (MHD) without being deterred. We take a less mathematical path and like the above mentioned pioneers limit ourselves to an energetic consideration as we did for thermal-based acceleration. We further restrict ourselves to flight-proven and widely used *electrostatic* thruster technology. Figure 5.12 shows the basic principle of this type. It consists of a charged particle $q_i$ in a homogeneous

---

[14] Konstantin Tsiolkovsky (1935$^{\dagger}$), a pioneering Russian rocket scientist of modest origin and who largely self-educated himself. From all known early space pioneers it was Tsiolkovsky who achieved incredibly much from very little, despite personal tragedies, disabilities, low social esteem and limited access to higher education. He is rightfully regarded as one of the founding fathers of modern aeronautics and astronautics. He formulated the famous rocket rocket equation that is named after him, envisioned the use of liquid propellants and build the first wind-tunnel in Russia. Additionally, Tsiolkovsky proposed advanced space travel concepts, including space stations, multi-stage rockets, and the colonisation of the solar system to name a few [3].

[15] Robert Goddard (1945$^{\dagger}$) was an American professor and inventor and a leading researcher in rocketry related disciplines in the US [4].

[16] Hermann Oberth (1989$^{\dagger}$) was a Romanian-German scientist who has inspired many, including Wernher von Braun, with is work about space travel in the early 20 s of the 20th century. Furthermore, he laid the theoretical foundation for human space exploration [5].

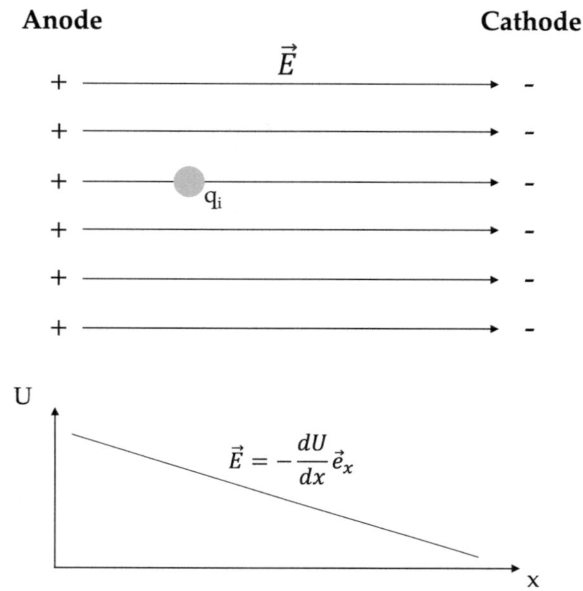


**Fig. 5.12** Electrostatic acceleration of a charged particle in an electric field

electrical field **E**, which in turn is the result of a gradient in the electrical potential field $V(x)$. A positively charged particle will move in direction of decreasing potential value, that is in direction of the electrical field lines **E**, while a negatively charged particle will move in the opposite direction. We are again interested in a pure energetic view.

Equation 5.9 establishes the relationship between electrical potential energy and the kinetic energy of particles (i.e. ions), assuming a full and loss free conversion of energy states.

$$E_{\text{el}} = E_{\text{kin,i}}, \tag{5.8}$$

$$qV = \frac{1}{2} m_i v_i^2, \tag{5.9}$$

| | |
|---|---|
| $E_{\text{el}}$ | energy of the homogeneous electric field, |
| $E_{\text{kin,q}}$ | kinetic energy of ion, |
| $q$ | ion charge, |
| $V$ | electric voltage, |
| $m_i$ | mass of single ion, |
| $v_i$ | velocity of a single ion. |

Obviously, electric propulsion systems require charged particles and a high voltage to accelerate the ions to high velocities. We will discuss in Sect. 7.2 different means of propellant ionisation and acceleration. Electric propulsion engineers refer to the exhaust of these thruster as *ion beam* or simply *beam* instead of jet stream. Under the assumption of a uniform velocity profile, the single ion velocity is then equivalent to the *ion beam velocity*: $v_i = v_\mathrm{b}$. Solving Eq. 5.9 for $v_i$ yields:

$$v_i = \sqrt{2\frac{q}{m_i}U}. \tag{5.10}$$

Note that this equation does not differ much in structure from Eq. 5.7 and there is in fact an electrothermal analogy: electric voltage is the equivalent to temperature and the ratio ion mass per charge $(m_i/q)$ the equivalent to molecular mass.

The process of ionising an element is energy-intensive, and as a result, the quantity of gas that can be sufficiently ionised is limited by the available energy source. The ionised mass flow is, in fact, very small, in the order of $10^{-3}$ g/s, and consequently also the thrust, which is measured in millinewton (mN). This is a fundamental difference to the thermal-based acceleration principles.

Propellant Properties

Analysing Eq. 5.10 with respect to the desired propellant properties reveals the following:

$$(q \uparrow, m_i \downarrow) \Rightarrow v_i \uparrow \tag{5.11}$$

A small atomic mass and a high degree of ionisation favour a high particle velocity. Given the mechanisms discussed above this is not a surprise and analog to thermal propulsion. The picture is not complete yet as we need to take into account the thrust relation for electric propulsion systems. We start with the above derived formulation for the thrust (5.2) that is generated by the ion beam:

$$T = \dot{m}_\mathrm{b} \cdot v_\mathrm{b}, \tag{5.12}$$

| | |
|---|---|
| $\dot{m}_\mathrm{b}$ | mass flow of the ion beam, |
| $v_\mathrm{b}$ | bulk velocity of the ion beam. |

We then make use of the relation between the ion beam flow $\dot{m}_\mathrm{b}$ and the electrical current of the beam $I_b$:

$$\dot{m}_\mathrm{b} = I_b \cdot \frac{m_i}{q_i}. \tag{5.13}$$

Inserting this relation as well as Eq. 5.10 into Eq. 5.12 gives us:

$$T = I_b\sqrt{2\frac{m_i}{q_i}U}. \tag{5.14}$$

For Xenon, a popular propellant for electric propulsion systems, the factor $\sqrt{2m_i/q_i}$ equals $1.65 \times 10^{-3}(kg/C)^{1/2}$, which gives:

$$T_{Xe} = 1.65 \times 10^{-3} I_b \sqrt{U} \quad [\mathrm{N}]. \tag{5.15}$$

Comparing both Eqs., 5.10 with 5.14, reveals that beam velocity and thrust do favour different propellant properties:

$$(q_i \uparrow, m_i \downarrow) \Rightarrow v_i \uparrow, \tag{5.16}$$

$$(q_i \downarrow, m_i \uparrow) \Rightarrow T \uparrow . \tag{5.17}$$

Typically, such a scenario would necessitate a trade-off. However, given the already staggeringly high bulk velocity of the ion beam and the low thrust level, the need for thrust a priori outweighs any potential trade-offs. Moreover, electric power is a valuable resource in a spacecraft, which is generally limited in mass, cost and accommodation volume. Given these limitations, the remaining parameter to enhance thrust is the mass of the ions. Hence, the popularity of the inert gas xenon for high performance electric thruster.

### *5.3.1 Examples of Electromagnetic Acceleration*

During the Cold War, the two superpowers, the Soviet Union and the United States of America investigated two technical realisations of this principle: Gridded Ion Thruster (GIT) and Hall Effect Thruster (HET). While the USA pursued GIT, the Sovjet Union opted for HETs [6]. They belong to the group of *electrostatic* thruster following the same basic principle discussed above. They will be discussed in detail in Sect. 7.2.

The striking feature of this acceleration type is the high jet velocity that can be generated. The ion beam consists of charged particles with a velocity well in excess of 100,000 m/s. The high exhaust velocity comes with a caveat, the accelerated mass flow is small, leading to low thrust levels ranging from below 5 mN up to 5 N—a factor of merely 1000 compared to 100 million for chemical engines. Therefore, unlike chemical engines, electrostatic thruster do not show the same *scalability* in thrust. This limitation is not a fundamental issue but rather related to the power source. We will see in Sect. 10.3 how this drives the power subsystem of a spaceship. Fortunately, in-space manoeuvre and interplanetary travel, do not have in general high acceleration requirements, and due to their high propellant efficiency, electrostatic thruster are often considered the preferred choice.

A famous example in which an electrostatic propulsion system made a mission possible was the ESA mission GOCE, depicted in Fig. 5.13, which was powered by the T5, a gridded ion thruster of the British company Qinetiq.

GOCE's prime objective was to precisely measure Earth's gravitational field, which is non-homogeneous, see Fig. 2.5. To do so, the spacecraft needed to fly at an extremely low altitude, in average between 250 and 270 km over a nominal time span of 20 months. There, the rest atmosphere is still dense and without active orbit control, an object would spiral down to Earth within weeks. In fact, from fuel depletion (21.10.2013) till re-entry (11.11.2013) passed only three weeks [10]. Consequently,

**Fig. 5.13** GOCE spacecraft propelled by one T5 ion thruster with back-up thruster in cold redundancy. *Credit* European Space Agency—ESA/ATG medialab

**Table 5.3** Nominal performance figures of electric thruster according to manufacturer specification

| Type | | T6 GIT (20 cm) | NEXT GIT (40 cm) | PPS®5000 HET | HERMeS HET | X3 HET |
|---|---|---|---|---|---|---|
| Exhaust velocity | m s | 36,395 | 41,398 | 19,620 | 27,272 | 25,310 |
| Thrust | mN | 73 | 235 | 300 | 590 | 1770 |
| Efficiency | % | 66 | 70 | 50 | 67 | 66 |
| Discharge power | kW | 2.4 | 6.9 | 5 | 12.5 | 33.7 |

the primary objective of the propulsion system was to compensate for the meticulous but detrimental drag (5–20 mN) in order to create a quasi-force-free environment for the gravimetric instruments. The mission objective could only be achieved through the use of electric propulsion. It is important to note, however, that this comes with a high power demand, which drives the system layout. Table 5.3 shows the specifications of five electric thruster. In pursue of higher thrust, high power GITs, like NEXIS with 25 kW [7] and HiPEP with 10–40 kW [8], have been studied intensively but the effort slowly waned in favour of high power HETs: partially due to limitations caused by space charging and grid erosion [11]—a phenomena detailed in Sect. 7.2—and partially because of the high power demand. HET technology has proven to be easier to scale and, as a result, has made significant strides. The thrust-to-power ratio is a key parameter specific to electric propulsion systems. Gridded Ion Thrusters feature a lower thrust-to-power ratio than Hall Effect Thrusters, thus requiring more power to reach the same thrust level. On the positive note, GITs generally boast higher beam velocities and thruster efficiencies. This underscores the inherent interconnection between the propulsion system and the overall spacecraft design, necessitating system-level trade-offs to achieve an optimal solution tailored to the specific mission.

Finally a last type of thruster shall be mentioned, the magnetoplasmadynamic thruster (MPD or MPDT). This technology accelerates the complete plasma and not only positively charged particles as for the electrostatic thruster previously discussed. To achieve that, it uses either an external or a self-induced magnetic field in combination with an arc-generated electrical field.

The technology offers a very high thrust density, and a specific impulse in the range of 1500–8000 s. A thrust level of 12.5 N in combination with a specific impulse of 4000 s and an efficiency of 48% could be demonstrated in tests [6], which corresponds to a power demand of 500 kW. Although still at experimental stage, these characteristics render this propulsion technology the first choice for future large scale interplanetary travel in combination with a nuclear-electric power source (NEP).

In conclusion to this section, we will examine the previously established taxonomy scheme, depicted in Fig. 5.14. We have seen that the energy demand is extremely high and it is no surprise that chemical energy can be clearly ruled out as a viable energy source. Not only because of its low specific energy content but also due to the associated conversion losses (40–60%). A continuous power supply would be ideal. This leaves only two options, nuclear energy and solar power. The first, known as nuclear-electric propulsion (NEP), is subject of intense research and promises a profound leap in space exploration capabilities. The second option, that is based on

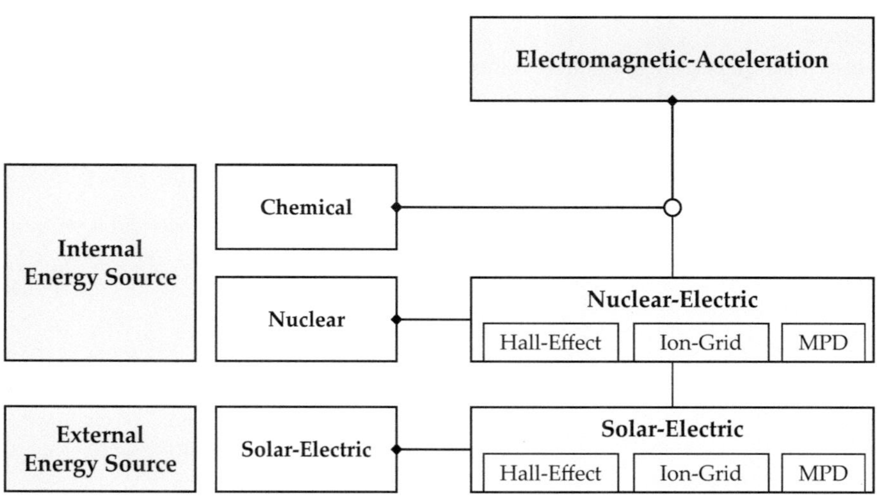


**Fig. 5.14** Categorization of the EM-based propulsion technologies

solar power is called Solar Electric Propulsion (SEP) and is currently the first choice for electric thruster.

For the sake of completeness, Fig. 5.15 depicts the generic power train options for an electric propulsion system. Note that the power source can be either external or internal in line with the established taxonomy.

## 5.4 Performance Parameter

There are several characteristic performance parameters of propulsion systems that are important in system-level trade-offs. In this section a selection of seven distinct parameters will be discussed:

1. Thrust,
2. Specific Impulse,
3. Total Propulsion Efficiency,
4. Total Impulse,
5. Thrust-to-Weight Ratio,
6. Thrust Density,
7. Propulsion System Mass Fraction.

The list is intentionally kept general, omitting technology-specific parameters such as the aforementioned thrust-to-power ratio for the sake of broad applicability. Moreover, it shall be noted that not all of these parameters are equally relevant for a mission. It is therefore part of the tasks of the systems engineering team to identify the key parameters for the system and drive the trade-off and selection process accordingly.

Thrust
From mission perspective, thrust a measure of how fast a mission can be conducted and the prime function of a propulsion system is to provide *thrust*. As discussed above, thrust, $T$, is the (re-action) of the exhaust force, $F_{\text{jet}}$ (action) and therefore points into the opposite direction: $\mathbf{T} = -\mathbf{F}_{\text{jet}}$. We will ignore in the following this vector relation and concentrate on the physics.

The thrust of a jet stream has two contributors, an *impulsive term* and a *pressure term* [12]:

$$T = \dot{m} u_{\text{ex}} + (p_{\text{ex}} - p_{\text{amb}}) A_{\text{ex}}, \tag{5.18}$$

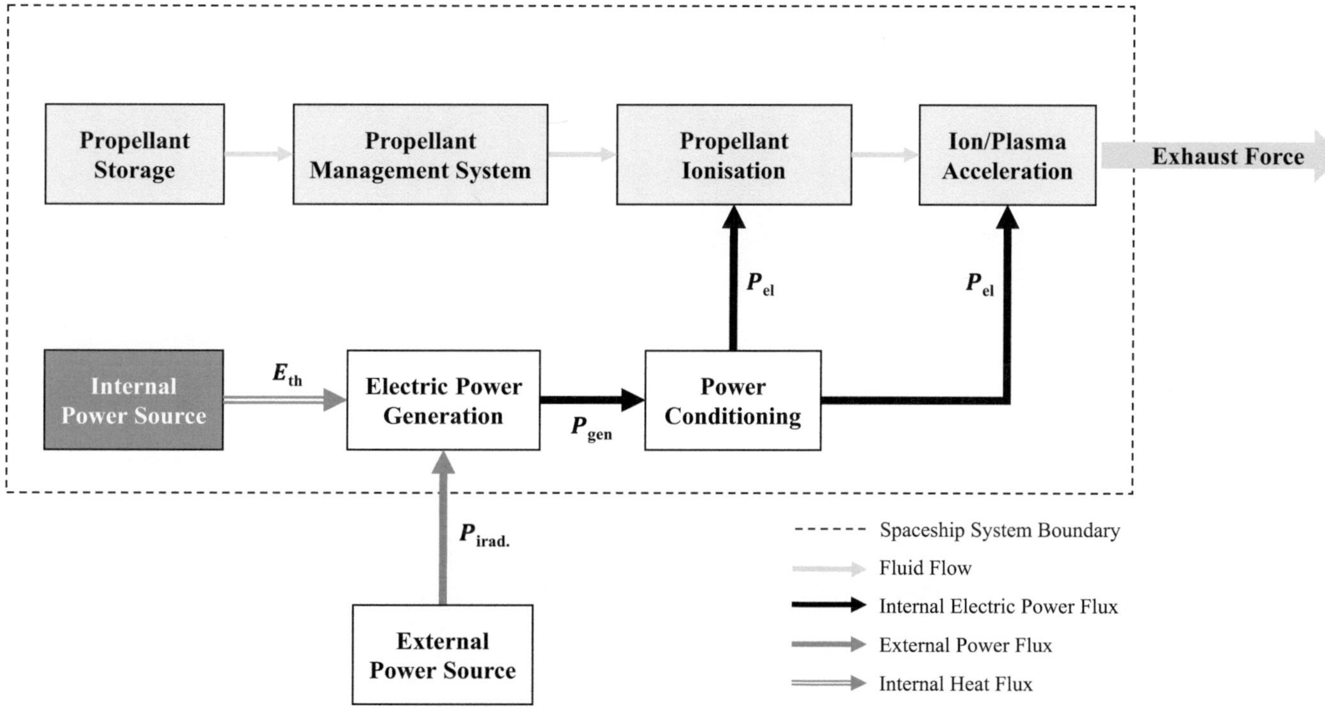


**Fig. 5.15** Generic functional block diagram of an EMED power-train based on electric acceleration for two cases: external and internal power source

| | |
|---|---|
| $T$ | measured engine thrust, |
| $\dot{m}$ | propellant mass flow, |
| $u_{ex}$ | mean jet velocity at the nozzle's exit plane, |
| $p_{ex}$ | pressure at the nozzle's exit plane, |
| $p_{amb}$ | ambient pressure, zero in vacuum, |
| $A_{ex}$ | nozzle's cross section at the exit plane. |

We have already encountered the first term, see Eq. 5.2. The second term is specific to engines that utilise the thermal acceleration principle, see Sect. 7.1. The equation can be rearranged to show that there is an engine intrinsic term and a term that depends on the external environment:

$$\begin{aligned} T &= \dot{m}u_{ex} + (p_{ex} - p_{amb})A_{ex}, \\ &= \underbrace{\dot{m}u_{ex} + p_{ex}A_{ex}}_{T_{eng}} - \underbrace{p_{amb}A_{ex}}_{F_{amb}} . \end{aligned} \tag{5.19}$$

| | |
|---|---|
| $T_{eng}$ | intrinsic engine thrust, i.e. thrust measured in vacuum, |
| $F_{amb}$ | negative force produced by ambient pressure, zero in vacuum. |

The force produced by the ambient atmosphere is negative and reduces the actual thrust level as it points in exhaust velocity direction. The term, $F_{amb}$, reduces with increasing height and is non-existing in the vacuum of space. The actual thrust of a rocket engine varies with height and thus time, which is the reason why there are two thrust figures for first stage rocket motors: one for sea-level and one for vacuum-level. Chemical engines are, therefore, either designed for atmospheric flight or for in-space application as will be detailed in Sect. 7.1.

Electromagnetic engines on the other hand can only be operated (meaningfully) in vacuum condition. The reason is the very low density of the plasma, known as *rarefied flow* which does not stand a chance of overcoming Earth's dense atmosphere at sea-level.[17] The contribution by the pressure term is indeed low ($p_{ex}A_{ex} \ll \dot{m}u_{ex}$) and can be ignored for electric thruster without loss of accuracy [12]. Introducing the electrostatic specific beam variables, yields the previously established Eq. 5.12:

$$T = \dot{m}u_{ex} = \dot{m}_b v_b.$$

[17] In fluid dynamics, there is a distinction between two different flow regimes: the *continuous* regime and the *free molecular* regime. The distinguishing parameter is the ratio of the particle's *mean free path*, $\lambda$, to the dimension of interest, $L$. The first is the average distance a particle travels before colliding with another particle. The latter is the subject of interest, such as the diameter of a blood vain, a nozzle length, or the diameter of a discharge chamber of an electric thruster. The continuous flow regime is applicable to all phenomena in which the ratio is very small, that is, when $\lambda/L \ll 1$. This is always the case for thermal-based engines. Conversely, we speak of a rarefied flow when $\lambda/L \gg 1$, which is the case in electric thruster. The physics of the two are different and so is the mathematical description.

Finally it shall be emphasised that this is a linear consideration. Collimation is in reality not perfect, which causes divergence losses of the exhaust gas. The consequences will be discussed in the context of electric thruster since this propulsion type is prone to divergence losses, Sect. 7.2.3.

Specific Impulse
The *specific impulse*, $I_{sp}$, is probably the most famous performance parameter, next to thrust, in astronautical engineering. To define the $I_{sp}$, it is necessary to start with another term, the *effective velocity*, $c_e$. It originates from the wish to avoid the use of two terms, that is $u_{ex}$ and $p_{ex}$ from Eq. 5.18 by introducing a new velocity to characterize the efficiency of an engine, defined as:

$$c_e := \frac{T}{\dot{m}} = u_{ex} + \frac{(p_{ex} - p_{amb})}{\dot{m}} A_{ex}. \tag{5.20}$$

The effective velocity, $c_e$, encompasses both terms the pressure as well as the impulsive term of the thrust, see Eq. 5.19. For electric thruster, effective velocity and actual average jet velocity are practically identical due to the low pressure term.

An even more convenient figure is the specific impulse, defined as the effective velocity normalized with the gravitational acceleration[18] $g_0 = 9.81$ m/s$^2$:

$$I_{sp} := \frac{c_e}{g_0} \tag{5.21}$$

$$= \frac{u_{ex}}{g_0} + \frac{(p_{ex} - p_{amb})}{\dot{m} g_0} A_{ex}. \tag{5.22}$$

Since the exit pressure is still present in the expression for the $I_{sp}$, a distinction is made between sea-level impulse ($I_{sp,sea}$) and vacuum impulse ($I_{sp,vac}$). In intermediate heights the following relation holds:

$$I_{sp} = I_{sp,vac} - \frac{p_{amb}}{\dot{m} g_0} A_{ex}. \tag{5.23}$$

The latter term vanishes for in-space propulsion systems like for upper-stage engines and electric thruster due to the zero ambient pressure in vacuum:

$$I_{sp,el} = \frac{c_e}{g_0} \approx \frac{u_{ex}}{g_0}. \tag{5.24}$$

We will explore in Sect. 6.2 that, from a system-level perspective, high specific impulse systems are not always desirable. This is in particular the case for solar electric propulsion (SEP) and nuclear electric propulsion (NEP) systems.

[18] A further advantage of this normalisation is that the resulting unit, the 'second,' is used in both the imperial and international (SI) systems.

The specific impulse serves as a measure of the efficient utilisation of propellant by the engine. The higher the specific impulse of an engine the less propellant mass is required for the same thrust to accomplish a specified velocity demand $\Delta v$.

Total Engine Efficiency

The specific impulse makes a statement on how efficient the propellant is used by the propulsion technology. It does, however, not make any statement on the efficiency of the *entire* thrust-train, Fig. 5.5. Precisely speaking, on how efficient the induced or released power, $P_{\text{in}}$, is converted into power of the exhaust jet $P_{jet}$. The parameter that characterises the efficiency of this conversion process is the *total efficiency*, defined as:

$$\eta_{\mathrm{T}} := \frac{P_{jet}}{P_{\text{in}}}. \tag{5.25}$$

The power of the exhaust jet can be derived from the energy of the jet stream:

$$P_{jet} = \lim_{\Delta t \to 0} \frac{\Delta E_{jet}}{\Delta t} = \lim_{\Delta t \to 0} \frac{\Delta m}{\Delta t} \frac{c_{\mathrm{e}}^2}{2} = \dot{m} \frac{c_{\mathrm{e}}^2}{2}. \tag{5.26}$$

The higher $\eta_{\mathrm{T}}$ the better the energy conversion. If the energy source is internal, we speak of released power, e.g. chemical-reaction and nuclear fission. If it is external, we speak of induced power, e.g. solar cells, solar-thermal generator. Collectively we refer to all forms as input power, $P_{\text{in}}$. The above derived expression for $P_{jet}$ can be expressed as function of thrust:

$$P_{jet} = \dot{m} \frac{c_{\mathrm{e}}^2}{2} = \frac{1}{2} T c_{\mathrm{e}}. \tag{5.27}$$

Introducing this expression into Eq. 5.25 and solving for $P_{\text{in}}$ yields:

$$P_{\text{in}} = \frac{1}{2} \frac{T^2}{\dot{m} \eta_{\mathrm{T}}}. \tag{5.28}$$

Introducing further the definition of the specific impulse gives:

$$P_{\text{in}} = \frac{1}{2} \frac{T I_{\text{sp}} g_0}{\eta_{\mathrm{T}}}, \tag{5.29}$$

or

$$\eta_{\mathrm{T}} = \frac{1}{2} \frac{T I_{\text{sp}} g_0}{P_{\text{in}}}. \tag{5.30}$$

These are fundamental relationships and highly relevant in the understanding of propulsion systems based on internal momentum exchange (IMEDs). It means that thrust and specific impulse are not independent in these propulsion systems and that the available power cannot be arbitrarily divided between thrust and specific impulse. This is independent of whether the power source is internal or external. The designer must choose between a propulsion system with either high thrust or high specific impulse, as these variables are inversely proportional to each other:

$$T = \left(\frac{2P_{\mathrm{in}}\eta_{\mathrm{T}}}{g_0}\right) \cdot \frac{1}{I_{\mathrm{sp}}} = C \cdot \frac{1}{I_{\mathrm{sp}}} \approx \frac{1}{I_{\mathrm{sp}}}. \tag{5.31}$$

To increase both parameters, it is necessary to raise the proportionality factor, $C$. This can be accomplished by either increasing the input power, $P_{\mathrm{in}}$ or by enhancing the efficiency factor, $\eta_{\mathrm{T}}$. It is important to note that this crucial equation is often falsely attributed solely to electric propulsion systems. Quite the opposite, this equation is applicable to all mass-ejection-based propulsion systems, i.e. to all IMEDs.

Total Impulse
The parameters presented above are either specific to the engine or to the power conversion. The *total impulse*, $I_{\mathrm{tot}}$, refers to a capability of the entire propulsion system including propellant. It is defined as the integral of the thrust over mission time, encompassing all propellant mass:

$$I_{\mathrm{tot}} = \int_0^{t_{\mathrm{b}}} T(t)\, dt. \tag{5.32}$$

The total impulse is a measure of the momentum that can be provided to the spaceship by the propulsion system. Or from the perspective of the mission analyst, the momentum that needs to be exerted by the propulsion system.[19] The value can be increased by loading more propellant. Hence, a system with a lower thrust level could have the same total impulse compared to a system with higher thrust, if the propellant mass is higher.

We will see in Sect. 7.2 that electric thrusters suffer from erosion processes that limit their lifetimes. This means that the limitation in total impulse is not due to the available propellant mass but due to thruster degradation.

Thrust to Weight Ratio
Mass is an important design parameter in space engineering, mainly due to the high launch cost. It is therefore justified to ask what is the required mass investment for a Newton of thrust. The respective performance is called *thrust-to-weight ratio* and defines as:

$$\Gamma := \frac{T_{tot}}{m_{\mathrm{eng,t}}g_0}, \tag{5.33}$$

[19] A technical specification represents a capability from the owner's perspective (e.g. propulsion engineer) and a requirement from the requester (here mission analyst). When these interfaces align, we say that the design *converges*. To balance the process and decide on necessary compromises is the task of the system engineer.

| | |
|---|---|
| $T_{\mathrm{tot}}$ | total thrust delivered by the engine cluster, |
| $m_{\mathrm{eng,t}}$ | total mass of engine cluster. |

The term stems from rocketry, where mass is crucial. Every saved 'kg' in the design increases the payload capability of the launcher. Note that, $\Gamma$, considers only the engines not the entire propulsion system or the entire vehicle (i.e. vessel or launcher). This is because $\Gamma$ is meant to compare engines among each other. Secondary parameter need to be derived to characterise the entire system.

Thrust Density
A parameter that describes the compactness of the engine cluster's base area is the *thrust density*. This figure is of particular importance for a multi engine configuration, i.e. engine cluster. It is defined as:

$$\delta = \frac{T_{cluster}}{A_{base}}, \tag{5.34}$$

| | |
|---|---|
| $T_{\mathrm{cluster}}$ | total thrust in case of multi engine configuration, |
| $A_{\mathrm{base}}$ | total base area enveloping all engines. |

The relevance of the thrust density is due to a very profane reason: engine accommodation space is ultimately limited and spaceships need to be launched in a fairing of limited diameter. This constrain in diameter may demand compromises on the type and number of engines to be used or might necessitate the need of a nozzle extension mechanism, see Sect. 7.1. We will discuss the launcher compatibility check in Chap. 10, in the context of an exemplified stage design. The launcher itself is also driven by several factors. Launchers that are manufactured at different places and transported over some distance for integration, are often limited by infrastructure, e.g. tunnel diameter as is the case for the Russian Soyuz and others. In consequence, if the core stage is not capable of accommodating all engines, additional boosters are required. Modern launch systems, like SpaceX' Starship, are manufactured at the launch site. In this case, launcher design is fully determined by technology and almost not at all by logistics.

Propulsion System Mass Fraction
Since mass is a valuable resource in spaceship design, it is reasonable to introduce a parameter that measures how much of this resource is utilised by the propulsion system. This is achieved by introducing a dedicated performance parameter, the *propulsion system mass fraction*:

$$\mu_{\mathrm{ps}} := \frac{m_{\mathrm{ps}}}{m_0}, \tag{5.35}$$

with

$$m_{\mathrm{ps}} = m_{\mathrm{pms}} + m_{\mathrm{eng,t}}, \tag{5.36}$$

$$m_{\text{pms}} = m_{\text{tank}} + m_{\text{fps}}, \tag{5.37}$$

$m_{\text{ps}}$ propulsion system mass,
$m_{\text{pms}}$ propellant management system mass,
$m_{\text{eng,t}}$ total engine mass, i.e. of the cluster,
$m_{\text{tank}}$ propellant tank mass,
$m_{\text{fps}}$ feed and pressurization mass.

The propulsion system mass includes all propulsion related dry masses, i.e. without propellant. It is, therefore, part of the vessel's dry mass. Ambiguity exists regarding propulsion-related structures, such as the thrust frame, on which the engines are mounted, and the tank support structure. It's crucial to acknowledge that there is no generally agreed *standard* that addresses this topic; each company and agency approach the matter differently and sometimes inconsistently. Hence, standard clarification and agreement are of utmost importance, particularly in multinational programmes.

## 5.5 Comparison of IMEDs and Trade-Off Criteria

Among the listed parameters, thrust and specific impulse are arguably the most decisive features of a propulsion technology. Figure 5.16 depicts the domains occupied by the aforementioned technologies in this two parameter space. The achievable thrust magnitude is essentially a measure of how fast a mission can be carried out, while the specific impulse is a measure of how efficiently the fuel is utilised.

Note the decrease in thrust with increasing specific impulse; this is directly related to the inverse proportional relationship between the two properties discussed above, Eq. 5.31.

*Thermo-chemical propulsion* (TCP) systems span an incredible range in terms of thrust, eight orders of magnitude as we have seen. This renders TCPs the most versatile propulsion technology, boasting a wide range of applications from rocket propulsion (7,900 kN) to satellite fine-pointing (0.02 N). The maximum achievable $I_{sp}$ with non-toxic propellant is about 465 s. We will see later that some chemical substances provide higher values, well above 500 s. Since we are only interested in exothermic chemical reactions, this propulsion system type is generally referred to as chemical propulsion system and abbreviated as *CPS* or *CPPS*.

*Nuclear-thermal propulsion* (NTP) systems promise a higher specific impulse than a TCP, about 900 s which is twice as high but feature a lower thrust magnitude. The experimental NERVA engine achieved 30 kN and upscaling to 100 kN seems possible. The development of this technology was halted in the early 1970s, but has recently regained significant interest from space and defence agencies worldwide. It is rightly regarded an enabler for interplanetary travel, starting in cis-lunar space.

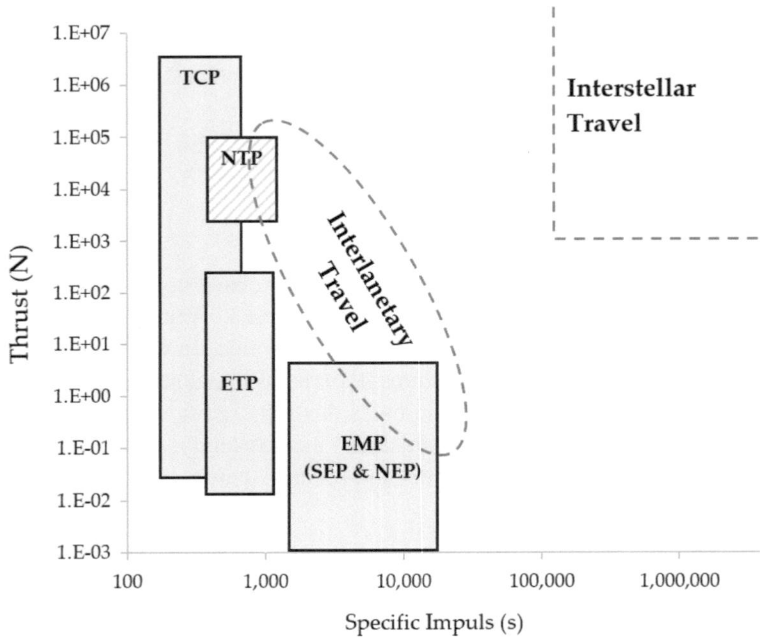


**Fig. 5.16** Propulsion system domains within the thrust to specific impulse domain

*Electro-thermal propulsion* (ETP) systems cater currently a niche market in between TCP and electric propulsion systems. Scalability of this technology towards higher thrust is debatable. Especially the Arcjet faces severe challenges due to the high temperature material requirements. Research on Megawatt actively cooled arcjets is currently pursued aiming to elevate this weakness [14]. The indicative performance figures are impressive: a thrust magnitude of up to 90 N and an $I_{sp}$ of 1100 s if hydrogen is used and 170 N and an $I_{sp}$ of 600 s in case of ammonia as propellant.

*Electromagnetic propulsion* (EMP) systems represented by Gridded Ion Thruster (GIT) and Hall Effect Thruster (HET) have become a veritable alternative to chemical propulsion systems as it is currently the case for geostationary satellites. Furthermore, they show compelling performances for $\Delta v$ demanding missions with high payload mass requirements, as required in interplanetary missions. Within this category are also magnetoplasmadynamic (MPD) propulsion systems. Though, still in its infancy, the concept promises a specific impulse of up to 10,000 s [15]. This domain is shared by solar electric and nuclear electric power sources (SEP, NEP). Both systems produce large amounts of excess heat that needs to be effectively radiated into space via large radiators. The amount of waste heat is, however, much larger in case of NEPs due to the lower efficiency, see Sect. 11.3.

All propulsion systems that utilise electric thruster are referred to as electric propulsion systems, abbreviated as EPS or EPPS. This naming is often interpreted as a reference to the energy source, and electrothermal propulsion systems are therefore almost always subsumed under this naming convention, even though the acceleration principle is purely thermal.

To memorise the different propulsion types, it is helpful to draw an analogy to athletes. The high thrust TCP systems can be compared to a sprinter. They consume their energy very quickly and reach their top speed very fast. The acceleration phase is accordingly short. The second group consists of NTP and ETP systems. These are middle distance runner, not that fast but they can go further reaching their top speed later. The last group, EMP systems are the marathon runner. They are the most efficient group and can go far with little thereby reaching their top speed late. Every comparison with other disciplines has its limits. In contrast to human athletes, the velocity of spaceships propelled with EMPs accumulates over time. This implies that although acceleration is initially significantly lower—by orders of magnitude—it will eventually become the fastest.

The graph features also two further domains: interplanetary and interstellar travel. While the first seems within reach with technology that is readily available within the next two decades, the latter requires a fundamental breakthrough in propulsion technology, like nuclear fusion.

## 5.6 Generic Design Configuration of Space Propulsion

As highlighted in the previous chapter, all currently used propulsion systems are based on the internal momentum exchange drive principle, creating forward motion by cannibalising on their onboard mass. We have identified two applied acceleration mechanisms based on different energy sources: *thermal* and *electric* propulsion systems.[20]

This section aims to provide an overview and guide to the subsequent chapters, building on the understanding gained so far. The upcoming chapters will outline the structure and functions of propulsion systems, using generalisations and specific examples to make the subject more accessible.

Thermal Propulsion Systems

From a technological point of view, thermal propulsion systems differ strongly in many respects but not in their basic structure. Figure 5.17 shows a generic schematic of this propulsion type.

Since the entire system has the sole task of mass ejection, the discussion commences with the physics and technical realisation of the engine nozzle where flow acceleration and collimation takes place, this is subject of Sect. 7.1. The process of energy release is subject of Sect. 8.2 for chemical propulsion systems (CPPS) and Chap. 11 for nuclear thermal propulsion systems (NTP). The supply of propellant to the engine(s) is a crucial task managed by the *Propellant Management System* (PMS), which is subject of Chap. 9. We will see in Sect. 7 how engine thrust and

[20] Solid-fuel is rarely used for propulsion in space. Exceptions are kick-stages like, Magellan or Dawn.

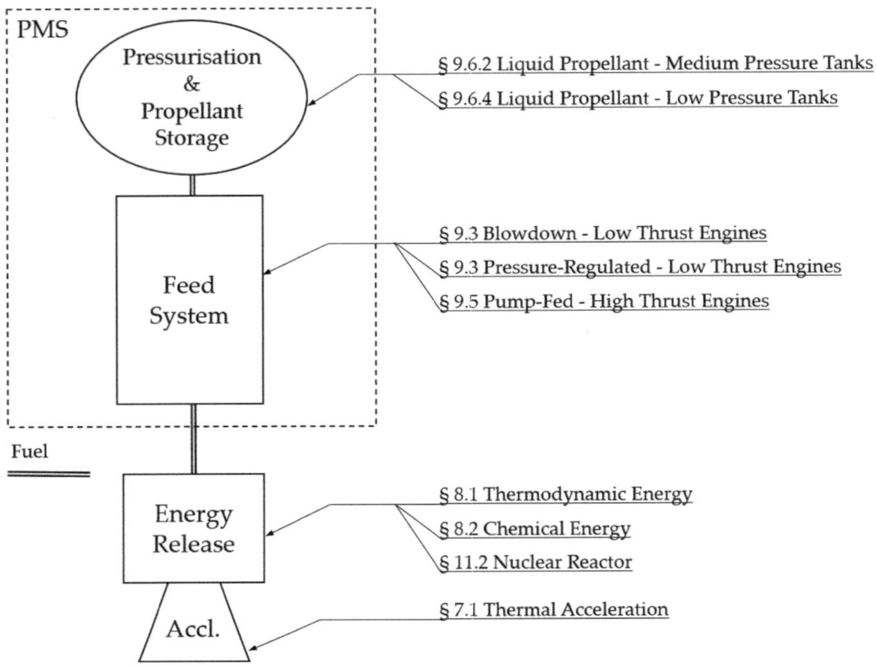


**Fig. 5.17** Functional set-up of thermal rockets with internal energy source

engine chamber pressure are related to each other, which is a prime example of requirement flow-down, since the pressure needs to be provided by the PMS, see Chap. 9, as reflected in the figure. The PMS has also the task to store the liquid or gatehouse propellant in either low and or high pressure tanks as we will see. Tank pressurisation during depletion of liquid tanks is subject of the same sections. Special considerations will be given to cryogenic fuels, especially when it comes to long-term storage.

Electric Propulsion Systems

Like thermal engines, electric propulsion systems share a similar functional design layout, depicted in Fig. 5.18.

A striking difference is that the functional block *Energy Release* needs to be replaced with the string of blocks containing *Power Generation* and *Power Processing*. The discussion starts with the physics of propellant mass acceleration in electrostatic propulsion systems, Sect. 7.2. In contrast to thermal engines, electric thruster need two physical properties: propellant and electric power. While understanding chemical propulsion systems requires basic knowledge of chemistry, electric engines require a basic understanding of power generation in space. This is subject of Sect. 8.3, which discusses solar power generators, currently the only feasible power source for electric thruster. A further difference to thermal propulsion systems is the fact that EPPS need their propellant in gaseous form, which is stored under very

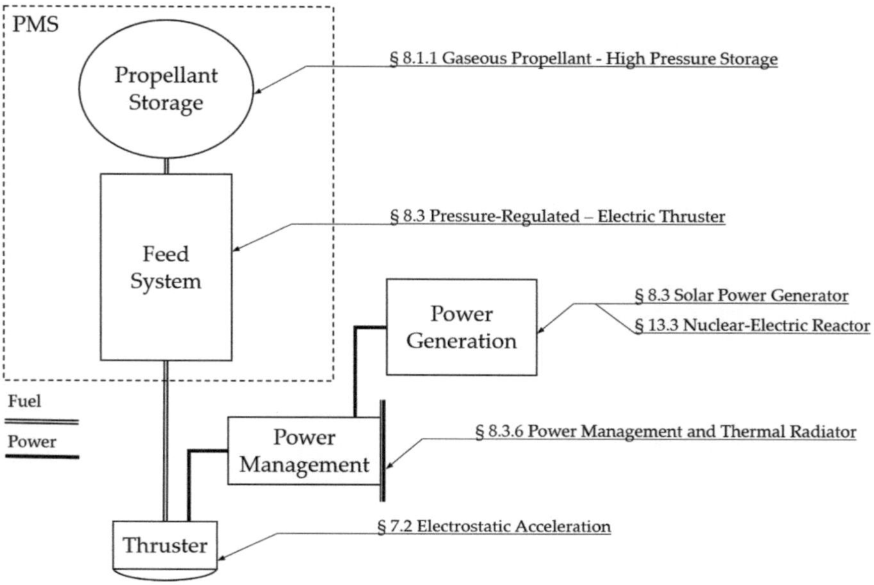


**Fig. 5.18** Functional set-up of electric thruster with external energy source

high pressure of up to 310 bar, in a supercritical state. This renders an additional pressurisation system obsolete. Propellant storage and the feed system of EPPS is subject of Sect. 9.4.

## 5.7 External Momentum Exchange Drive EMED

We conclude this chapter by a brief review of concepts that do *not* consume their onboard mass and are referred here as *External Momentum Exchange Drive*. To create forward motion in an inertial reference frame these systems need to be pushed by another object or medium.

### *5.7.1 Solar Sail*

The concept of *solar sail* is based on the fact that even mass-less photons carry a momentum. The physical theory of describing this field is called *radiation transfer*. We will take a heuristic approach to introduce the matter. The momentum of a single photon is:

$$p_{ph,i} = \frac{E_{ph,i}}{c} = \frac{hc}{\lambda}\frac{1}{c} = \frac{h}{\lambda}, \tag{5.38}$$

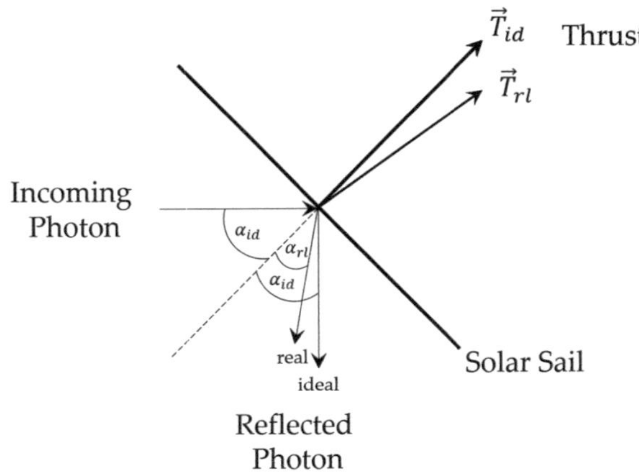


**Fig. 5.19** Photon reflection at solar sail, ideal versus real case

$p_{ph,i}$ momentum of a single photon in Ns,
$E_{ph,i}$ energy of a single photon in J,
$c$ speed of light 299, 792, 458 m/s,
$h$ Planck's constant $6.62607015 \cdot 10^{-34}$ J/Hz,
$\lambda$ wavelength of photon $i$.

The momentum, $p_{ph,i}$, is the energy divided by the speed of light. The energy of a single photon is a function of its wavelength and so is the momentum. The impact of a single photon on a macroscopic objects is small but our Sun submits myriads of photons per second into space. Figure 5.19 shows the principle idea. A stream of photons hits the surface of a solar sail and experiences a momentum exchange. The exerted momentum is largest if the impact is perfectly elastic (i.e. ideal). The photons are then perfectly reflected (angle of incidence equals angle of reflection). The resulting force is the thrust and it points in this ideal case in direction of the angle bisector. In reality, however, reflection is not ideal and about 10% of the irradiance is absorbed leading to a slight tilt in the thrust vector.

In order to quantify the achievable thrust, we need to have an expression for the radiation pressure over the entire solar spectrum. This expression can be derived by taking into account that the unit of pressure is 'force per unit area' and that force is the time derivative of momentum, we then arrive quickly to an expression for the radiation pressure:

$$P_{\text{rad}} = \frac{F(r)}{c}, \tag{5.39}$$

$$= \frac{1}{c}\frac{L_{\text{s}}}{4\pi r^2}, \tag{5.40}$$

$P_{\text{rad}}$ radiation pressure,
$F(r)$ solar flux at distance $r$,
$L_s$ solar luminosity $3.846 \times 10^{26}$ W,

We have already determined the expression of the flux density $F(r)$, Eq. 3.5. This yields a pressure of $4.50312 \cdot 10^{-6}$ N/m$^2$. This is in fact a tiny value. To illustrate this; a space probe of a mass of 1 kg and a solar sail of 1 m$^2$ needs 257 days to accelerate to a velocity 100 m/s. The obvious conclusion is that a larger solar sail is needed, which highlights the key feature of solar sail systems, large surface-to-mass ratios. This comes with the challenge to stabilise and steer such large structures in a controlled way. A further challenge comes from the inverse law of the flux density,

$$F \propto 1/r^2. \tag{5.41}$$

As evaluated in Table 3.1, the flux decrease rapidly with distance from the Sun. In consequence, solar sails are most effective the closer the object is to the Sun, that is if the Sun is the only radiation source. Key challenges remain for the sail design [16]:

- high-efficiency packing,
- three axis stabilisation of large solar sail structure,
- membrane with very good reflectivity properties as well as resilience against space environment, especially against ultraviolet radiation exposure,
- self-deployment mechanisms,
- booms with high specific stiffness and strength,
- effective attitude control.

Despite these challenges, the technology promises to be an enabler for very specific scenarios. Generally speaking, missions into the inner part of the solar system are more costly in terms of $\Delta v$ than into the outer solar system. Solar sails could help to improve access to this region. If carefully conducted, solar sails can be used for aerobreaking manoeuvre in which the rest atmosphere of the planet is used to achieve rendezvous.

Finally, it should be noted that there are concepts that propose using a laser beam to push the solar sail, either as a complementary means or to enhance the solar flux [17].

Implementation and Technological Status

The first solar sail experimental mission, Znamya 2, was conducted by the soviet union in 1993. The recent Japanese IKAROS mission, launched May 2010 demonstrated further key capabilities:

- unfolding of a 173 m$^2$ solar sail of $14 \times 14$ m,
- power generation through ultra thin solar cells printed on solar sail, called flexible solar array (FSA),
- active attitude control via liquid crystal device (LCD).

The membrane, consisting of polyimide, measures only 7.5 $\mu$m in thickness and is covered with a 80 nm evaporated aluminium layer. The probably most innovative part

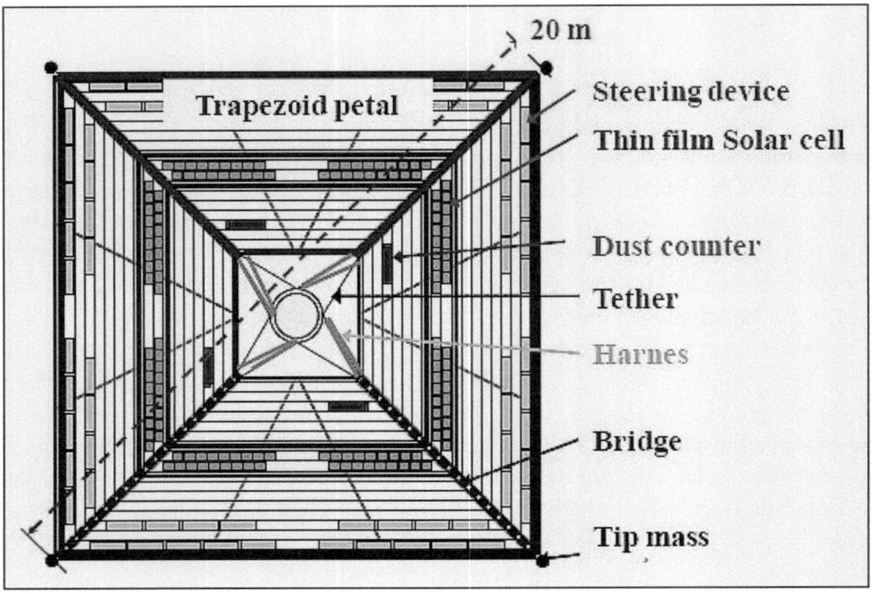


**Fig. 5.20** Solar sail set-up of IKAROS [19]. *Credit* JAXA

of the experiment was the third point in the list. The liquid crystal device was part of a Reflectivity Control Device (RCD). As the name suggests, this device could alter its reflectivity by applying a voltage, thus changing the momentum transfer with the sun. Differential actuation of RCDs, that are distributed over the solar sail, yields to a torque in the desired direction [18]. This is an important step towards a three-axis stabilised solar sail propelled spacecraft. Since IKAROS several other in-space experiments have been conducted mainly on a small scale (Fig. 5.20). Solar cruiser, a promising large scale NASA mission with a solar sail of $1200\,m^2$ has been cancelled in 2023. It would have been the first solar sail mission with an actual science case, beyond technology demonstration. The primary mission objective was to image and characterise a near-Earth asteroid (NEA) during a slow flyby. The solar sail was supposed to have a total mass of 90 kg.

## References

1. Vance, L., Espinoza, A., Dominguez, J., Rabade, S., Liu, G., & Thangavelautham, J. (2024). A solar thermal steam propulsion system using disassociated steam for interplanetary exploration. *Aerospace, 11*, 84. https://doi.org/10.3390/aerospace11010084
2. Britannica, T. Editors of Encyclopaedia. (2024). "Wernher von Braun." Encyclopedia Britannica. https://www.britannica.com/biography/Wernher-von-Braun
3. Arlazorov, M. S. (2023). Konstantin Tsiolkovsky. Encyclopedia Britannica. Retrieved March 29, 2024, from https://www.britannica.com/biography/Konstantin-Eduardovich-Tsiolkovsky

4. Lehman, M., & Lehman, M. K. (2024). Robert Goddard. Encyclopedia Britannica. Retrieved March 6, 2024, from https://www.britannica.com/biography/Robert-Goddard
5. Britannica, T. Editors of Encyclopaedia. (2024). Hermann Oberth. Encyclopedia Britannica. Retrieved March 29, 2024, from https://www.britannica.com/biography/Hermann-Julius-Oberth
6. Jahn, R., & Choueiri, E. (2003). Electric propulsion. In *Encyclopedia of physical science and technology* (pp. 125–141). https://doi.org/10.1016/B0-12-227410-5/00201-5
7. Goebel, D., Brophy, J., Polk, J., Katz, I., & Anderson, J. (n.d.). Variable specific impulse high power ion thruster.
8. Foster, J., Haag, T., Patterson, M., Williams, J, G., Sovey, J., Carpenter, C., Kamhawi, H., Malone, S., & Elliot, F. (2004). The high power electric propulsion (HiPEP) ion thruster. https://doi.org/10.2514/6.2004-3812
9. Burns, D. E., Lenox, K. E., O'Brien, R. C., Rieco, I., Palomares, K. B., Searight, W., Todosow, M., & Werner, J. Options for subscale maturation of advanced reactor technologies testing for nuclear thermal propulsion. United States. https://doi.org/10.2172/1844192
10. Steiger, C., Romanazzo, M., Emanuelli, P. P., Floberghagen, R., & Fehringer, M. (2014). The deorbiting of ESA's gravity mission GOCE—Spacecraft operations in extreme drag conditions. https://api.semanticscholar.org/CorpusID:123881052
11. Dale, E., Jorns, B., & Gallimore, A. (2020). Future directions for electric propulsion research. *Aerospace, 7*, 120. https://doi.org/10.3390/aerospace7090120
12. Sutton, G. P., & Biblarz, O. (2017). *Rocket propulsion elements*, 9th ed. Wiley (US).
13. Wollenhaupt, B. L. (2017). Die Entwicklung thermischer Lichtbogentriebwerkssysteme, Thesis DUniversity Stuttgart. https://doi.org/10.18419/opus-10059
14. Litchford, R. (2011). *High power hydrogen arcjet performance characterization*. https://doi.org/10.2514/6.2011-4013
15. LaPointe, M., Strzempkowski, E., & Pencil, E. (2004). High power MPD thruster performance measurements. In *40th AIAA/ASME/SAE/ASEE Joint Propulsion Conference and Exhibit*. https://doi.org/10.2514/6.2004-3467
16. Zhao, P., Wu, C., & Li, Y. (2023). Design and application of solar sailing: A review on key technologies. *Chinese Journal of Aeronautics, 36*(5), 125–144. https://doi.org/10.1016/j.cja.2022.11.002
17. Carzana, L., Dachwald, B., & Noomen, R. (2017). Model and trajectory optimization for an ideal laser-enhanced solar sail.
18. JAXA. Press Release. (2010). Small solar power sail demonstrator 'IKAROS' successful solar sail deployment. https://www.jaxa.jp/press/2010/06/20100611_ikaros_e.html
19. eoPortal. (2012). IKAROS—Interplanetary kite-craft accelerated by radiation of the sun. https://www.eoportal.org/satellite-missions/ikaros#solar-sail

# Chapter 6
# Rocket Equations and Spaceship Design

**Abstract** The rocket equation is rightfully the most known and most significant equation in space engineering. In this paragraph, we will show that to fully grasp its essence it is important to distinguish between two of its forms: the classical form, known as the Tsiolkovsky equation, in which the energy is *propellant intrinsic* (e.g. chemical propulsion) and the form in which the energy is *propellant extrinsic* (e.g. solar electric and nuclear). We will discuss the strengths and limits of this equation in the context of spaceship design as well at its key parameter.

## 6.1 The Classical Tsiolkovsky Equation

The principle of rocket motion was presumably derived and published first by the British Mathematician Sir William Moore in 1813[1] in 'A Treatise on the Motion of Rockets' [1]. But, it was not until the remarkable Konstantin Tsiolkovsky, a Russian mathematician and self-educated aerospace scientist that this equation was brought into its rightful context, space travel. Tsiolkovsky has derived the rocket equation independently from Moore, so did the US rocket pioneer Robert Goddard and the Austro-Romanian Physicist Hermann Oberth two decades later.

In Sect. 5.1 the internal momentum exchange principle was introduced; spaceships utilising this principle propel themselves by ejecting board mass, i.e. propellant, into space. For all those spaceships and spacecrafts the rocket equation applies and is essential in assessing the vessel's capabilities in terms of achievable velocity change, $\Delta v$. The equation is not applicable to vehicles with a propulsion system that does not consume and expel board mass, e.g. solar sails.

The Tsiolkovsky equation is derived from Newton's second law, which states that in an inertial reference frame, the rate of change of an object's linear momentum over time is equal to the sum of the external forces acting on it. In the language of mathematics this is:

[1] Sir William Moore (1813$^{\dagger}$) visited the Royal Military Academy at Woolwich, England.



F. Gamgami, *Space Propulsion and Spaceship Design*, Springer Praxis Books,
https://doi.org/10.1007/978-3-031-71336-1_6

$$\frac{d\mathbf{I}}{dt} = \sum \mathbf{F}_{\text{ext,i}}, \tag{6.1}$$

| | |
|---|---|
| $\mathbf{I}$ | total linear momentum of the spaceship, |
| $t$ | time variable, |
| $\mathbf{F}_{\text{ext,i}}$ | external force of contributor '$i$' acting on the spaceship. |

Assuming that the rocket is in free space with perfect vacuum and no external forces acting on it, the right hand side in Eq. 6.1 is zero and the spaceship can only start moving by redistributing its internal momentum, as discussed in Sect. 5.1. This achieved by ejecting mass, more precisely by ejecting propellant $\Delta m_\text{p}$ with a high velocity. The rocket then moves into the opposite direction of the ejected mass as illustrated in Fig. 5.2. This is the starting point for the derivation of the rocket equation. Two derivations, a heuristic and a formal one, are provided in Appendix C. We will take a 'fast track' at this point and make use of the equation of motion introduced in Sect. 2.2.3:

$$\dot{v} = -\frac{\mu}{r^2}\sin\gamma + \frac{T_\text{T}}{m_2} + f_{\text{p,T}}. \tag{2.38}$$

The mass $m_2$ is in this context the mass of the spaceship. We replace in the following the subscription '2' with 'sc' for spaceship. The subscription 'T', which stands for 'tangential' is redundant since we consider only tangential forces in this 1 dimensional analysis. Furthermore, we consider a force free environment meaning that no gravity nor perturbing forces act on the spaceship. This reduces the equation above to:

$$\dot{v} = \frac{T}{m_\text{sc}}. \tag{6.2}$$

Introducing the compact definition of thrust with the effective velocity $c_\text{e}$, $T = \dot{m}c_\text{e}$ (Eq. 5.20) thereby considering that the exhaust mass flow by the propellant $\dot{m}$ reduces the spacecraft mass, $\dot{m}_\text{sc} = -\dot{m}$, yields:

$$\dot{v} = -\frac{\dot{m}_\text{sc}c_\text{e}}{m_\text{sc}}, \tag{6.3}$$

$$\frac{\text{d}v}{\text{d}t} = -c_\text{e}\frac{1}{m_\text{sc}}\frac{\text{d}m_\text{sc}}{\text{d}t}. \tag{6.4}$$

The infinitesimal time duration d$t$ appears on both sides and it says that both sides of the equation experience the same change over d$t$, namely:

$$\text{d}v = -c_\text{e}\frac{\text{d}m_\text{sc}}{m_\text{sc}}. \tag{6.5}$$

Integration from 0 to the end of manoeuvre (EOM):

$$\int_{0}^{\mathrm{EOM}} dv = -c_e \int_{0}^{\mathrm{EOM}} \frac{1}{m_{sc}} dm_{sc}, \tag{6.6}$$

finally yields the famous Tsiolkovsky equation:

$$\Delta v = c_e \ln\left(\frac{m_0}{m_{\mathrm{EOM}}}\right), \tag{6.7}$$

$\Delta v$ achieved velocity change of the spaceship from begin to end of manoeuvre: $\Delta v = v_{\mathrm{EOM}} - v_0$,
$c_e$ effective velocity of the exhaust gas observed from the spaceship's reference frame,
$m_0$ mass of fully loaded spaceship at begin of the manoeuvre, called wet mass
$m_{\mathrm{EOM}}$ mass of the spaceship at the end of manoeuvre, $m_{\mathrm{EOM}} = m_0 - \Delta m_p$, with consumed propellant mass $\Delta m_p$.

The Tsiolkovsky equation, tells us immediately the velocity increment, $\Delta v$, attainable by a spaceship with an initial mass $m_0$, when the propulsion system ejects the propellant mass $\Delta m_p$ at an effective velocity of $c_e$ into space. It can be reformulated to determine the necessary propellant mass for given $c_e$ and $m_0$:

$$\Delta m_p = m_0 \left[1 - e^{-\frac{\Delta v}{c_e}}\right]. \tag{6.8}$$

Both Eqs., 6.7 and 6.8, are important in the design process of the mission architecture and the spaceship itself. Equation 6.7 is used to compute the achievable $\Delta v$ for a given propellant mass, whereas Eq. 6.8 is used to compute the required propellant mass for a given $\Delta v$, see process description Fig. 6.1.

### *6.1.1 Performance Parameters and Limits of the Rocket Equation*

There are two performance parameters hidden in the Tsiolkovsky equation: the specific impulse and the dry mass index. We have already made acquaintance with the first parameter in Sect. 5.4, it quantifies the efficiency of propellant utilisation by the propulsion system. The latter parameter measures the structural efficiency of the overall spaceship. The two parameter shall be discussed in the context of the rocket equation and their implications for spaceship design.

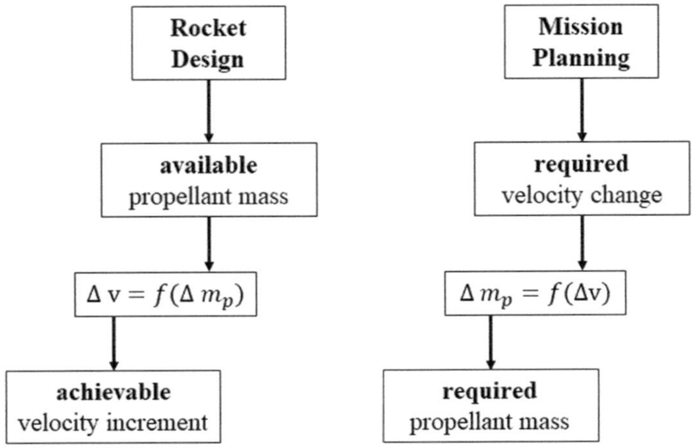


**Fig. 6.1** The two potential design processes using the Tsiolkovsky equation

Specific impulse and the rocket equation

Inserting the definition of the specific impulse, Eq. 5.21, into the two forms of the rocket equation yields:

$$\Delta v = I_{\mathrm{sp}} g_0 \ln\left(\frac{m_0}{m_0 - \Delta m_{\mathrm{p}}}\right), \tag{6.9}$$

and

$$\Delta m_{\mathrm{p}} = m_0 \left[1 - e^{-\dfrac{\Delta v}{I_{\mathrm{sp}} g_0}}\right]. \tag{6.10}$$

The relation between $\Delta v$ and $I_{\mathrm{sp}}$ is obviously linear whereas the relation with $m_{\mathrm{p}}$ is more intricate. It is instructive to plot the two relations for five spaceships of 100 t wet mass each but different specific impulses. Figure 6.2 depicts the achievable velocity increment as function of the available propellant mass and the specific impulse as a performance parameter. Let us assume all spaceships have loaded the same amount of propellant, namely 50 t, which results in a propellant mass fraction of 50 %. A spaceship with a specific impulse of 4500 s could then achieve more than 30 km/s when consuming all its propellant, whereas a spaceship with a specific impulse of 450 s achieves only 3 km/s. This disparity becomes larger the more propellant is consumed relative to the spaceship's wet mass. Figure 6.3 depicts the required propellant mass as function of the specified velocity demand (from mission analysis, Sect. 2.2.3), with the same $I_{\mathrm{sp}}$ variation as before. The logarithmic nature of the relation becomes more apparent for low specific impulse values. Consider that all spaceships are supposed to achieve a velocity change of 5 km/s. The spaceship with an $I_{\mathrm{sp}}$ of 4500 s requires for this feat only 11 t of propellant while the spaceship with

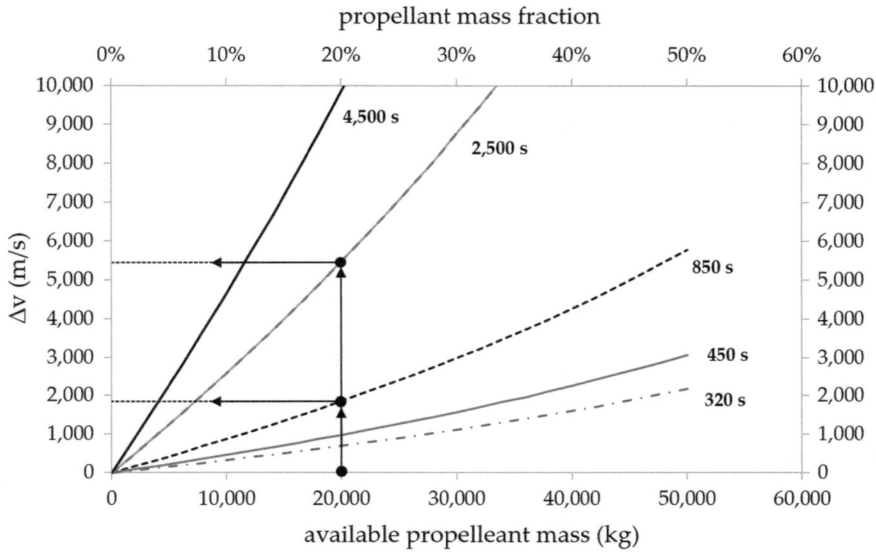


**Fig. 6.2** Effect of specific impulse on delta v capability as function of available propellant mass

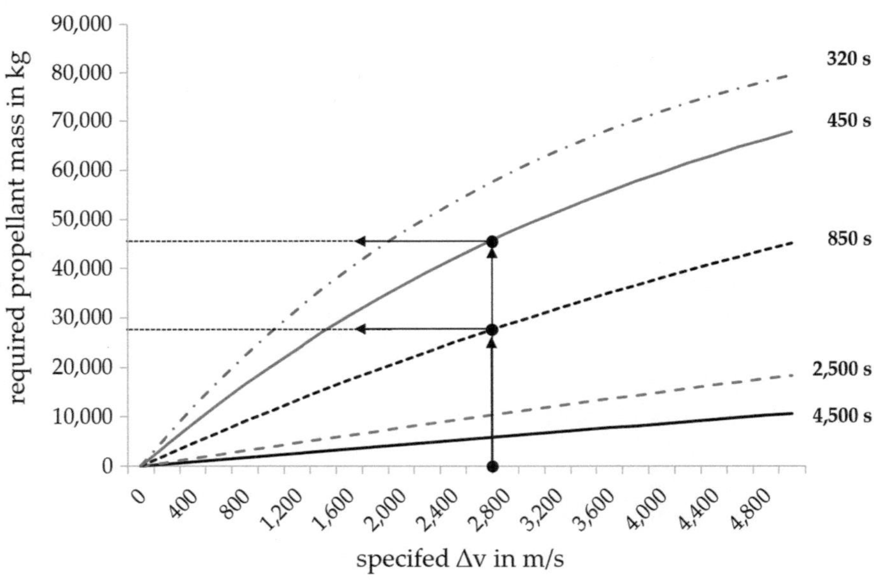


**Fig. 6.3** Comparison of the effect of specific impulse on required propellant mass as function of specified delta v demand

an $I_{sp}$ of 450 s needs 68 t of propellant leaving only 32 t for other subsystems and payload. The superiority of high $I_{sp}$ propulsion systems is indeed compelling.

On the other hand, if the velocity demand is small as is the case for typical LEO satellites, the specific impulse plays a subordinate role in the trade-off. Other criteria such as reliability, cost efficiency and technical integration effort weight more heavily in the decision process.

Dry Mass Index

The second hidden performance parameter in the Tsiolkovsky equation, the dry mass index $\sigma$, is defined as the ratio of all dry mass of the spaceship to the total loaded propellant mass:

$$\sigma = \frac{m_{\mathrm{dry}}}{m_{\mathrm{p}}}. \tag{6.11}$$

This definition goes back to classic rocketry where it is called *structural index* instead. This is understandable since structural mass $m_s$ is the dominant item in the dry mass budget of a rocket stage, hence $m_s \simeq m_{\mathrm{dry}}$. This is in general not the case for satellites and planetary landing systems. We will therefore adopt the more general term *dry mass index* throughout this textbook. Special attention is required in other contexts.

The aim of introducing this dimensionless parameter was to have a figure of merit at hand with which the structural performance of a design can be characterised and measured. The engineers wanted to know how much physical structure is needed to carry a certain amount of propellant mass and than be able to compare different designs. The dry mass index is, therefore, a function of clever design on the one hand and material selection on the other. The lower $\sigma$ the better the structural design and the less dead[2] mass needs to be accelerated. We will discuss in Sect. 9.6.4 some design methods to save structural mass.

Table 6.1 shows the structural indices of all three stages of the Saturn V rocket as an instructive example. It is important to note that the index increases with decreasing propellant mass, or in other words, the smaller the stage the larger the fraction of structural mass. This is in fact a direct consequence of the surface to volume ratio of hollow bodies.

The relation of the dry mass index to the rocket equation can be derived by re-examining the spaceship's wet mass constituents:

$$m_0 = m_{\mathrm{dry}} + m_{\mathrm{p}} + m_{\mathrm{pl}}, \tag{6.12}$$

$m_0$ wet mass, i.e. fully loaded spaceship,
$m_{\mathrm{p}}$ loaded propellant mass,
$m_{\mathrm{pl}}$ carried payload mass.

[2] Dead mass is an expression from rocketry that refers to mass that is not propellant but needs to be accelerate. Another expression is *inert mass*. The dead mass problem will lead us to the need of staging, see Sect. 6.1.2.

**Table 6.1** Specification of Saturn V stages

| Stage | Gross mass (t) | Empty mass (t) | Prop mass (t) | $\sigma$ |
|---|---|---|---|---|
| 1 | 2,286,217 | 135,218 | 2,150,999 | 0.063 |
| 2 | 490,778 | 39,048 | 451,730 | 0.086 |
| 3 | 119,900 | 13,300 | 106,600 | 0.125 |

We shall ignore in the following the payload mass and assume that the spaceship has to carry only itself. Inserting the mass definition into the rocket equation and assuming that all propellant is consumed to achieve the maximum possible velocity increment, yields:

$$\Delta v = I_{sp} g_0 \ln\left(\frac{m_{dry} + m_p}{m_{dry}}\right). \tag{6.13}$$

Rearranging the mass ratio in brackets

$$\left(\frac{m_{dry} + m_p}{m_{dry}}\right) = \left(1 + \frac{m_p}{m_{dry}}\right) = \left(1 + \frac{1}{\sigma}\right) = \left(\frac{\sigma + 1}{\sigma}\right),$$

leads to an expression of the Tsiolkovsky equation with the dry mass index:

$$\Delta v = I_{sp} g_0 \ln\left(\frac{\sigma + 1}{\sigma}\right). \tag{6.14}$$

What makes this expression remarkable is the lack of any information about the absolute mass of the rocket.

Limits of the Tsiolkovsky Equation

As shown above the $\Delta v$ performance of a spaceship depends only on the two performance parameter, $I_{sp}$ and $\sigma$:

$$\Delta v = f(I_{sp}, \sigma). \tag{6.15}$$

Remarkably, there is no absolute information about the spaceship in this form. The Total mass could be 50,000 t or merely 500 kg. A closer look at the Tsiolkovsky equation in form of Eq. 6.14, reveals a further fascinating feature: there is no limit in the achievable $\Delta v$ as long as the dry mass index can be decreased to arbitrarily low values. In the limit of $\sigma \to 0$ the equation predicts an infinite velocity increase:

$$\lim_{\sigma \to 0} \Delta v(\sigma) = \infty. \tag{6.16}$$

This is certainly not correct and we have encountered a limitation of the Tsiolkovsky equation. Though a trivial statement, engineers in early design phases are often tempted to over-tweak the dry mass index in order to achieve the required $\Delta v$ within the given constraints, for instance available wet mass $m_0$. It is therefore mandatory to check with independent analysis tools these important parameters $I_{sp}$ and $\sigma$ for technical feasibility.

We will see later in this chapter that for electric propulsion systems, the two parameters are not fully decoupled from each other and that it takes a significant amount of energy and thus mass to generate the $I_{sp}$.[3] A detailed discussion is provided in Sect. 6.2. Furthermore it is crucial to note that time information is not a variable in the Tsiolkovsky equation, it was 'lost' in the integration,[4] see Eq. 6.6. This means that even if a high $I_{sp}$ and low $\sigma$ can be realised, it could still take years to reach the required velocities. As anticipated in Fig. 5.16 and discussed in detail in Sect. 7.2, it happens that propulsion systems that offer very high $I_{sp}$ offer only low thrust-levels. This is why a launcher with electric propulsion is not feasible.

In addition to these technical limitations, the Tsiolkovsky equation also faces a physical limitation imposed by Einstein's theory of special relativity. With the hope that this will one day be relevant in spaceship design, the relativistic form of the Tsiolkovsky equation is derived and discussed in the Appendix C.2.

In conclusion to this section, two Tables, 6.2 and 6.3 are provided, that list the dry mass indices of launcher stages and the related masses. The first lists only those that loaded cryogenic propellant (hydrolox) and the second those that loaded storable propellant. It is noteworthy that $\sigma$ is in general lower for stages with storable propellant. The reasons for these effects of the fuel on the design are discussed in Sect. 8.2.2. It can be anticipated that the dry mass index does not only correlate with the propellant mass but also with the density of the propellant combination. This makes perfectly sense because denser propellant combinations lead to compact tanks and consequentially less surface. The surface is in first order a measure for the amount of material needed, i.e. mass.

### 6.1.2 *The Concept of Staging*

We will explore the design process highlighted in Fig. 6.1 and focus on the right string in which the velocity increment $\Delta v$ is given and spaceship specification needs to be determined. There are several ways to formulate the problem. We will assume that the payload mass is a requirement, and thus known. We further assume that the dry mass index is known either from a database or from a structural analysis. With these two assumptions and the following three relations:

[3] The currency in aerospace is mass, and indeed in aerospace projects you will often be asked by systems engineers and project managers how much mass your solution costs.

[4] The time dependency was given by the physical quantity thrust.

**Table 6.2** Database of launcher stages utilising hydrolox, sorted by dry mass index [6]

| Rocket stage | $m_0$ (kg) | $m_{dry}$ (kg) | $m_p$ (kg) | $F$ (kN) | $I_{sp}$ (s) | $\sigma$ (−) |
|---|---|---|---|---|---|---|
| Ariane 5 EPC | 186000 | 12700 | 173300 | 1114 | 434 | 0.068 |
| Ariane H155 | 170800 | 12700 | 158100 | 1114 | 430 | 0.074 |
| Saturn Stage 2 | 490778 | 39048 | 451730 | 5165.79 | 421 | 0.080 |
| Centaur V1 | 22825 | 2026 | 20799 | 99.19 | 451 | 0.089 |
| Energia EUS | 77000 | 7000 | 70000 | 1962.03 | 455 | 0.091 |
| Centaur IIIB | 22960 | 2130 | 20830 | 198.32 | 451 | 0.093 |
| Centaur IIIA | 18710 | 1905 | 16805 | 99.16 | 451 | 0.102 |
| Energia M | 272000 | 28000 | 244000 | 1960 | 455 | 0.103 |
| Centaur I | 15600 | 1700 | 13900 | 146.80 | 444 | 0.109 |
| Centaur II | 18833 | 2053 | 16780 | 146.80 | 444 | 0.109 |
| Centaur I | 15600 | 1700 | 13900 | 146.80 | 444 | 0.109 |
| Saturn Stage 3 | 119900 | 13300 | 106600 | 1031.6 | 421 | 0.111 |
| Delta 4H-2 | 30710 | 3490 | 27220 | 110.05 | 462 | 0.114 |
| Centaur G | 23880 | 2775 | 21105 | 146.80 | 444 | 0.116 |
| Delta 4-2 | 24170 | 2850 | 21320 | 110.05 | 462 | 0.118 |
| Delta RS-68 | 226400 | 26760 | 199640 | 3312.76 | 420 | 0.118 |
| Centaur IIA | 19073 | 2293 | 16780 | 185.01 | 449 | 0.120 |
| H-2A-1 | 113600 | 13600 | 100000 | 1098 | 440 | 0.120 |
| H-2-1 | 98100 | 11900 | 86200 | 1078 | 446 | 0.121 |
| Delta 3-2 | 19300 | 2476 | 16824 | 110.03 | 462 | 0.128 |
| Centaur C | 15600 | 1996 | 13604 | 133.45 | 425 | 0.128 |
| CZ-3A-3 | 21000 | 2800 | 18200 | 156.00 | 440 | 0.133 |
| GSLV-3 | 14600 | 2200 | 12400 | 75.05 | 460 | 0.151 |
| Centaur D/E | 16258 | 2631 | 13627 | 131.22 | 444 | 0.162 |
| H-2A LRB | 117000 | 17800 | 99200 | 2196 | 440 | 0.152 |

$$m_0 = m_{pl} + m_p + m_{dry}, \tag{6.17}$$

$$m_p = m_0 \cdot \left(1 - e^{-\frac{\Delta v}{c_e}}\right), \tag{6.18}$$

$$m_{dry} = \sigma m_p, \tag{6.19}$$

$m_0$ wet mass, i.e. fully loaded spaceship,
$m_{pl}$ carried payload mass,
$m_p$ loaded propellant mass,
$m_{dry}$ carried payload mass,

**Table 6.3** Database of storable-propellant stages [6]

| Rocket stage | $m_0$ (kg) | $m_{dry}$ (kg) | $m_p$ (kg) | $F$ (kN) | $I_{sp}$ (s) | $\sigma$ (−) |
|---|---|---|---|---|---|---|
| Titan 3A-1‡ | 116573 | 5443 | 111130 | 2339 | 302 | 0.049 |
| Titan 4-1‡ | 163000 | 8000 | 155000 | 2428 | 302 | 0.052 |
| Titan 3B-1‡ | 139935 | 7000 | 132935 | 2413 | 302 | 0.053 |
| Tsyklon 2-†1 | 122300 | 6400 | 115900 | 2640 | 301 | 0.055 |
| Titan 2-1‡ | 117866 | 6736 | 111130 | 2172 | 296 | 0.061 |
| CZ-4A-1† | 192700 | 9500 | 183200 | 3000 | 289 | 0.063 |
| CZ-3-1† | 151000 | 9000 | 142000 | 3000 | 289 | 0.063 |
| Kosmos-3M-1* | 87200 | 5300 | 81900 | 1740 | 292 | 0.065 |
| Tsyklon 3-1† | 127000 | 8300 | 118700 | 3032 | 301 | 0.070 |
| Proton K-2† | 167828 | 11715 | 156113 | 2399 | 327 | 0.075 |
| Ariane 4-1† | 245900 | 17900 | 228000 | 3034 | 278 | 0.079 |
| Tsyklon 2-2† | 49300 | 3700 | 45600 | 940.4 | 317 | 0.081 |
| Titan 3B-2‡ | 37560 | 2900 | 34660 | 460.3 | 316 | 0.084 |
| Proton K-3† | 50747 | 4185 | 46562 | 630.2 | 325 | 0.090 |
| Titan 2-2‡ | 28939 | 2404 | 26535 | 444.8 | 316 | 0.091 |
| Tsyklon 3-2† | 53300 | 4800 | 48500 | 941 | 318 | 0.099 |
| Titan 3A-2‡ | 29188 | 2653 | 26535 | 453.7 | 316 | 0.100 |
| Ariane 4-2† | 37130 | 3625 | 33505 | 805 | 296 | 0.108 |
| Ariane 2-2† | 37130 | 3625 | 33505 | 805 | 296 | 0.108 |
| Agena D* | 6821 | 673 | 6148 | 71.17 | 300 | 0.110 |
| CZ-4A-2† | 39000 | 4000 | 35550 | 831 | 295 | 0.113 |
| CZ-3-2† | 39000 | 4000 | 35000 | 761.9 | 295 | 0.114 |
| Proton KM-4† | 22170 | 2370 | 19800 | 19.6 | 326 | 0.120 |
| Ariane 2-1† | 160030 | 13750 | 146280 | 2880 | 281 | 0.120 |
| Titan 4-2‡ | 39500 | 4500 | 35000 | 459.5 | 316 | 0.129 |
| PSLV-2† | 45800 | 5300 | 40500 | 725 | 293 | 0.131 |
| Agena B* | 7167 | 867 | 6300 | 71.17 | 285 | 0.138 |
| GSLV-0† | 45600 | 5600 | 40000 | 735 | 281 | 0.140 |
| GSLV-2† | 42900 | 5400 | 37500 | 725 | 295 | 0.144 |
| Tsyklon 2-3† | 3200 | 400 | 2800 | 77.96 | 317 | 0.143 |
| Fregat† | 6535 | 1100 | 5435 | 19.6 | 327 | 0.202 |

* HNO3/UDMH
† N2O4/UDMH
‡ N2O4/A-50

**Table 6.4** Comparison of $\Delta v$ capacity of three space vessels with different staging architecture

| | Spaceship A | Spaceship B | | Spaceship C | | |
|---|---|---|---|---|---|---|
| | Single stage | Stage 1 | Stage 2 | Stage 1 | Stage 2 | Stage 3 |
| $m_{\mathrm{p}}$ | 83,333 | 41,667 | 4,1667 | 27,778 | 27,778 | 27,778 |
| $m_{\mathrm{dry}}$ | 21,667 | 63,333 | 8,333 | 5,556 | 5,556 | 5,556 |
| $m_{st}$ | 100000 | 50000 | 50000 | 33333 | 33333 | 33333 |
| $m_{\mathrm{pl}}$ | 5,000 | 55,000 | 5,000 | 71,667 | 38,333 | 5,000 |
| $m_0$ | 105,000 | 105,000 | 55,000 | 105,000 | 71,667 | 38,333 |
| $\Delta v_{st}$ | 5,378 | 1,984 | 5,561 | 1,206 | 1,924 | 5,061 |
| $\Delta v_{tot}$ | 5,378 | 7,544 | | 8,191 | | |

we can derive a useful equation that relates the total mass and the payload mass

$$\frac{m_{\mathrm{pl}}}{m_0} = \mu_{\mathrm{pl}} = \left[1 - (1+\sigma)\left(1 - e^{-\frac{\Delta v}{ce}}\right)\right]. \tag{6.20}$$

This equation says that the *payload mass fraction* $\mu_{\mathrm{pl}}$ is only determined by three parameter: $\Delta v$, $\sigma$ and $I_{\mathrm{sp}}$ and is a another form of the classical Tsiolkovsky equation. Again, no statement on absolute masses is involved. We will revisit this equation below and examine its full implication.

In the previous section the term dead mass was used to indicate structural mass that needs to be accelerated even though it is in principle no longer needed. To understand the dead mass problem we will examine the total $\Delta v$ capacity of three spaceships.[5] Spaceship A consists of a single stage, spaceship B consists of two equal stages and spaceship C consists of three equal stages. All spaceships and their stages have the same propulsion systems with an $I_{\mathrm{sp}}$ of 400 s and the same dry mass index,[6] $\sigma = 0.2$. The total mass of each spaceship shall be 100 t. All three space vessels shall carry a payload of 5 t such that the initial wet mass is 105 t for each.

Table 6.4 illustrates the striking advantage of staging even under the simplified assumption that all stages within a rocket have equal wet masses. The performance of a two stage vehicle is 40% larger compared to a single stage. This figure is further exceeded by a three stage vehicle for which it reaches 52%. This is why rockets for space transportation are all made up by at least 2 stages. In fact, detailed optimisation for space transportation systems show that launchers with moderate to high specific impulse that target mainly Low Earth Orbit should consist of two stages and those that shall deliver payload to more energetic orbits beyond LEO, such as

[5] The term 'total' shall highlight that we are interested on the stack performance. Stack means the sum of all stages. Instead of total, the reader may also find the term 'stack performance' in the literature.

[6] It would be legitimate to raise the objection that larger stages have lower structural indices as discussed before. A more detailed consideration with more realistic structural indices would, however, not alter the results drastically since the effect is of this correction is of secondary order.

the Geo Synchronous Orbits should have three stages. As a rule of thumb, the lower the specific impulse, the more stages (and absolute propellant mass) are needed to achieve comparable performances in terms of velocity increment and payload.

The Tsiolkovsky equation clearly motivated the introduction of staging but it does not make any statements on the physical architecture and layout of the rocket design. Engineers came up with two distinct concepts for staging: a), tandem staging (like Delta IV Heavy) and serial staging (like Falcon 9 and Starship). A thorough discussion of staging optimisation methods would require too much space and is beyond the objective of this textbook. It shall be emphasized that the mathematical complexity of this optimisation problem is due to different structural indices among the single stages, different specific impulses and the atmospheric ascent in the specific case of launchers. Furthermore, the objective function which in our simple case consists merely of $\Delta v$, is in general more complex as it contains several weighted parameter, e.g. $\Delta v$, transfer time and production cost to name but a few.

### 6.1.3 Extended Tsiolkovsky Equation

For the derivation of the Tsiolkovsky Eq. 6.7 at the beginning of this section we assumed that the spaceship is in a force-free environment. We will drop this restriction and consider the full equation:

$$\dot{v} = -\frac{\mu}{r^2}\sin\gamma + \frac{T_\mathrm{T}}{m_2} + f_{\mathrm{p,T}}, \tag{2.38}$$

We again drop the subscription 'T' for the thrust and replace '2' by 'sc' for reasons explained above. Integration over time yields:

$$\Delta v = \underbrace{-\int_0^{\mathrm{EOM}} \frac{\mu}{r^2}\sin\gamma \,\mathrm{d}t}_{\text{gravity loss}} \underbrace{-c_\mathrm{e}\int_0^{\mathrm{EOM}} \frac{1}{m_\mathrm{sc}} dm_\mathrm{sc}}_{\text{thrust}} + \underbrace{\int_0^{\mathrm{EOM}} f_\mathrm{p}\,\mathrm{d}t}_{\text{perturbation}}\,. \tag{6.21}$$

All three terms cause a velocity change of the spaceship:

$$\Delta v = \Delta v_\mathrm{gl} + \Delta v_\mathrm{thrust} + \Delta v_\mathrm{p}, \tag{6.22}$$

$\Delta v_\mathrm{gl}$ $\Delta v$ loss due to gravity,
$\Delta v_\mathrm{thrust}$ $\Delta v$ achieved by the propulsion system,
$\Delta v_\mathrm{p}$ $\Delta v$ perturbation, in case of atmospheric drag it is a loss.

The middle term, $\Delta v_{\text{thrust}}$, is familiar, it was derived at the start of this section. It did not receive the subscription 'thrust' because it was the sole term and it was derived directly from the acceleration caused by thrust. The first and the third term are new.

The first term, $\Delta v_{\text{gl}}$, expresses the velocity demand of the spaceship to work against gravity. It is zero, if the flight path angle, $\gamma$, is zero and maximum, if the flight path angle is 90°. Consequentially the term is negative and represents a loss and since the loss is due to gravity, it is called *gravity losses*. Launcher for instance lift off vertically, $\gamma = 90°$, and therefore experience a significant amount of gravity losses. An extreme example of gravity losses would be a vehicle that hovers over the ground with a constant acceleration of $1g_0$. All of its propellant would be required to act against gravity without gaining velocity relative to ground. We will encounter gravity losses in Chap. 10 again.

The third term, $\Delta v_{\text{p}}$, is the velocity increment created by any kind of perturbative force acting on the spaceship. The result is very diverse and depends strongly on the direction of the force. Launcher again need to traverse the dense atmosphere thereby experiencing a *aerodynamic drag* which acts against the thrust vector and reduces the exerted velocity increment $\Delta v_{\text{thrust}}$. Another example is *aerobreaking* whereby a spaceship uses the upper atmosphere of a planetary body to reduce its velocity. This has been frequently applied for missions to Mars. Solar sails on the other hand rely fully on the third term and use the solar pressure to achieve a net velocity change, $\Delta v$.

## 6.2 Tsiolkovsky Equation for Electric Propulsion Systems

As anticipated earlier the Tsiolkovsky equation is different for vessels that are not propelled by chemical propellant or any other form in which the energy source is intrinsic to the propellant. The generalised form was first derived for solar electric propulsion (SEP) systems but is in fact valid for all propulsion forms in which propellant and energy source are separate, like nuclear thermal and nuclear electric propulsion systems.

The objective of this section is to derive a form of the Tsiolkovsky equation that is equivalent to Eq. 6.20 in the context of electric propulsion systems. We remember that the right hand side of Eq. 6.20 does not contain absolute masses but only the dimensionless dry mass index $\sigma$. The starting point for this case should be the mass budget equation and the introduction of suitable indices. We therefore re-examine Eq. 6.12 and extract the mass required for the power subsystem, $m_{\text{eps}}$, out of the dry mass. We call the new dry mass simply 'remaining dry mass' $m_{\text{r,dry}}$. This yields:

$$m_0 = m_{\text{r,dry}} + m_{\text{eps}} + m_{\text{p}} + m_{\text{pl}}, \tag{6.23}$$

$$1 = \frac{m_{\text{r,dry}}}{m_0} + \frac{m_{\text{eps}}}{m_0} + \frac{m_{\text{p}}}{m_0} + \frac{m_{\text{pl}}}{m_0},$$

$$1 = \mu_{\text{rd}} + \mu_{\text{eps}} + \mu_{\text{p}} + \mu_{\text{pl}}, \tag{6.24}$$

$\mu_{\text{rd}}$ remaining dry mass fraction after subtraction of the power subsystem system,
$\mu_{\text{eps}}$ mass fraction of the power subsystem system,
$\mu_{\text{p}}$ propellant mass fraction,
$\mu_{\text{pl}}$ payload mass fraction.

With the equations introduced in Sect. 5.4, it is possible to establish relations between these single masses and the propulsion system characteristics by introducing meaningful parameters. We start with the power subsystem and its relation to the propellant mass. Starting point is the expression for the internal propulsion efficiency, Eq. 5.25. We already derived the equation for $P_{\text{jet}}$, Eq. 5.26 but we are missing one for $P_{\text{in}}$. The means to generate the required power depend on the precise method used, e.g. solar electric or nuclear electric. Without knowledge of the method, it is reasonable to assume that the power output will be proportional to the invested mass. The proportionality coefficient is called *specific mass* $\alpha$ and has the unit kg/W:

$$\alpha = m_{\text{eps}}/P_{\text{in}}. \tag{6.25}$$

Introducing this into Eq. 5.30 for $\eta_{\text{T}}$ yields:

$$\eta_{\text{T}} = \frac{1}{2}\dot{m}c_{\text{e}}^2 \cdot \frac{\alpha}{m_{\text{eps}}}. \tag{6.26}$$

In addition, the propellant mass flow $\dot{m}$ can be replaced by the total propellant mass, $m_{\text{p}}$, considering the total manoeuvre time $\tau$:

$$\eta_{\text{T}} = \frac{1}{2}\frac{m_{\text{p}}}{\tau}c_{\text{e}}^2 \cdot \frac{\alpha}{m_{\text{eps}}}. \tag{6.27}$$

We can now use the definitions made above for the single mass fractions to eliminate $m_{\text{p}}$ and $m_{\text{eps}}$ by considering that $m_{\text{p}}/m_{\text{eps}} = \mu_{\text{p}}/\mu_{\text{eps}}$:

$$\eta_{\text{T}} = \frac{1}{2}\frac{\mu_{\text{p}}}{\tau}c_{\text{e}}^2 \cdot \frac{\alpha}{\mu_{\text{eps}}}. \tag{6.28}$$

Solving for $\mu_{\text{eps}}$ yields finally:

$$\mu_{\text{eps}} = \mu_{\text{p}} \cdot \frac{\alpha}{2\tau\eta_{\text{T}}}c_{\text{e}}^2, \tag{6.29}$$

$\tau$ thrust duration (s),
$\alpha$ ratio of power subsystem mass to EP power demand, (kg/kW).

Note that since no other efficiency factor is involved, we drop the index 'T' for $\eta$ in the following. The relation between the propellant mass is given by the classical Tsiolkovsky Eq., 6.8, and divided by the wet mass yields the dimensionless form:

$$\mu_{\mathrm{p}} = \left[1 - e^{-\frac{\Delta v}{c_{\mathrm{e}}}}\right]. \tag{6.30}$$

Finally, the relation between the remaining dry mass and the propellant mass is assumed to be constant and characterised by the dry mass index, $\mu_{\mathrm{rd}} = \sigma\mu_{\mathrm{p}}$. Inserting the three relations for $mu_{\mathrm{eps}}$, $\mu_{\mathrm{p}}$ and $\mu_{\mathrm{rd}}$ into Eq. 6.17 yields the formula for the payload mass fraction in case of an electric propulsion system:

$$\begin{aligned}
\mu_{\mathrm{pl}} &= 1 - \frac{\alpha}{2\tau\eta}c_{\mathrm{e}}^2\left[1 - e^{-\frac{\Delta v}{c_{\mathrm{e}}}}\right] - \left[1 - e^{-\frac{\Delta v}{c_{\mathrm{e}}}}\right]\cdot(1+\sigma), \\
&= 1 - \left[1 - e^{-\frac{\Delta v}{c_{\mathrm{e}}}}\right]\cdot\left(1 + \sigma + \frac{\alpha}{2\tau\eta}c_{\mathrm{e}}^2\right), \\
&= 1 - \left(1 + \sigma + \frac{c_{\mathrm{e}}^2}{v_{\mathrm{ch}}^2}\right)\cdot\left[1 - e^{-\frac{\Delta v}{c_{\mathrm{e}}}}\right].
\end{aligned} \tag{6.31}$$

The term $2\tau\eta/\alpha$ has the dimension of a velocity squared and is called characteristic or *Stuhlinger* velocity, to honour the scientist who contributed to these findings [3]. Comparing Eq. 6.20 with Eq. 6.31 reveals the fundamental difference between these two propulsion systems. Formally speaking, the ratio $c_{\mathrm{e}}^2/v_{ch}^2$ increases the dry mass index and consequentially reduces the payload mass fraction. We therefore introduce a second 'dry mass index', $\sigma_{fep}$, the subscription 'fep' indicates flight with EP (electric propulsion). It is sometimes also referred to as 'external' dry mass. The general form of the Tsiolkovsky equation is then:

$$\mu_{\mathrm{pl}} = 1 - \left(1 + \sigma + \sigma_{fep}\right)\cdot\left[1 - e^{-\frac{\Delta v}{c_{\mathrm{e}}}}\right]. \tag{6.32}$$

It is important to emphasis that this index is by nature *not* a dry mass index but contains information about the engine performance, $c_{\mathrm{e}}$, the mission duration, $\tau$, the engine efficiency, $\eta$, and specific mass $\alpha$:

$$\sigma_{\text{fep}} \propto \frac{\alpha}{\tau\eta} \cdot I_{\text{sp}}^2, \tag{6.33}$$

and it behaves as follows:

$$(\tau \uparrow, \eta \uparrow, \alpha \downarrow) \Rightarrow \sigma_{\text{fep}} \downarrow . \tag{6.34}$$

For chemical propulsion systems $\alpha$ is zero and so is the index and we return to the classical form, Eq. 6.20. The influence of the specific impulse is twofold in the generalised form: on the one hand it increases the external dry mass index, $\sigma_{\text{fep}}$, on the other hand it reduces the propellant demand. To understand the derived function better, it is helpful to take a step back and to examine its constituents. The dependency of the single mass contributors is depicted in Fig. 6.4 for reference parameters. As we already know, the propellant mass decreases with specific impulse, as does the dry mass fraction. However, in case of an electric propulsion system, the mass required to power the propulsion system increases with specific impulse, Eq. 6.29. This contributor to $\mu_{\text{pl}}$ does not exist for chemical propulsion systems, due to the simple fact that the fuel *is* the energy source. Consequently the competing relations lead to an *optimum* specific impulse for which the payload mass fraction has a maximum. This is in striking contrast to thermal rockets, where $\mu_{\text{pl}}$ grows monotonic with $I_{\text{sp}}$. The directive, the higher, the better, is not anymore valid for electric propulsion systems and a case by case analysis is needed.

In order to deepen the understand of the impact of these four parameter ($\alpha$, $\Delta v$, $\eta$, $\tau$) on the payload mass fraction, it is helpful to perform a *sensitivity*

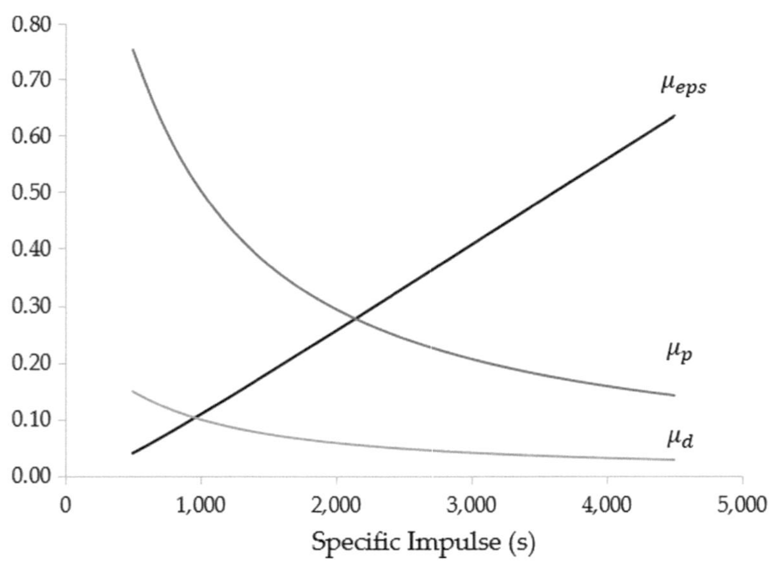


**Fig. 6.4** Mass fraction as function of $I_{\text{sp}}$

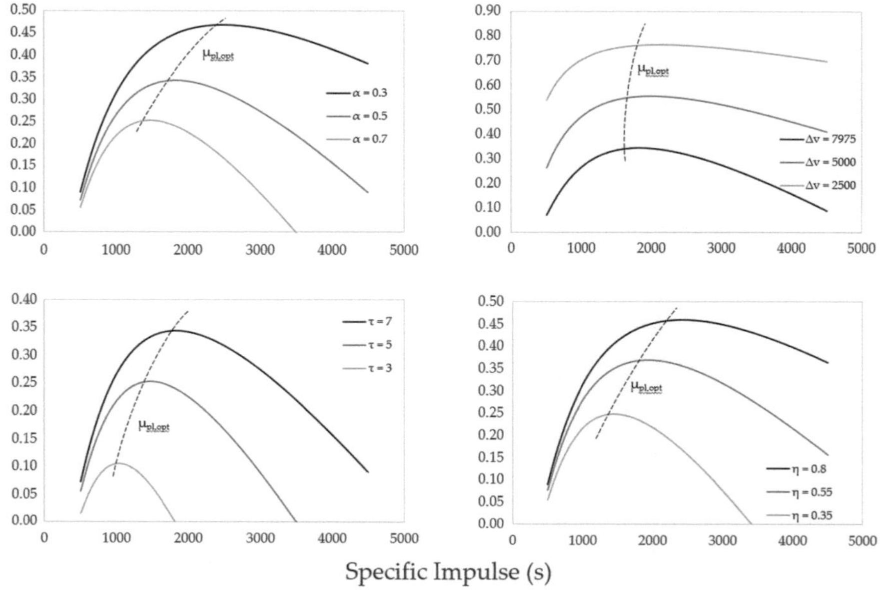


**Fig. 6.5** Parameter trend analysis for the payload mass fraction

*analysis*. The results are depicted in Fig. 6.5 and the dependencies can be explained as follows:

*dependency on* $\Delta v$: the higher the velocity increment that needs to be achieved the more propellant is needed. This increases firstly $\mu_\mathrm{p}$ and via Eq. 6.29 also $\mu_\mathrm{eps}$

*dependency on* $\alpha$: the less mass is required to produce 1 kW the more mass is available for the payload. Given a constant total mass, this means a higher payload mass fraction $\mu_\mathrm{pl}$ the lower $\alpha$.

*dependency on* $\tau$: if more time is available to achieve a velocity change $\Delta v$, less thrust needs to be exerted and consequently less power to be generated, see Eq. 5.29, for constant $I_\mathrm{sp}$

*dependency on* $\eta$: an efficient energy conversion means that less power generation capacity is needed to supply the requested power. In consequence the EPS mass can be lower in favour of a higher payload mass.

While velocity change and flight time are practically determined by the mission requirements, the other two parameters, as discussed in Appendix B, are of a purely technical nature. Although both important, one may argue that the most decisive parameter in current research shall be $\alpha$, i.e. the capability to generate electricity with a small amount of mass. This is because engines with efficiencies up to 0.8 do already exist as we will see in Sect. 7.2. Hence, as tempting as the idea of an ultra high specific impulse system may seem, it is only meaningful, if there is a power generation system available with a low $\alpha$ value.

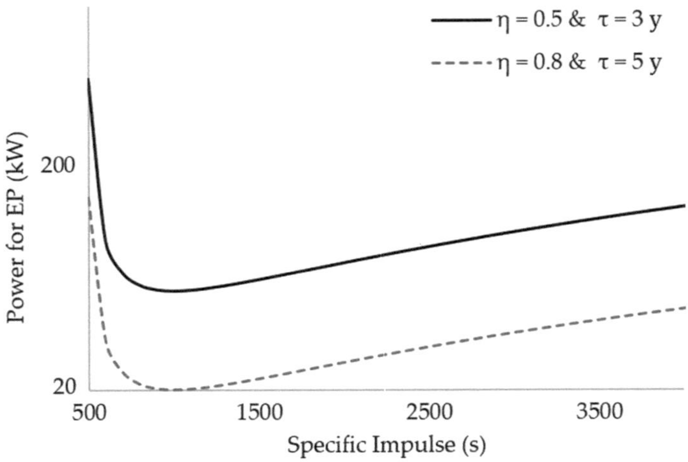


**Fig. 6.6** Parameter trend analysis for the power to $I_{sp}$ relation

Flight time $\tau$ and propulsion efficiency $\eta$ have a considerable effect on the power demand of the propulsion system, while their direct impact on the wet mass is small, see discussion in Sect. 10.3. The propulsion power demand on the other hand is driven by a short flight requirement and low engine efficiencies:

$$(\tau \uparrow, \eta \uparrow) \Rightarrow P_{EP} \downarrow . \tag{6.35}$$

The combined effect is depicted in Fig. 6.6.

Increasing the engine's efficiency from 50 to 80 % leads to a reduction of 40 % in power demand. Equally, allowing the spaceship a longer time of flight (3 yr → 5 yr) leads to a similar saving in power demand. Gridded ion Thruster exhibit in general a higher efficiency factor than Hall Effect Thruster, Sect. 7.2. These relations will be revisited in the context of the conceptual design of *C-One* in Sect. 10.3.

In conclusion the here established set of equations for the generalised Tsiolkovsky equation will be referred in the following as the 'EP model' for conciseness.

PPE—Power and Propulsion Element

Now that we have analysed the basic relationships, it is time to apply them to a real case study. NASA's PPE (Power and Propulsion Element) for the lunar gateway is a suitable example, as it is a *space tug* with an electric propulsion, Fig. 6.7. It has a power generation capacity of 50 kW averaged over its lifetime [5].

The task of a space tug is to transport a payload from an initial obit to a specified destination and often back. It is basically a space logistic vehicle dominated by its propulsion and solar electric subsystems. According to the lunar gateway schedule, the PPE will launch together with HALO, the Habitation and Logistics Outpost, on a Falcon 9 Heavy.

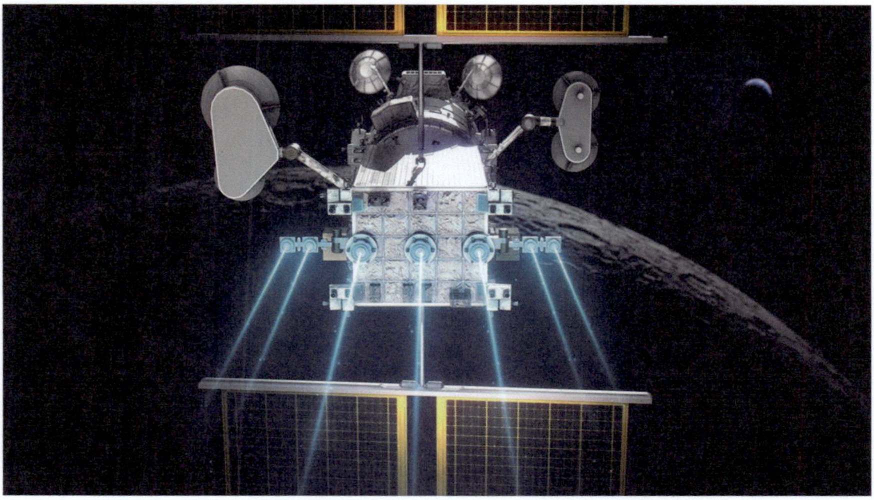

**Fig. 6.7** Power and propulsion element (PPE) with HALO module. Credit: NASA

**Table 6.5** Mass specifications of stack and PPE

| | | |
|---|---|---|
| $m_\mathrm{p}$ | 2.77 t | xenon |
| $m_\mathrm{dry}$ | 2.23 t | Dry mass |
| $m_\mathrm{ppe}$ | 5.0 t | Wet mass |
| $m_\mathrm{pl}$ | 10.0 t | HALO |
| $m_0$ | 15.0 | Stack mass |

Dual launches like this are called *co-manifested*. In this case, the cost for a second launch is saved and the risk of failure during in-orbit docking is eliminated [5]—at the expense of having two eggs in a single basket. Table 6.5 lists the publicly available masses from which we can compute the required mass fractions: $\mu_\mathrm{pl} = 0.667$, $\mu_\mathrm{p} = 0.185$ and $\mu_\mathrm{d} = \mu_\mathrm{eps} + \mu_\mathrm{rd} = 0.15$. Table 6.6 lists the specifications of PPE's electric propulsion system, which is a HET. Information on the EPS mass (i.e. solar generator, solar array drive mechanism, power management and distribution) is not publicly available, neither is the dry mass index $\sigma$ given. Reverse engineering necessitates to narrow down the missing parameters by (a) bottom up analysis (b) by analogy based on a rigorous literature search or (c) best engineering guess (BEG). The EP model parameter for the PPE are listed in Table 6.7.

For practical reasons, we will adopt the latter approach and compute the $m_0 - I_\mathrm{sp}$ diagram for the PPE space tug. Reasonable values are $\sigma = 0.3$ leading to $\alpha = 0.029\,\mathrm{kg}\backslash\mathrm{W}$. The result shows that the actual design of the PPE although close, does not coincide with the optimum (given our assumptions). This is because the rate of increase of the payload mass fraction with specific impulse reduces towards the optimum (the curve becomes flatter towards the optimum).

**Table 6.6** HERMeS nominal specifications [?]

| | | |
|---|---|---|
| $P_{st}$ | 12.5 kW | Per thruster string |
| $V_{dc}$ | 600 V | Max |
| $I_{dc}$ | 20.8 A | Max |
| $F$ | 590 mN | Max per thruster |
| $I_{sp}$ | 2780 s | Max |
| $\eta$ | 0.67 | Max |

**Table 6.7** Parameter of EP model for the PPE

| | |
|---|---|
| $\tau$ | $3.2 \cdot 10^7$ s |
| $\sigma^*$ | 0.3 |
| $\alpha^*$ | 0.029 kg/W |
| $v_{ch}$ | 32.4 km/s |
| $\Delta v$ | 5510 m/s |

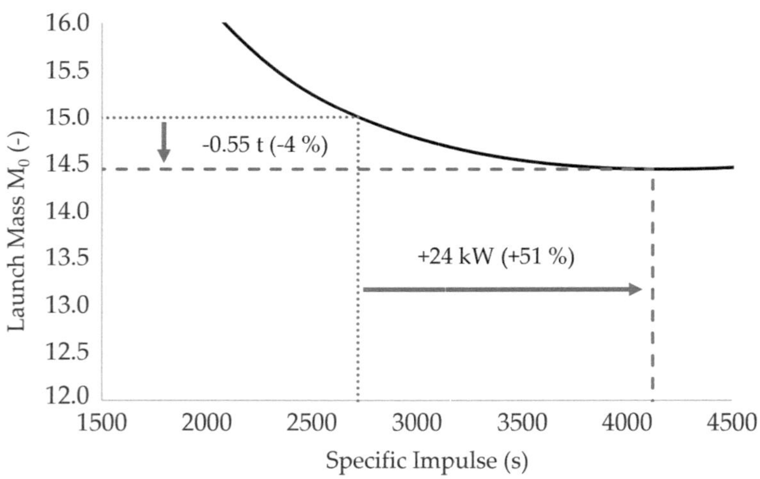


**Fig. 6.8** Ep model with PPE system and propulsion parameters

As an result, the power demand increases considerably while the gain in launch mass reduction is marginal. This effect is illustrated in Fig. 6.8. Clearly, the marginal launch mass saving of 4% does not justify a 51% increase in solar array, particularly considering that it is the most costly subsystem. Due to the plateau behaviour (typical for low $\alpha$ values) it is obviously not worth to utilise an electric propulsion system with a higher specific impulse. Instead, the choice of an engine with an $I_{sp}$ in the

**Table 6.8** Estimate of mass fraction of PPE & HALO stack

| | $\mu_{pl}$ | P (kW) | $m_0$ (t) | $m_{eps}$ (t) | $m_p$ (t) | $m_{dry}$ (t) |
|---|---|---|---|---|---|---|
| PPE | 0.667 | 47 | 15.00 | 1.4 | 2.8 | 0.8 |
| Optimum | 0.692 | 71 | 14.44 | 2.0 | 1.8 | 0,6 |

range of 2600–2800 s seems the right choice of the design team for this cislunar mission (Table 6.8).[7]

The derived operating point fits very well with the average power demand of the PPE, as reported in the literature [5].

## 6.3 Caveat of the Rocket Equation

Though powerful and mighty, both forms of the rocket equation, Eqs. 6.9 and 6.23, do have limitations, since both depict a *static* picture but spaceflight is *dynamic*. Although rooted in the rocket equation, the equation of motion is not explicitly solved. As mentioned before, information is lost while integrating over time. In other words, the rocket equation does not know how long it takes to achieve the desired $\Delta v$. This is in particular relevant for electric propulsion systems. The main physical quantity that is lost in the integration is the mean acceleration $\bar{a}$, put in technical terms, the required thrust level and its time dependence. The Tsiolkovsky equation does not tell whether 1000 kN or 500 mN are needed. In free space, i.e. without any gravity source, this lack of information does not change the resulting $\Delta v$ but it does impact the time of flight:

$$\tau = \frac{\Delta v}{\bar{a}} = \frac{\Delta v \bar{m}_{sc}}{\bar{F}}. \tag{6.36}$$

For constant thrust and thus constant board mass consumption, the mean spaceship mass is simply the arithmetic average of the initial and the mass after the manoeuvre. If gravity and other sources of force (e.g. rest-atmosphere) are involved, the trajectory will be subject to varying influences of forces. In a conceptual mission design, the rocket equation must be therefore supplemented by a dynamic analysis that takes into account all the forces acting, see Sect. 2.2.3 'Continuous Orbital Transfer'.

[7] A through analysis needs to consider that thruster with higher $I_{sp}$ (e.g. GITs) do have higher efficiency values.

## References

1. Johnson, W. (1995). Contents and commentary on William Moore's A treatise on the motion of rockets and an essay on naval gunnery. *International Journal of Impact Engineering, 16*(3), 499–521. https://doi.org/10.1016/0734-743X(94)00052-X
2. Arlazorov, M. S. (2023). "Konstantin Tsiolkovsky". Encyclopedia Britannica. Retrieved March 29, 2024, from https://www.britannica.com/biography/Konstantin-Eduardovich-Tsiolkovsky
3. Martinez-Sanchez, M., & Lozan, P. (2015). Lecture notes in space propulsion. Session 3-4. MIT Open Course Ware.
4. Scholze, F., Tartz, M., Neumann, H., Leiter, H., Kukies, R., Feili, D., & Weis, S. (2007). Ion analytical characterization of the RIT 22 ion thruster. In *Collection of Technical Papers—43rd AIAA/ASME/SAE/ASEE Joint Propulsion Conference*, vol. 2. https://doi.org/10.2514/6.2007-5216
5. Herman, D. A., Gray, T., Johnson, I., Hussein, S., & Winkelmann, T. (2022). Development and qualification status of the electric propulsion systems for the NASA PPE mission and gateway program. In *Presented at the 37th International Electric Propulsion Conference, IEPC-2022-465*. Cambridge, MA.
6. Rohrbeck, M. (2012). Conceptual study of large transfer stages. Diplomarbeit R 1221 D. Technical University Braunschweig, Germany

# Chapter 7
# Acceleration Principles and Technologies

**Abstract** Current propulsion technology relies on the ejection of mass, known as reaction mass, to generate forward movement or thrust. The objective of this chapter is to present and discuss the physical laws and technological means required to accelerate and *collimate* the reaction mass. According to the previously established taxonomy, we will distinguish between thermal and electrostatic acceleration. The former pertains to chemical and nuclear thermal propulsion systems, while the latter is relevant for solar and nuclear electric propulsion. Both technologies will be discussed separately.

## 7.1 Thermal Acceleration

We have already outlined the core idea of thermal acceleration in Fig. 5.5 of Sect. 5.1, it is the conversion of *internal energy* of a fluid into *directed kinetic energy*. In space propulsion, this fluid is called reaction mass or simply propellant. Since the laws we will discuss in this section are agnostic to the technical context, the term *fluid* is preferred.[1]

From the law of energy conservation, we derived the following relation for an isentropic flow:

$$u_{\max} \propto \sqrt{\frac{T_0}{\mathcal{M}}}. \tag{5.7}$$

We concluded that a high initial temperature and a low molecular mass of the fluid lead to high exhaust velocities. But as long as this energy is not collimated to form a directed gas stream (i.e. jet), no thrust can be generated.

[1] Fluid dynamics is the study of fluid motion, which encompasses gases, liquids, and critical states where it is impossible to distinguish between the two physical states. It is even possible to mathematically model the movement of Earth's crust and the emergence of mountain ranges using the same laws.

 
F. Gamgami, *Space Propulsion and Spaceship Design*, Springer Praxis Books,
https://doi.org/10.1007/978-3-031-71336-1_7

### 7.1.1 Laval Nozzle

The important device that converts the enthalpy of a liquid into a high-speed jet is called a nozzle. It is interesting to note that it is simply a rigid device with no moving parts, and that the magic lies in its particular geometry. In the context of space propulsion (and jet-engines), it is the task of the nozzle to perform the above mentioned energy transformation. The generated thrust has two constituents as introduced in Sect. 5.4:

$$T = \dot{m}u_{\text{ex}} + (p_{\text{ex}} - p_{\text{amb}})A_{\text{ex}}. \tag{5.18}$$

We will see below that both jet properties, $u_{\text{ex}}$ and $p_{\text{ex}}$, are not independent and that the nozzle's geometry is the connecting and determining element.

The nozzle geometry with which the highest jet velocities can be achieved is named after its inventor, *Laval-Nozzle*.[2] The basic working principle of a nozzle was well-known to scientists and engineers since at least the 18th century. A nozzle *increases* the velocity of a fluid while a diffuser *decreases* the velocity. Whether the cone acts as a nozzle or diffuser depends simply on the flow direction through the cone, as depicted in Fig. 7.1 assuming radial-symmetry. This is in accordance with our everyday experience and intuition: a narrowed exit of a garden hose creates an improved water fountain. This behaviour can be explained by the law of mass conservation, which states that for a stationary flow, the same amount of mass $\Delta m$ enters the device as leaves it in the time span $\Delta t$. Translating this prose into mathematical language gives:

$$\dot{m}_{\text{in}} = \dot{m}_{\text{ex}} = const. \tag{7.1}$$

with the mass flow defined as [1]:

$$\dot{m} = \rho u A = const.$$

Since for an *incompressible* flow, density does not change, we can go even one step further and write:

$$uA = \dot{V} = const, \tag{7.2}$$

$\rho$ density of the fluid,
$u$ velocity of the fluid,
$A$ cross section of the duct/nozzle,
$\dot{V}$ volume flow.

The relation states that the volume flow is constant, even if the cross section or the velocity changes along the nozzle. With help of this simple equation we can

---

[2] Gustav de Laval (1913†) was a resourceful Swedish engineer and entrepreneur of French-Huguenot origin. He was the first to come up with a nozzle design that is able to accelerate the fluid (steam) to supersonic velocity. However, it took another resourceful engineer to fully grasp the science behind this peculiarity, the Hungarian born Aurel Boleslav Stodala [1].

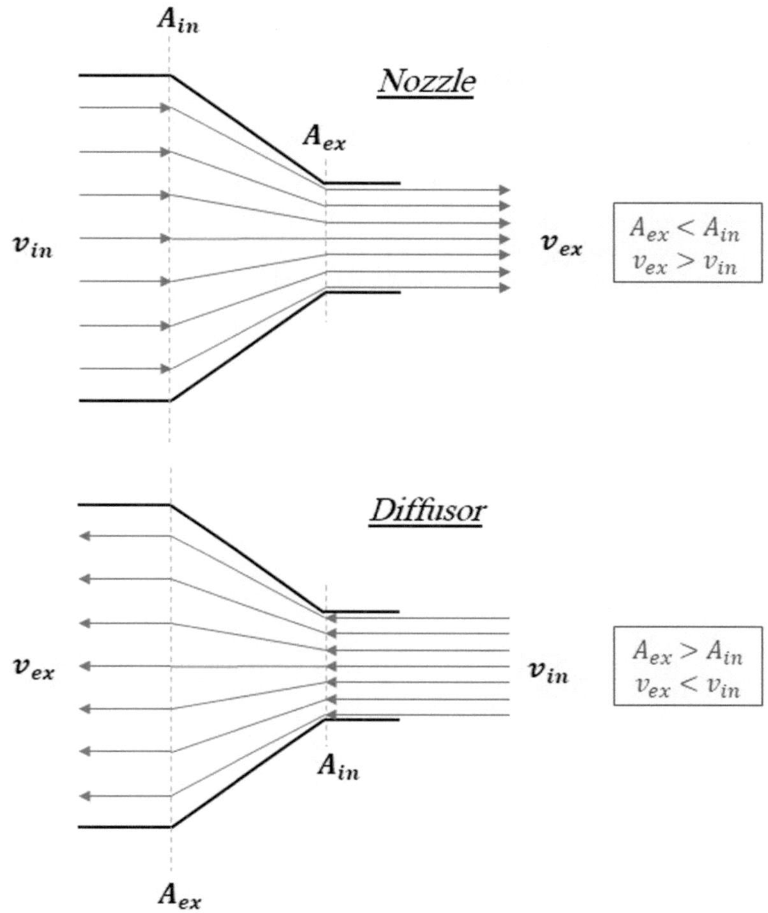


**Fig. 7.1** Simplified *subsonic* ($v_{\text{in}} < a$) flow through nozzle (top) and through diffuser (bottom)

confirm our intuition about the behavior of a flow through a nozzle by considering the *differential* form of Eq. 7.2:

$$
\begin{aligned}
d(uA) &= 0, \\
du A + u dA &= 0, \\
\frac{dA}{A} &= -\frac{du}{u}.
\end{aligned}
\tag{7.3}
$$

The important conclusion of this brief derivation is that the relation between the *change* of duct area, $dA$, and the *change* of flow velocity, $du$, is *negative* proportional, meaning that a decreasing cross-section in flow direction, $A_{\text{ex}} < A_{in} \Rightarrow dA < 0$, leads to an increasing velocity $u_{\mathrm{ex}} > u_{in} \Rightarrow du > 0$ and vice-versa. This is due to the different signs of the two sides of the equation:

$$dA \propto -du. \tag{7.4}$$

As the relatively simple math suggests, this behavior was known since some time. What modern engineers discovered at the turn of the former century, though, was that this obvious relation is turned around for flow velocities that are larger than the speed of sound, so called *supersonic* flows, depicted in Fig. 7.2. The reason for this

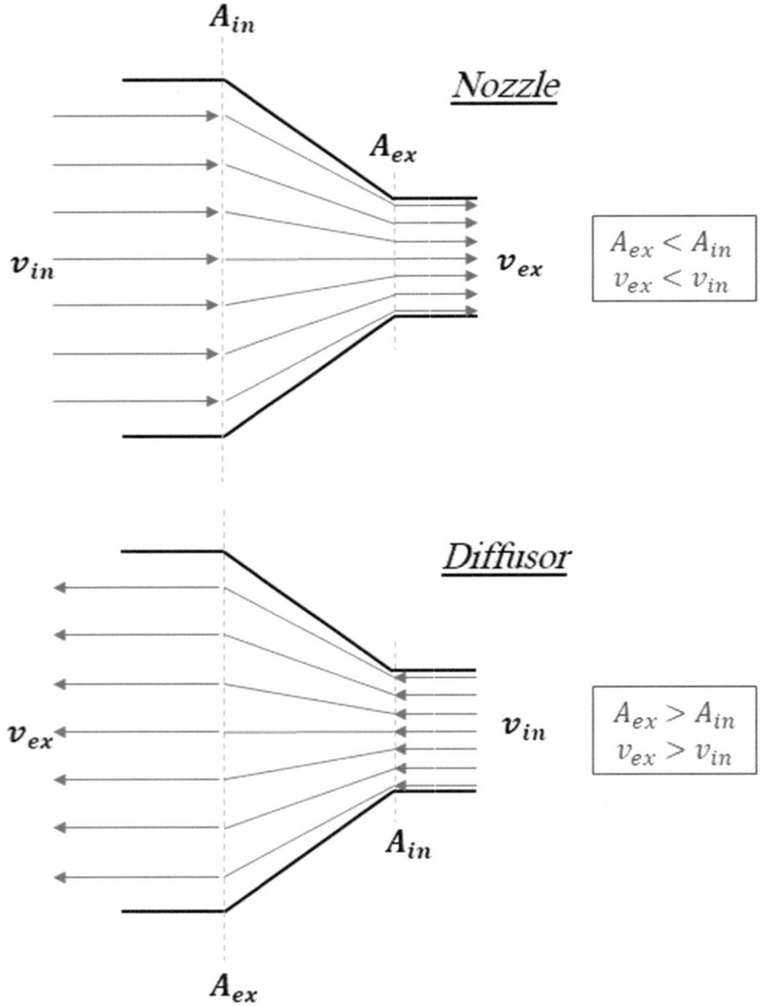


**Fig. 7.2** *Supersonic* ($v_{\mathrm{in}} > a$) flow through nozzle (top) and through diffuser (bottom)

is that unlike in Eq. 7.3 changes in density need to be considered the faster the fluid flows. This is because the larger the velocity, the stronger the compressibility effects. The relation that characterises the compressibility of a flow is the Mach number,[3] defined as the ratio of flow velocity to the speed of sound[4]:

$$M = \frac{u}{a}. \tag{7.5}$$

$M$ Mach number, dimensionless number for the flow's degree of compressibility
$a$ speed of sound

Given this definition, we can derive a generalized version of Eq. 7.3:

$$\begin{aligned} d(\rho u A) &= 0, \\ d\rho u A + \rho du A + \rho u dA &= 0. \end{aligned} \tag{7.6}$$

Applying the required thermodynamic relations yields after some arrangement, [1]:

$$\frac{dA}{A} = (M^2 - 1)\frac{du}{u}. \tag{7.7}$$

We highlight what matters by omitting the normalisation with $A$ and $u$ respectively:

$$dA \propto (M^2 - 1)du. \tag{7.8}$$

This equation is similar in structure to Eq. 7.4 except the fact that the sign of $du$ depends now on the Mach number. For Mach numbers below 1 the sign is negative like in Eq. 7.3, meaning that the flow behaves as discussed above and shown in Fig. 7.1. In fact for very low inlet Mach numbers, $M \ll 1$, Eq. 7.7 approaches Eq. 7.3. The sign, however, is reversed for inlet Mach numbers larger than 1 ($v_{\text{in}} > a$): the velocity then can only be increased to $du > 0$ by increasing the cross section $dA > 0$. The roles of the nozzle and diffuser are therefore reversed when the flow is supersonic.

To make the fluid flow in the first place, it is mandatory to apply a *pressure gradient*. The fluid then flows in the direction in which the pressure decreases, $-\nabla p$. For a subsonic flow the maximum achievable velocity at the nozzle's exit is limited to the speed of sound. Irrespective of the magnitude of the pressure gradient, this limit

---

[3] Ernst Mach (1916†) was an Austrian physicists and philosopher who made significant contributions in the understanding of shock waves in the field of gas dynamics. His contributions in the philosophy of science, in particular *empiricism*, influenced many thinkers of the 20th century including Albert Einstein [2].

[4] The advantage of dimensionless figures such as the Mach number is that they allow the comparison of flow physics irrespective of model size. This significantly advanced the development of aeronautics at the beginning of the 20th century through the use of wind tunnels.

cannot be exceeded, if the flow is isentropic. [5] Thus, starting subsonic and becoming supersonic is only possible by combining the nozzle in Fig. 7.1 with a diffuser in Fig. 7.2. The result is a convergent-divergent nozzle, called *Laval-Nozzle* in honor of its inventor. This idea works, if the pressure gradient along the nozzle is sufficiently high to keep the fluid accelerating against ambient pressure.

The shape of the Laval-Nozzle has become iconic for rocket engines, and for good reason. It is the best device to convert the high energy state of a fluid at rest into a high velocity jet stream. Since we only consider Laval-Nozzles in this book, we will henceforth refer to this special design simply as 'nozzle'.

We should be interested in a quantitative formula that shows the dependence of the fluid velocity on the geometric parameters of the nozzle in order to design a suitable nozzle for the engine. This formula and others can be derived from first principle laws of physics, known as governing equations of fluid dynamics. The differential form of these equations for an *isentropic* and *quasi one-dimensional* flow are:

$$\text{continuity equation:} \quad d(\rho u A) = 0, \tag{7.9}$$

$$\text{momentum equation:} \quad dp + \rho u du = 0, \tag{7.10}$$

$$\text{energy equation: :} \quad dh + u du = 0. \tag{7.11}$$

$p$ static pressure of the fluid (measured in a co-moving reference frame)

The first two equations are known since the end of the 18th century [3]. These are the Euler equations that describe a frictionless flow without heat transfer.[6] However, their power to describe the flow trough a Laval-nozzle was not unleashed before the end of the 19th century.

A set of five equations can be derived from the governing equations that help us to grasp the essential physics of the nozzle flow, also referred to as nozzle-theory, [12]. The assumption of an isentropic flow, where heat exchange between the fluid and wall as well as friction are neglected, does not alter the conclusion drawn here but helps to focus on the essence without complicating the picture with secondary phenomena. Having said that, a real nozzle design needs to take these processes into account.[7] The first equation gives the relation between the area ratio $\varepsilon$ and the Mach number, [1]:

[5] To be precise, this could be enabled, if heat exchange were allowed, and the flow not isentropic.

[6] Leonhard Euler (1783$^{\dagger}$) was a genius Swiss mathematician, physicist, and engineer who made groundbreaking contributions to various fields of physics and mathematics, including calculus, number theory, and graph theory. He spent most of his life in St Petersburg, Russia, where he made a name for himself and occupied high positions. [4].

[7] The governing equations are the *Navier-Stokes equations* in combination with a chemistry model. They cannot be solved analytically but require numerical models.

$$\varepsilon = \frac{A}{A_t} = \frac{1}{M^2}\left\{\frac{2}{\kappa+1}\left(1+\frac{\kappa-1}{\kappa}M^2\right)\right\}^{\frac{\kappa+1}{\kappa-1}}, \tag{7.12}$$

$$\varepsilon = f(\kappa, M), \tag{7.13}$$

$\kappa$ isentropic exponent of fluids defined as the ratio of heat capacities $c_p \backslash c_v$,
$A_t$ area of nozzle throat, the smallest cross section of the nozzle,
$\varepsilon$ expansion ratio.

Although $\varepsilon$ is given as a function of the Mach number, it is in fact $\varepsilon$ that is *the* decisive design parameter: the engineer controls the expansion ratio $\varepsilon$ by design of the nozzle, which in turn leads to a Mach number profile within the nozzle and most importantly at the nozzle's exit.

The following three equations give the change of the fluid's thermodynamic properties: temperature $T$, pressure $p$ and density $\rho$ as function of the Mach number along the nozzle axis [1]:

$$\frac{T_0}{T} = \left(1+\frac{\kappa-1}{2}M^2\right), \tag{7.14}$$

$$\frac{p_0}{p} = \left(1+\frac{\kappa-1}{2}M^2\right)^{\frac{\kappa}{\kappa-1}}, \tag{7.15}$$

$$\frac{\rho_0}{\rho} = \left(1+\frac{\kappa-1}{2}M^2\right)^{\frac{1}{\kappa-1}}, \tag{7.16}$$

$T_0$ temperature of the fluid at rest, i.e. upstream the nozzle,
$p_0$ pressure of the fluid at rest, i.e. upstream the nozzle,
$\rho_0$ density of the fluid at rest, i.e. upstream the nozzle.

The equations above are depicted in Fig. 7.3. Note that the thermodynamic properties are normalised to their respective value at rest, e.g. $p/p_0$ local pressure relative to the pressure at rest. Once the fluid enters the nozzle, it starts to accelerate and the pressure starts to drop.

The fluid is pushed outside of the nozzle due to the lower ambient pressure compared to the pressure at rest. This is why it is said, that the fluid *expands* and the area ratio $\varepsilon$ is called *expansion ratio*. The Mach number at the nozzle's throat is 1 and the area ratio there is referred to as $\epsilon_t$. The area ratio at the nozzle's exit is simply referred to as $\varepsilon$ since the ratio in between is not relevant for these considerations. It is important to note that the temperature does not drop as rapidly as the other thermodynamic variables. At the throat of the nozzle, where $M = 1$, there is still a significant temperature of $0.88 \cdot T_0$. We will see in Sect. 8.2.1, when discussing cooling mechanisms, why this is in fact an issue to care about in rocket engine design.

The last of the five equations gives the velocity $u$ along the nozzle, [1]:

$$u = \sqrt{\frac{2\kappa}{\kappa - 1} \frac{R^* T_0}{\mathcal{M}} \left\{ 1 - \left( \frac{p}{p_0} \right)^{\frac{\kappa}{\kappa - 1}} \right\}}, \tag{7.17}$$

$R^*$ universal gas constant, 8314 J/K/mol,
$\mathcal{M}$ mean molecular mass of the fluid.

In order to achieve a high nozzle exit velocity, $u_{ex}$, it is necessary to expand the fluid sufficiently. This is achieved by reducing the ratio $p_{ex}/p_0$ in Eq. 7.17. In the best case, the ratio becomes zero. This however, requires (mathematically speaking) an *infinitely* large expansion ratio $\varepsilon$. Practically sufficiently low pressure ratios can be achieved for finite expansion ratios due to the steep drop of the static pressure with Mach number as depicted in Fig. 7.3.

We assume for a moment that $\varepsilon \gg 1$ such that $p_{ex} \ll p_0$. Eq. 7.17 reduces then to:

$$u_{\max} = \sqrt{\frac{2\kappa}{\kappa - 1} \frac{R^* T_0}{\mathcal{M}}} \propto \sqrt{\frac{T_0}{\mathcal{M}}}. \tag{7.18}$$

We have found the origin of Eq. 5.7. The importance of this equation cannot be overestimated. It compellingly shows that the maximum velocity generated by a Laval-nozzle is a function of three parameters:

1. the expansion ratio of the nozzle $\varepsilon$,
2. the initial temperature of the fluid $T_0$ prior to expansion,
3. the mean molecular mass $\mathcal{M}$ of the expanding fluid.

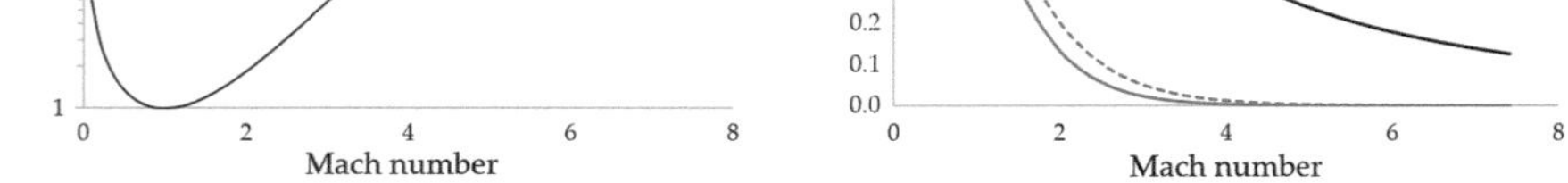


**Fig. 7.3** Area ratio $A/A_t$ (left) and ratio of static value to value at rest of temperature, density and pressure (right) as function of Mach number

**Table 7.1** Nozzle properties of selected upper stage engines

| Engine | $\varepsilon$ | $I_{sp}$ (s) | Thrust (kN) | Length (m) | Weight (kg) |
|---|---|---|---|---|---|
| RL-10B-2 | 280 | 465.5 | 110 | 4.1 | 301 |
| Vinci | 240 | 457.2 | 180 | 3.2 | 550 |
| LE-5A | 130 | 452.0 | 121 | 2.9 | 248 |

A nozzle with a large $\varepsilon$ leads to high exit velocities and therefore to a high specific impulse. So do rocket engines that can heat up the fluid to a high temperature before expansion. Interestingly, we derived this relation in Sect. 5.2 without any knowledge about Laval Nozzles.

Table 7.1 shows three upper stage engines with expansion ratio.

### 7.1.2 Generating Thrust

We have seen in Sect. 5.4 that the thrust generated by an engine has two components:

$$T = \dot{m}u_{ex} + (p_{ex} - p_{amb})A_{ex}. \tag{5.18}$$

Depending on the relation between the fluid exit pressure and ambient pressure different cases can be distinguished, as shown in Table 7.2.

The pressure term in the thrust equation is in general negative for low altitudes where ambient pressure is high, and positive in high altitudes when ambient pressure has decreased significantly. The first case is called *over-expansion*, meaning that the expansion in the nozzle is too large, thereby reducing the flows exit pressure below that of the ambient pressure: $p_{ex} < p_{amb}$. This is typically the case for 1st stage engines. The second case is called *under-expansion* and occurs in high altitudes where atmospheric pressure is low. The situation here is reversed, $p_{ex} > p_{amb}$. This is where upper stage engines operate. The special case when exit pressure matches

**Table 7.2** Cases of fluid exit conditions

| Environment | Relations | Case | Thrust |
|---|---|---|---|
| Lower altitude | $p_{ex} < p_{amb}$ | Over-expansion | $\dot{m}u_{ex} + (p_{ex} - p_{amb})A_{ex}$ |
| Intermediate altitude | $p_{ex} \approx p_{amb}$ | Adapted nozzle | $\dot{m}u_{ex}$ |
| Upper altitude | $p_{ex} > p_{amb}$ | Under-expansion | $\dot{m}u_{ex} + (p_{ex} - p_{amb})A_{ex}$ |
| Vacuum | $p_{ex} \gg p_{amb}$ | Under-expansion | $\dot{m}u_{ex} + p_{ex}A_{ex}$ |

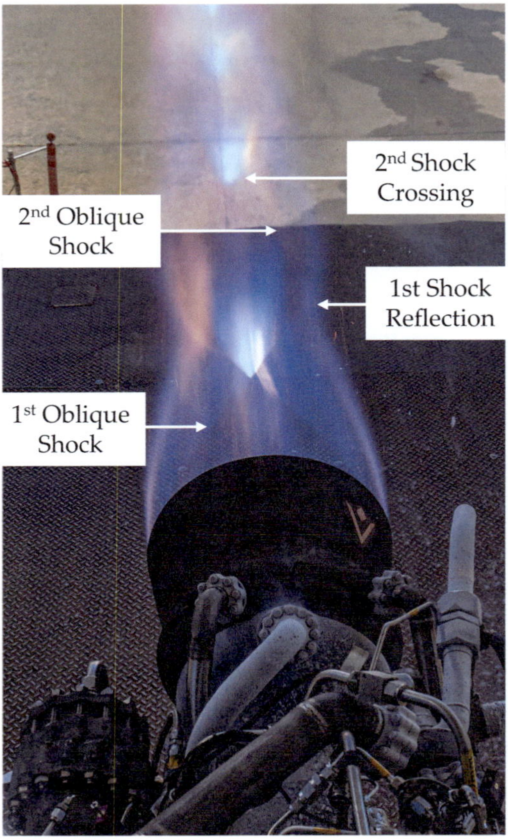


**Fig. 7.4** Hot firing test of 'E2' of the US company Launcher. Courtesy: John Kraus

ambient pressure is called *adapted nozzle*. It can be shown, that an engines with an adapted nozzle, exhibit the highest thrust level [12].

If the engine is operated in the atmosphere like for all rocket engines, adaption can only be achieved for a specific altitude due to the nozzle's rigidity. In that sense, these 1st stage engines run inefficiently most of the time and are only optimized for a specific moment of the flight altitude. The choice of that moment is a system design parameter in the overall launch vehicle optimisation task.

Figure 7.4 shows the 1st stage engine E2 developed by the company Launcher on the test-stand. It is clearly visible that the engine is slightly over-expanding. The ambient pressure confines the exhaust jet to a tube, not allowing the jet to widen up. This interaction creates a conical shock structure known as *rhombus*, which is typical for all 1st stage engines. The shocks are reflected by the ambient pressure, cross and are reflected again, thereby increasing the pressure magnitude inside the flow behind each shock. This continues until the pressures have equalised, $p_{\text{jet}} \simeq p_{\text{amb}}$.

It is important to note that under-expansion results in performance loss, while strong over-expansion can be detrimental to the engine. The reason for this is a gas

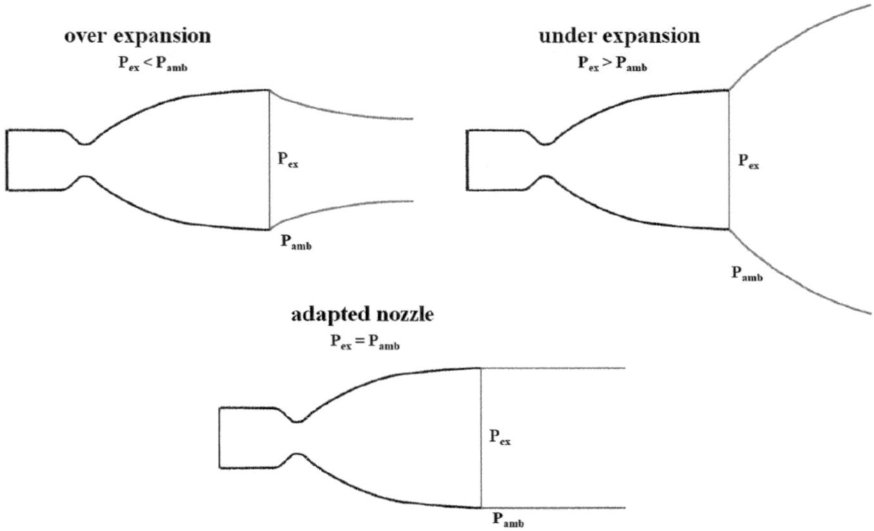


**Fig. 7.5** Plume topology for different backpressure/exit conditions. Note that a benign over-expansion case is shown

dynamic shock inside the nozzle, where the pressure jumps to the higher ambient pressure, causing flow separation[8] at the adjacent nozzle surface. This is a dynamic process that causes the shock position to oscillate, thereby increasing the mechanical loads on the nozzle, which could lead to a failure of its structural integrity.

The schematic of the exhaust jet topology for the three expansion cases discussed is depicted in Fig. 7.5. If operated in space, a nozzle with an infinitely large expansion ratio ($\epsilon$) would be required to reduce the exit pressure to near vacuum, creating an adapted nozzle. However, this is neither possible nor desirable because a larger $\epsilon$ results in a longer nozzle with more mass. There is an optimum $\epsilon$ for the engine, usually characterized by the *thrust-to-weight ratio*, see Sect. 5.4. Therefore, all in-space engines operate in under-expansion. Their exhaust jet is called *plume*.[9]

The plasma within electrostatic and electromagnetic thrusters has extremely low densities. In fact, electrostatic thruster as will be discussed in Sect. 7.2 do not possess a nozzle. This is because, due to the low densities, the flow is not a continuum flow but a *rarefied flow*, refer to footnote 17 in Sect. 5.4.

With this extremely low density, the pressure term, $p_{\text{ex}} A_{\text{ex}}$, is also negligible and the mean exhaust velocity matches the effective velocity $c_e$, such that without loss of accuracy, we can state: $T \approx \dot{m} u_{\text{ex}}$, Eq. 5.4.

[8] Flow separation is rarely desired by engineers. A famous example of flow separation with a detrimental effect is the stalling of an aircraft, which occurs when the angle of attack is too large and the flow cannot follow the wing profile.

[9] This term is only used by the community propulsion engineers that deal with in-space propulsion systems.

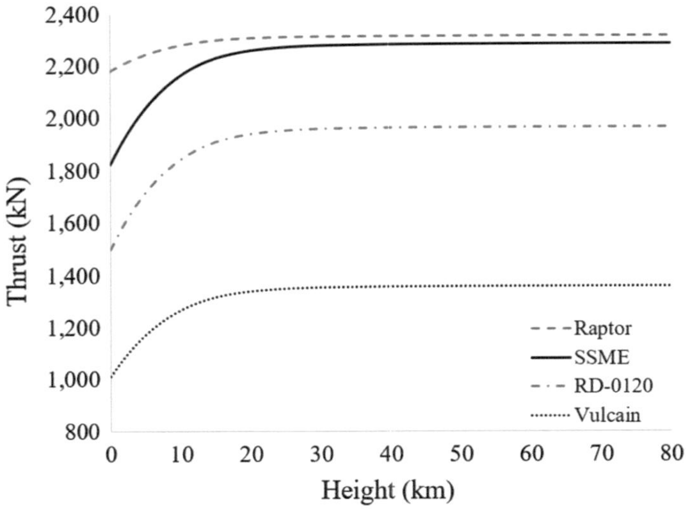


**Fig. 7.6** Thrust over height profile for selected engines for US standard atmosphere

Figure 7.6 shows the thrust profile of four 1st stage engines as function of height from ground till the edge of space. The US space shuttle main engine RS-25 and the Russian RD0120 belong to the few rocket engines to perform this feat, which explains the rather large hub in thrust. The graph shows a rapid change of the thrust profile; starting with a lower value at sea level it reaches 99 % of its asymptotic value, i.e. vacuum thrust level, already at 20 km altitude. This is why two thrust values are provided for 1st stage engines; sea-level thrust $T_{\text{sea}}$, and vacuum thrust $T_{\text{vac}}$.

To conclude this section we will explore the intriguing fact that despite the crucial role of temperature, $T_0$, for achieving high jet speeds the thrust itself is not a function of temperature. This seems counter intuitive, especially given the discussion above. The nature of thrust can be understood by reformulating the mathematical expression for it so that it is expressed as a function of quantities at the nozzle's throat. The fluid's velocity equals the speed of sound there and the mass flow is:

$$\dot{m} = \rho_t a A_{\text{t}}. \tag{7.19}$$

$\rho_t$ fluid density at the throat

Replacing the density with pressure and using basic thermodynamic relations, the following formulation can be derived, [12]:

$$\dot{m} = p_0 A_{\text{t}} \sqrt{\frac{\mathcal{M}}{T_0}} \mathcal{K}, \tag{7.20}$$

$\mathcal{K}$ propellant specific coefficient, $f(\kappa)$.

**Table 7.3** Comparison of Raptor engine variants (methalox) with RS25 (hydrolox)

| | Raptor 1 | Raptor 2 | Raptor 3 | RS25 |
|---|---|---|---|---|
| Chamber pressure (bar) | 250 | 300 | 350 | 206 |
| Sea level thrust (kN) | 1850 | 2300 | 2690 | 1860 |
| Mass (kg) | 2000 | 1600 | 1400 | 3177 |
| T/W (-) | 94 | 147 | 196 | 73 |
| Sea level $I_{sp}$ (s) | 330 | 327 | 327 | 366 |
| Length | 3 | 3 | 3 | 4.3 |
| Diameter | 1.3 | 1.3 | 1.3 | 2.4 |

This relation states that the maximum mass flow that can go through the nozzle's throat is directly proportional to

- the throat area: → the larger the area the more can go through,
- the chamber pressure: → the larger the pressure the stronger the push outside,
- the mean molecular mass: → the heavier the molecules the larger the mass flow.

Note that the mass flow is indirectly proportional to the temperature. This is because for a given chamber pressure $p_0$ the density decreases with increasing temperature, and so does the mass flow. Inserting Eqs. 7.20 and 7.18 into the definition of thrust and assuming an adapted nozzle yields:

$$\begin{aligned} F &= \dot{m} \cdot u_{\max}, \\ &= p_0 A_t \sqrt{\frac{\mathcal{M}}{T_0}} \mathcal{K} \cdot \sqrt{\frac{2\kappa}{\kappa - 1} \frac{R^* \cdot T_0}{\mathcal{M}}}, \\ &= p_0 A_t \mathcal{K} \cdot \sqrt{\frac{2\kappa}{\kappa - 1} R^*}, \\ &= p_0 A_t \mathcal{K}_2, \\ &\propto p_0 A_t, \end{aligned} \tag{7.21}$$

$\mathcal{K}_2$ propellant specific coefficient, $f(\kappa)$.

This equation is one of the most significant equations for thermal rockets. It shows clearly that there are only two design parameter for thrust; *chamber pressure* and *nozzle throat area* in absolute figures. The equation states, that in order to design a *compact* engine with a high *thrust-to-mass ratio* it is paramount to increase the chamber pressure. This leads to a mass and volume efficient design, which is of high interest for a clustered engine architecture. A negative accompanying circumstance of a high chamber pressure, is the increase of combustion instabilities as will be discussed in Sect. 8.2.

Table 7.3 compares the evolution of the Raptor engine variants towards higher chamber pressure with the RS25 (space shuttle main engine). The company SpaceX achieved a staggering increase of the thrust-to-weight ratio (T/W) by primarily increasing the pressure ratio, thus keeping the overall engine dimensions compact. In fact the volume envelope ratio of the RS25 to a Raptor engine is 2.64! In Sect. 8.2.2, we will discuss how the mean density of the propellant combination also influences engine compactness and in Sect. 9.5 we will explore engine cycles with which very high chamber pressures can be achieved.

## 7.2 Electrostatic Acceleration

Electric propulsion engines come in various technological shapes. This can lead to the perception that this type of engine is complex and difficult to understand. The intention of this section is to shed light into this domain following a systematic approach.

All electrostatic engines have two functions in common: the generation and the accelerations of ions in an electrostatic field. They convert electrical energy into thrust. This is the core idea formulated by the pioneers in this field, Konstantin Tsiolkovsky and 30 years later Robert Goddard and Hermann Oberth. The complexity of this propulsion type is due to the fact that these two functions can be achieved in various ways. Additionally, they are closely linked to the different technological implementations. Therefore, instead of structuring the section based on these functions, it is rather structure it according to their functional working principle. This book will focus on two major groups, which have been selected due to their scalability and thus their applicability to large space systems: *Gridded Ion Thruster* (GIT) and *Hall Effect Thruster* (HET). Both are currently used in a wide spectrum of space applications.

### *7.2.1 Gridded Ion Thruster*

Figure 7.7 depicts the functional set-up of Gridded Ion Thruster. It consists of five elements; propellant supply, electron supply, discharge chamber, acceleration grid and plume neutralisation. Electron supply is achieved by a hollow cathode, which, heated up, emits electrons. In this section we will focus on the heart of GITs; the discharge chamber and the acceleration grid.

#### 7.2.1.1 Discharge Chamber

The *discharge chamber* is the electrical counterpart to the combustion chamber in chemical engines. This is where the medium is conditioned for acceleration. While conditioning in thermal engines means heating up the propellant, conditioning in electrostatic propulsion systems means ionising the medium. For this reason, it is

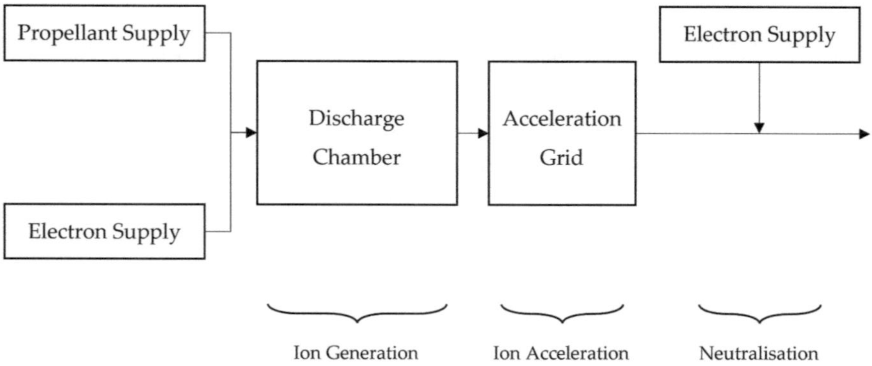


**Fig. 7.7** Functional block diagram of gridded ion thruster

also referred to as an *ionisation chamber* in order to emphasise the main process that takes place in the chamber. Ionisation is achieved by forced collision of free electrons with the neutral propellant gas. The result is a quasi-neutral plasma, consisting of ions and electrons. A well designed discharge chamber yields a high ionisation degree. Key to that is a high density of energetic electrons that spend a sufficient amount of time in the chamber to hit enough propellant atoms. We will discuss in the following two widely used mechanisms for electron acceleration: the Kaufman principle, named after the US physicist Harold R. Kaufman, and the radio frequency ion thruster (RIT), that was developed by Prof. H. Loeb at the university of Gießen, Germany, [5].

Discharge Chamber—Kaufman Type

Figure 7.8 shows the schematic design of a discharge chamber with magnetic ring cusps. There are various layouts but they all follow the same principle [6]. The propellant is injected into the chamber from the left together with the electrons that are generated in a hollow cathode. The efficiency of ionisation depends mainly on the trajectory and the dwell time of these electrons in the chamber.

The chamber wall is positively charged, around 1000 V, marked as 'Anode', and electrons are accelerated towards the walls while gaining kinetic energy. This kinetic energy is needed to ionise the propellant atoms by collision, which is why the principle is often referred to as 'electron bombardment ionisation'. The bad thing about this principle is the fact that electrons are lost too quickly to the walls, if no measures are taken. The consequence is a reduced dwell time in the chamber and thus reduced number of collisions. To increase the electron path length in the chamber, a magnetic field is applied that confines the electrons and forces them to follow a spiral trajectory according to the *Lorentz force*. This measure improves greatly the ionisation efficiency. The result is a *plasma* consisting of electrons, ions and neutral atoms. The exact location and number of the magnet cusp (i.e. sinks and sources of the magnetic field lines) determines greatly the topology of the field lines. The original Kaufman

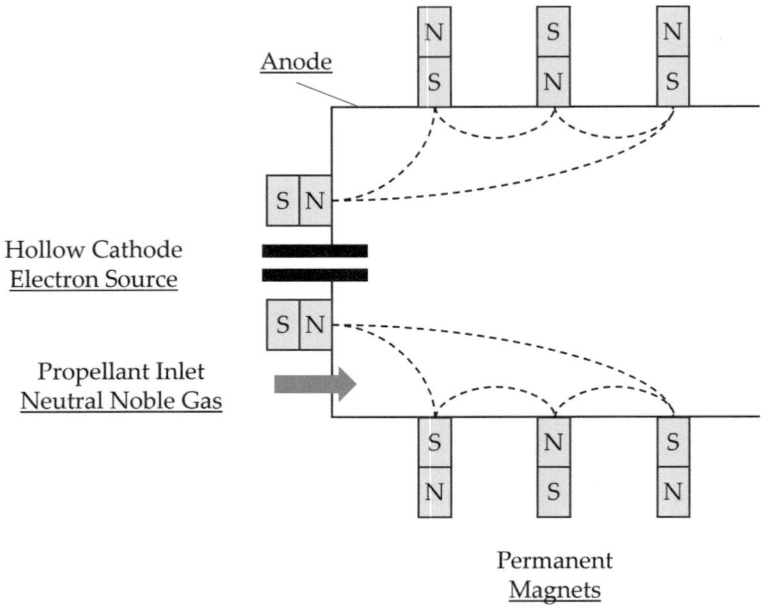


**Fig. 7.8** Schematic of an discharge chamber with magnetic ring-cusped design

design had only pair of magnetic cusps. Other layouts operate without a cusp-design but they all generate a magnetic field to trap the electrons and divert their path.

While electrons are attracted to the wall, ions are in principle repelled. Due to the thermal kinetic energy in the discharge chamber caused by collision among the particles, wall impingement by ions does occur often and is referred to as 'ion bombardment'. This bombardment can lead to serious damage of the chamber walls and is a limiting factor of the chamber's lifetime and total impulse capacity. In addition, ions that hit the wall are either absorbed by the wall or they recombine with free electrons at the wall thereby becoming neutral. In both cases, they can no longer be accelerated and are therefore lost to the thrust generation process. This effect contributes to the *discharge losses*, which means that although energy was used to ionise the atoms, these charged elements are lost for their actual purpose and so is the invested energy.

The Kaufman-type is currently the most common discharge chamber type and it has become standard practice to associate the Gridded Ion Thruster with this type of discharge chamber. However, there is yet another mechanism to ionise the propellant, which is based on *radio frequency excitement*.

Discharge Chamber—Radio Frequency Ionisation

As anticipated, the Radio frequency Ion Thruster (RIT) achieves ionisation in a radio frequency (RF) chamber without the need of electrodes. Instead, a mechanism known from plasma physics is utilised: *inductively coupled plasma* (ICP). The ICP is generated by an oscillating electro-magnetic field ($\dot{E} \neq 0$ and $\dot{B} \neq 0$) of about 1 MHz frequency. This is achieved by RF coils, that can be modeled as a solenoid with $N$ turns. In consequence the RF chamber has no magnets and no static field like the Kaufman type. A further difference is the fact that the walls of an RF chamber are insulated with a ceramic material [4].

Free electrons are accelerated by the field on closed-loop azimuthal trajectories thereby producing ions by collisions with neutral atoms. The fields are generated by an induction coil surrounding the discharge chamber. To create oscillating fields an alternating current (AC) is needed, which is supplied by a radio frequency generator (RFG). This is in contrast to the Kaufman type which runs with a direct current (DC). Figure 7.9 depicts the schematics of a RIT. To start this process free electrons are required that generate more free electrons and so on. These *seed electrons* are either free electrons within the propellant gas (a small fraction is naturally ionised), or additionally injected electrons from a (Tesla) spark generator. The latter is the predominant method in terrestrial plasma technology applications.

**Fig. 7.9** Schematic of RIT chamber

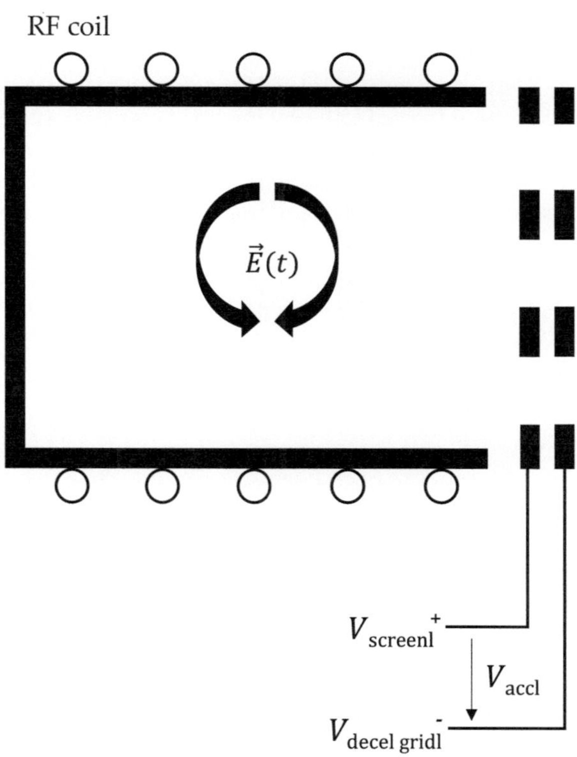

#### 7.2.1.2 Grid Assembly

The actual ion acceleration takes place in the grid assembly, where the electric potential drops significantly over a short distance. Figure 7.10 shows the structure of a three-grid arrangement and the corresponding voltage curve. We will first focus on the voltage curve before discussing the functions of the single grids and their geometric parameter.

Note that the voltage of the screen grid is very high (≈1000 V), and the question of why positively charged ions should not be repelled by it would be justified. This can be understood, when considering the voltage curve of the discharge chamber and acceleration grid combined. The walls of the discharge chamber are positively charged, as stated earlier. In fact the potential of the walls is commonly slightly higher than that of the screen grid (≈50 V), thus forming a potential gradient towards the screen grid. This gradient supports the *diffusion process* of the ions towards the screen in addition to the pressure gradient. Both are small in magnitude and the overall transportation process is slow. Once the ions have reached the screen grid, the coulomb force acting on them is able to accelerate them to towards the acceleration grid (≈-150 V) and thus to extremely high velocities of 20,000–80,000 m/s, which is 5–20 times higher than in chemical propulsion systems.

Without the screen grid and because of the steep voltage drop, the accelerating grid would be subjected to heavy ion bombardment, worse than that experienced by the walls due to the electrostatic acceleration, which would result in severe grid erosion. The screen grid serves as a protection shield of the acceleration grid and enhances the collimation of the ion beam. Due to its positive charge, the ion impacts are not very energetic and cause little damage such as *sputtering*. However, as with the chamber walls, ions that hit the screen grid are generally either absorbed or repelled as neutral atoms due to recombination processes at the grid surface, and are therefore lost for acceleration. A good screen grid design seeks to increase *grid transparency*, which is defined as the beam current that passes through the grid $I_b$ divided by the current that reaches the grid $I_i$:

$$T_s = \frac{I_b}{I_i}, \tag{7.22}$$

$T_s$ grid transparency,
$I_b$ beam current,
$I_i$ ion current that reaches the grid.

The third grid is the deceleration grid. Its function is to prevent the ion beam from diverging by slowing down the beam current and enhance collimation. A second function of the third grid is to repel ions from falling back into the discharge chamber. In summary, the grid assembly serves the following three purposes [8]:

*screen grid* extract Ions from the discharge chamber,
*acceleration grid* accelerate ions to create thrust and prevent electrons from entering the discharge chamber,

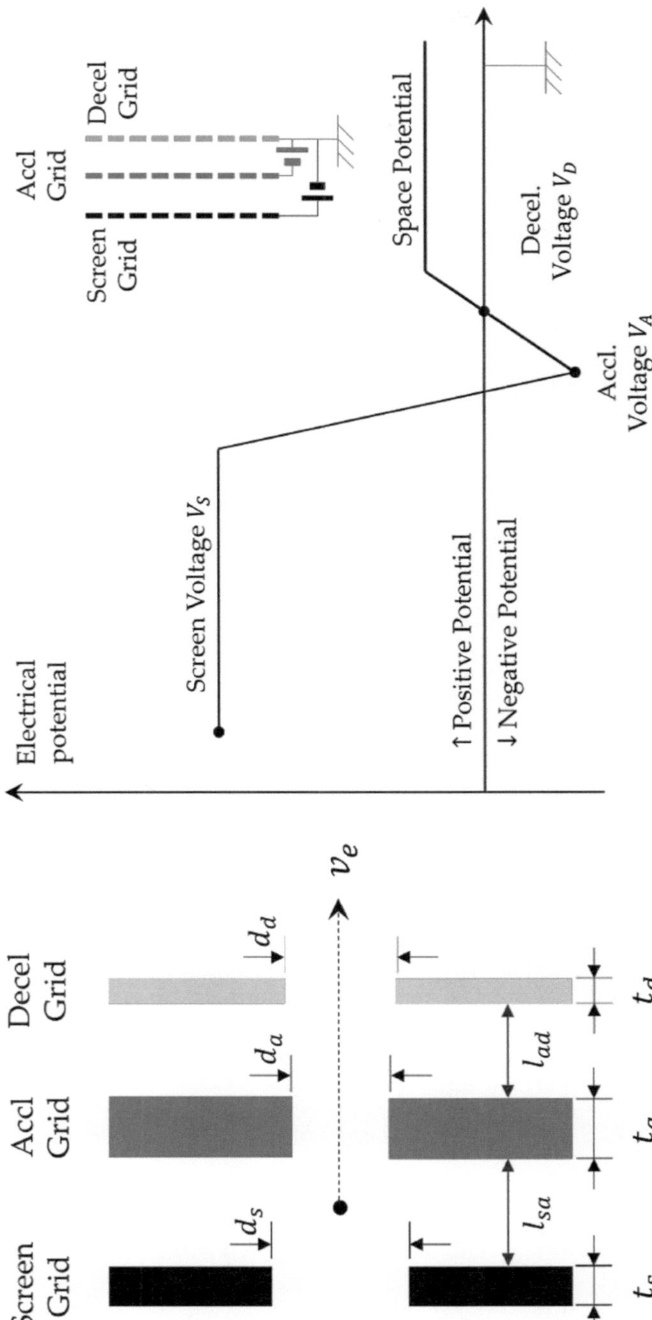


**Fig. 7.10** Left: Grid assembly and relevant geometric parameters. Right: Voltage drop

*deceleration grid* increase collimation, prevent ions from backstreaming into the discharge chamber.

In conclusion, an ideal grid assembly design extracts all ions, keeps all neutral atoms inside the chamber, does not erode (i.e. has an infinite lifetime) and creates a perfectly collimated and velocity-wise uniform ion beam profile. In the next paragraph, we will discuss why reality deviates from the ideal case and what adjustment parameters are available to the propulsion engineer

Ion Optics of Grid Assembly

The grid configuration is determined by *plasma physics* and the key parameter shall be explained briefly. We start with the famous *Child-Langmuir-Shottky* law, [10, 11]. It describes mathematically a phenomenon in plasma physics that is known as *space charge limit*. The name describes the phenomenon pretty well, there is a limit of how many charged particles can be stuffed into a finite space. The fact that there should be a limit should not wonder, considering that equally charged particles repel each other. The achievement of the equation's namesake lies in the fact that they asked themselves this question in the context of a plasma, which is by definition quasi-neutral. The result for a planar aperture is:

$$j_{\text{cls}} = \frac{I_{\text{cls}}}{A} = \frac{4}{9}\epsilon_0\sqrt{2\frac{q_i}{m_i}} \cdot \frac{U^{3/2}}{d^2}, \tag{7.23}$$

$I_{\text{cls}}$ ion current limited by space charging,
$j_{\text{cls}}$ ion current density limited by space charging,
$\epsilon_0$ permittivity of free space, $8.85 \times 10^{-12}$ F/m,
$q_i$ particle charge (single, dual etc.),
$m_i$ mass of ion,
$U$ electric Voltage drop between grids,
$d$ gap distance between screen and accl. grid.

The limit increases with voltage $V$ but decreases with longer acceleration distance $d$. A large charge over mass ratio, $q_i/m_i$, pushes the limit up leading to a higher ion throughput. In short:

$$(U \uparrow \; q_i/m_i \uparrow \; d \downarrow) \quad \Rightarrow \; j_{\text{cls}} \uparrow .$$

Evaluating Eq. 7.23 for electrons, ions and Xenon specifically leads to:

$$j_{\text{cls,e}} = 2.33 \cdot 10^{-6}\frac{U^{3/2}}{d^2}, \tag{7.24}$$

$$j_{\text{cls,ion}} = \frac{5.45 \cdot 10^{-8}}{\sqrt{M_a}}\frac{U^{3/2}}{d^2}, \tag{7.25}$$

$$j_{\text{cls,Xe}} = 4.75 \cdot 10^{-9}\frac{U^{3/2}}{d^2}. \tag{7.26}$$

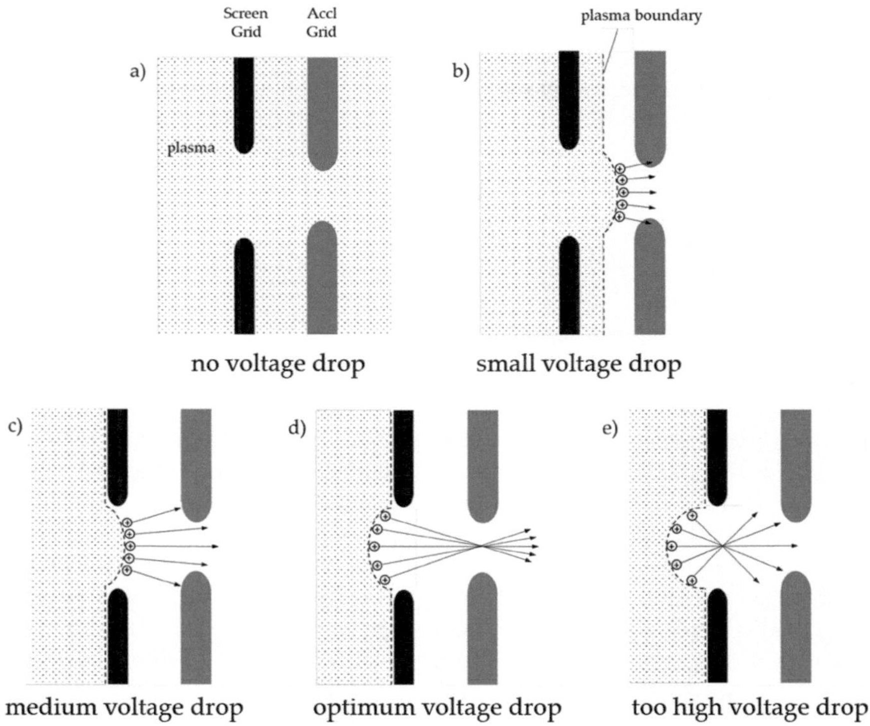


**Fig. 7.11** Ion optics as function of voltage drop

Assuming further an applied voltage drop of 1000 V and a grid gap of 1 mm, the resulting current density for Xenon GIT amounts to 15 mA/cm$^2$. The single aperture diameter is of the order of the gap distance, ranging from 1 to 5 mm for an accelerator grid [12]. Assuming a diameter of 3 mm, leads to a single aperture area of 0.07 cm$^2$, a maximum current of 1.06 mA and a maximum thrust of 0.056 mN. Hence, in order to achieve a total thrust[10] of 250 mN a net aperture area of 319 cm$^2$ (20 cm diameter) with a total of 4507 holes is needed resulting in a total beam current of 4.8 A.

We will now look at why electric propulsion engineers refer to the design of the grid as ion optics. Figure 7.11 shows the interaction of plasma with the screen and accelerator grid for an increasing voltage drop between the two.

A small voltage drop between the grids pushes the plasma back, that is away from the acceleration grid and towards the screen grid. Ions get extracted and accelerated moderately. Increasing the voltage drop between the two grids pushes the plasma further away from the acceleration grid. The shape of the plasma boundary opposite to the acceleration aperture is now concave; the generated ion beam diverges and hits the grid, leading to erosion. A further increase in voltage towards the optimum

---

[10] An an incredibly ridiculously small number compared to chemical propulsion systems. But we will see that in free space time is on the side of electric propulsion systems.

operating voltage results in the desired *convex* shape of the plasma boundary with the effect that the ions are focused in the center of the aperture causing no harm to the acceleration grid. An increase in voltage beyond the optimum operation voltage shifts the focal point towards the screen grid, which leads to strong ion impingement on the acceleration grid thereby eroding the grid and causing energy and thrust losses. These effect have an important system level implication: thruster voltage is kept constant during throttling because of this ion optical behaviour.

### 7.2.2 Hall Effect Thruster

Soviet engineers developed a technical design for an electrostatic thruster without the need for a grid assembly and the associated ion-optic physics. The result is the famous *Hall Effect Thruster* (HET), in which ion generation and acceleration occur in the same spatial area. Figure 7.12 shows the principle design of a HET. It is characterised by a strong radial magnetic field, that has its peak in strength at the exit of the tube, an anode in the back and a cathode outside the thruster for electron production. In stationary operation, a cloud of energetic electrons is formed inside the tube. This cloud is responsible for propellant ionisation and acceleration. Understanding the origin and mechanisms of this electron cloud is therefore of central importance for understanding HETs. Like for RF discharge chambers, the walls of a HET are insulated, which reduces ionisation losses to the wall but hits still produce sputtered material [8].

We start the discussion with the cathode that produces the free electrons. The emitted electrons are attracted by the anode, they start moving into the tube, where they interact with the magnetic field. The field forces the electrons into a classical *gyration* around the magnetic field lines, due to the *Lorentz force*:

$$\mathbf{F}_\mathrm{L} = q\mathbf{v} \times \mathbf{B}, \tag{7.27}$$

$\mathbf{F}_\mathrm{L}$ Lorentz force vector,
$q$ particle charge (here electron),
$\mathbf{v}$ velocity vector of charged particle,
$\mathbf{B}$ Magnetic field vector.

The electric field that attracted the electrons into this trap is still acting on them. The result is a peculiar motion in *crossed electric and magnetic fields*.[11] To understand this motion, we consider for simplicity a pure planar case as depicted in Fig. 7.13. An electron flying perpendicular to the magnetic field experiences the above mentioned Lorentz force that is responsible for trapping its trajectory into the *gyration*. Without an electric field this trajectory would be circular around the magnetic field lines with

[11] In his series of lectures, Richard Feynman provides the most comprehensive and insightful description of this motion [13].

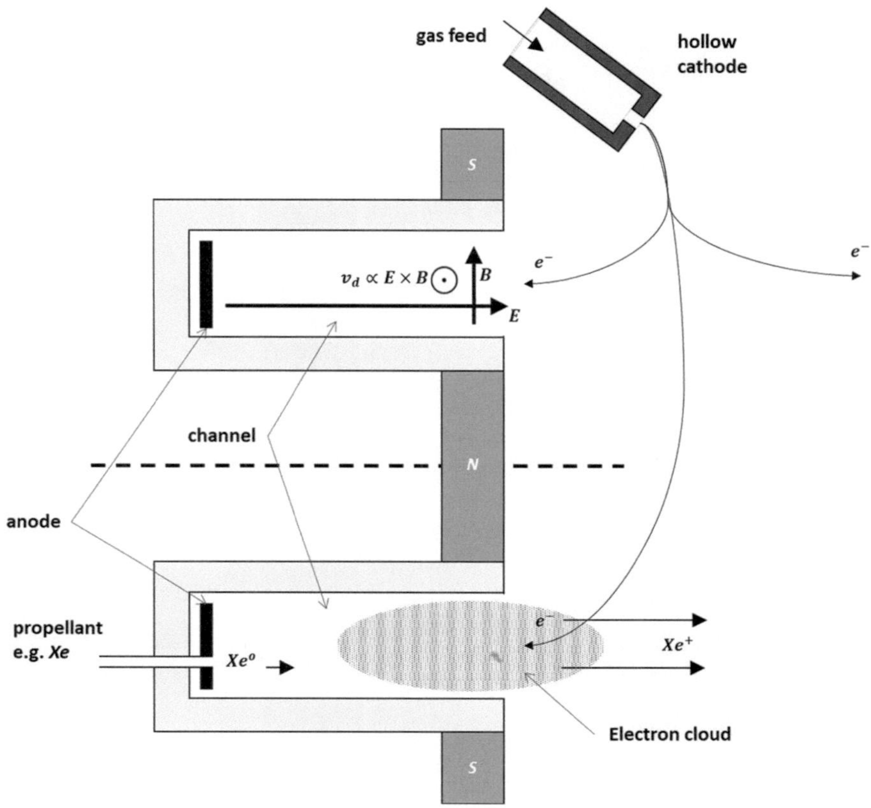


**Fig. 7.12** Schematic of a hall effect thruster with field lines, charged particles and plasma motion

a radius known as Larmor radius. Applying an electric field creates a bias in the electrons circular motion: when the electron flies parallel to the electric field lines, it is decelerated and accelerated when flying in opposite direction. Remember that the magnitude of the Lorentz force is directly proportional to the magnitude of the ion's velocity, leading to a gradient in this force and finally to a *drift* of the electron's motion perpendicular to the **B**- and **E**−fields, Fig. 7.13. The drift velocity is, [13]:

$$\mathbf{v}_\mathrm{d} = \frac{\mathbf{E} \times \mathbf{B}}{B^2}, \tag{7.28}$$

$\mathbf{v}_\mathrm{d}$ concentric velocity drift within the tube around the symmetry axis,
$\mathbf{E}$ electric field vector.

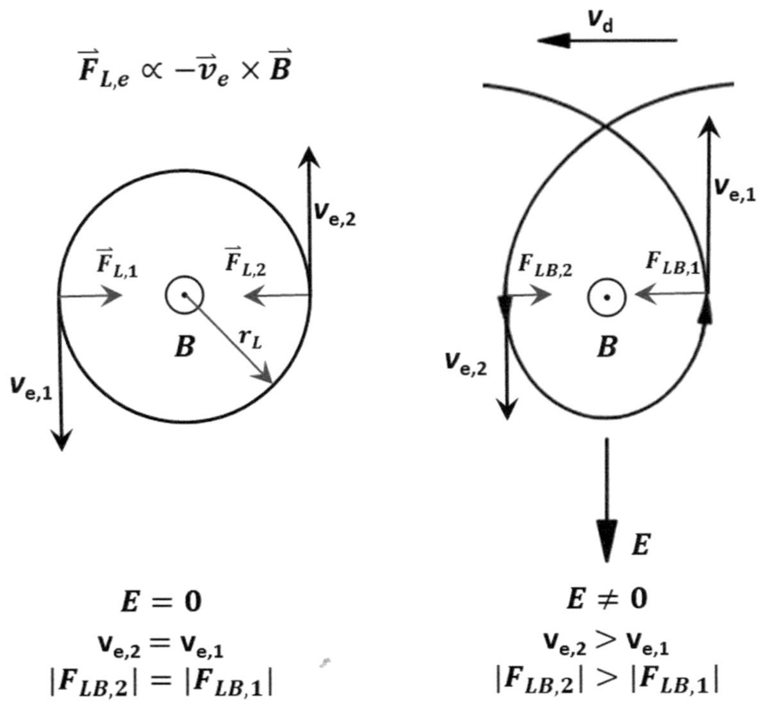


**Fig. 7.13** Electron drift in crossed electric and magnetic fields

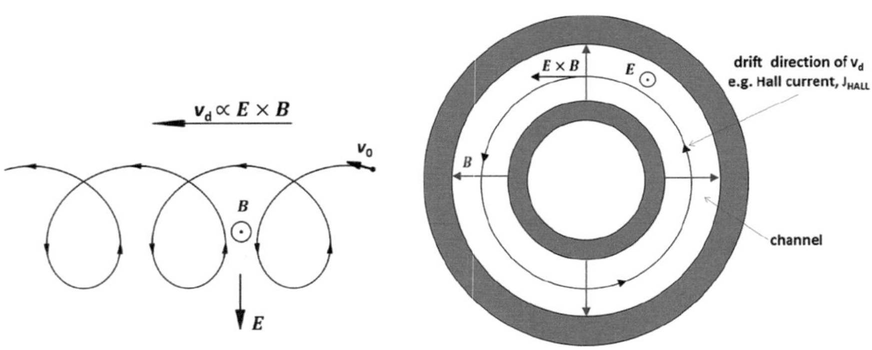


**Fig. 7.14** Schematic representation of circumferential drift of electrons within the discharge channel, called Hall current

The arguably most striking feature of this equation is the fact that it is not dependent on the particle's charge but purely on the strength of the applied fields. Figure 7.14 shows the extended electron drift. For a HET this movement takes place within the circular plasma channel in azimuthal direction. Hence, the drift acts around the inner coil.

The emergence of the drift results in an azimuthal current, which is commonly referred to as the *Hall current* or *cross field* diffusion.

Electrons that are trapped by the crossed electric and magnetic fields, form an energetic *electron cloud* that efficiently ionises the propellant it collides with. Once ionised, it experience the voltage drop and accelerate to high velocities. To make this process efficient, electrons need to be sufficiently *magnetized*. This is expressed in two ways. Firstly, the *Larmor radius* $r_{\mathrm{L}}$ (left in Fig. 7.13) shall be significantly smaller than the characteristic scale length of the thruster e.g. the channel width, $w$:

$$r_{\mathrm{L}} >> w, \tag{7.29}$$

with [8]:

$$r_{\mathrm{L}} = \frac{v_{\perp}}{\omega_{\mathrm{c}}} \tag{7.30}$$

$r_{\mathrm{L}}$ Larmor radius,
$w$ characteristic scale length of the thruster e.g. the channel width,
$v_{\perp}$ velocity component perpendicular to the magnetic field
$\omega_{\mathrm{c}}$ cyclotron resonance frequency $\omega_{\mathrm{c}} = eB/m_{\mathrm{e}}$

This criterion means that while the electrons circulate around the magnetic lines they shall be confined to the magnetic field lines such that the number of rotations is large and the likelihood to hit a xenon atom increased. The second expression of sufficiently magnetised electrons is reflected in a large electron Hall parameter, $\Omega_e^2$. This parameter is defined as the square of the ratio of cyclotron frequency to the total collision frequency, $\nu$, of an electron while circulating around the magnetic field line [8]:

$$\Omega_e^2 = \frac{\omega_c^2}{\nu^2} >> 1. \tag{7.31}$$

This criterion, too, has a tangible meaning: electrons shall perform several orbits around a magnetic field line before a collision with a neutral atom occurs. These two criteria ensure that the electrons are firmly bound to the magnetic field, which reduces their axial mobility towards the anode and continuously increases their kinetic energy for the successive impacts.

The following discussion will show why this thruster type is referred to as Hall thruster. We start by analysing a balance of the forces acting on the ionized propellant and the electrons. The ionised propellant is accelerated in the electric field that acts between the anode in the back and the circular electron cloud at the exit of the tube. The integral of this force over the entire channel volume is:

$$\mathbf{F}_{ion} = 2\pi \iint q n_i \mathbf{E}\, r dr dz, \tag{7.32}$$

$n_i$ ion charge density.

Note that the thrust, that pushes the spacecraft, is related to this force by Newton's second law: $\mathbf{T} = -\mathbf{F}_{ion}$. The electrons in the cloud are constrained not to move axially by the crossed fields as discussed above. Ideally, the acting forces on the electrons cancel each other, forming a quasi-stationary cloud:

$$\mathbf{F}_\mathrm{e} = \underbrace{2\pi \iint qn_\mathrm{e}\mathbf{E}\, r dr dz}_{\text{electrostatic force}} + \underbrace{2\pi \iint en_\mathrm{e}\mathbf{v}_\mathrm{d} \times \mathbf{B}\, r dr dz}_{\text{Lorenz force on Hall current}} = 0, \tag{7.33}$$

$n_\mathrm{e}$ electron charge density.

It is important to note that the second term contains the drift velocity $\mathbf{v}_\mathrm{d}$. This term describes the force that the magnetic field exerts on the Hall current! In summary, the magnetic field acts on the single electron forcing it into gyration around its field lines, it further acts together with the electric field on this circulating electron to form a Hall current by causing the electrons drift around the symmetry axis. Once the Hall current is formed, the magnetic field acts on it by the Lorenz force: $e\mathbf{v}_\mathrm{d} \times \mathbf{B}$.

In the following we will use the fact that the charge densities of ions and electrons are equal, forming a quasi neutral plasma, $qn_i = en_\mathrm{e}$. This implies that the first term in Eq. 7.33 equals in magnitude the right hand side of Eq. 7.32. We will also use the definition of the Hall current density $\mathbf{J}_\mathrm{Hall} = -en_\mathrm{e}\mathbf{v}_\mathrm{d}$ to derive a new expression for the thrust in a Hall Thruster:

$$\begin{aligned} T &= -\mathbf{F}_{ion}, \\ &= -2\pi \iint qn_i\mathbf{E}\, r dr dz, \\ &= 2\pi \iint en_\mathrm{e}\mathbf{v}_\mathrm{d} \times \mathbf{B}\, r dr dz, \\ &= 2\pi \iint \mathbf{J}_\mathrm{Hall} \times \mathbf{B}\, r dr dz. \end{aligned} \tag{7.34}$$

The resulting expression shows clearly that thrust in a HET is proportional to the strength of the magnetic field and the Hall current density. Hence the name Hall Effect Thruster.

Finally it shall be mentioned that the plume of HETs is approximately twice as large as for GITs, which means that the ion beam is not equally good collimated, [8]. This has not only an effect on the thrust vector, which is reduced in axial direction, but also on the remaining vehicle. Ions and, even worse, sputtered material hit the solar arrays or other sensitive parts of the spacecraft and payload thereby damaging them. Besides the immediate damage, the plume contributes to the *contamination* of the vehicle. As a consequence, optical equipment such as star tracker, which are required for attitude determination, degrade.[12]

[12] The matter is even more relevant, if the main payload is of optical nature. Contamination control (part of the discipline 'product assurance') becomes a vital subject during the design process.

## Nested Hall Effect Thruster

An intriguing variant of the standard HET design is the *nested* Hall Effect Thruster, jointly developed by: the private company Aerojet Rocketdyne (now L3Harris$^{TM}$), University of Michigan, NASA GRC, JPL and Air Force Office of Scientific Research [15]. The idea is as simple as brilliant: several discharge chambers are concentrically combined into a single thruster. The X3 of the University of Michigan is a prominent and advanced example, depicted in Fig. 7.15. It features three rings in a single engine and shows remarkable characteristics. The arguably most important feature is its high throttle range, since each ring can be throttled individually. This offers maximum flexibility to the mission analyst for the design of an optimal interplanetary trajectory. Table 7.4 demonstrates the versatility of this engine.

**Fig. 7.15** 80 cm X3 Nested HET [14]. Credit: NASA

**Table 7.4** Selected operation points of X3 as function of discharge voltage and power from min to max, [16]

| 300 V | | | | 400 V | | | | 500 V | | | |
|---|---|---|---|---|---|---|---|---|---|---|---|
| $P_{dc}$ | $I_{sp}$ | $F$ | $\eta_T$ | $P_{dc}$ | $I_{sp}$ | $F$ | $\eta_T$ | $P_{dc}$ | $I_{sp}$ | $F$ | $\eta_T$ |
| kW | s | mN | (–) | kW | s | mN | (–) | kW | s | mN | (–) |
| 4.9 | 1820 | 0.35 | 0.64 | 6.5 | 2050 | 0.39 | 0.59 | 8.2 | 2300 | 0.43 | 0.59 |
| 40.8 | 2030 | 2.58 | 0.62 | 42.7 | 2330 | 2.4 | 0.64 | 42.2 | 2560 | 2.19 | 0.64 |
| 74.9 | 2020 | 4.64 | 0.61 | 69.3 | 2300 | 3.94 | 0.64 | 67.8 | 2610 | 3.38 | 0.63 |
| – | – | – | | 98.6 | 2340 | 5.42 | 0.63 | 100.1 | 2570 | 5.03 | 0.63 |

### 7.2.3 *Efficiencies and Losses*

We have introduced in Sect. 5.4 the total engine efficiency $\eta_T$ in a general form irrespective of the acceleration principle. In the context of the discussion of the generalised form of the Tsiolkovsky equation, Sect. 6.2, it became clear that this parameter plays an important role in the design of electric propulsion based spacecrafts. We will approach the losses in the following from a physical perspective and concentrate on the main determining factors of this important system parameter. We take Eq. 5.28 as starting point and reformulate it to receive an expression for $\eta_T$:

$$\eta_T = \frac{1}{2}\frac{T^2}{\dot{m}_p P_{in}}. \tag{7.35}$$

The right side hand side of this formula can be interpreted as a combination of all losses that contribute to the total loss. This is a motivation to analyse the effects closer with the aim to identify the loss contributors.

First, we take into account that collimation is not perfect. Therefore, the ion beam diverges and forms the plume mentioned above. The thrust is reduced by the divergence factor $\gamma$ defined as:

$$\gamma = \cos\theta = \frac{T_{eff}}{T}. \tag{7.36}$$

A half-cone angle of 15° corresponds to a $\gamma = 0.966$. This is a loss in usable thrust of 3.4 % which needs to be considered in the propellant budget. This also means that the equation as formulated above for $\eta_T$ does not assume any thrust loss. The corrected form is:

$$\eta_T = \frac{1}{2}\frac{T_{eff}^2}{\dot{m}_p P_{in}} = \gamma^2 \frac{T^2}{2\dot{m}_p P_{in}}. \tag{7.37}$$

The updated equation reveals that the impact of the divergence loss on the total efficiency is squared and the system has to cope with an amplified detrimental effect: $0.966^2 = 0.933$.

We can gain more insight by introducing the expression of thrust for an electric thruster, Eq. 5.12, $T = \dot{m}_b v_i$:

$$\eta_T = \gamma^2 \cdot \frac{\dot{m}_b}{\dot{m}_p} \cdot \frac{T v_i}{2P_{in}}. \tag{7.38}$$

Finally we consider that the $T v_i/2$ is actually the beam power and the final expression is:

$$\eta_T = \gamma^2 \cdot \frac{\dot{m}_b}{\dot{m}_p} \cdot \frac{P_b}{P_{in}}. \tag{7.39}$$

This new expression consists of the divergence factor $\gamma$ and two ratios. The first is the *mass efficiency* factor, it is defined as the ratio of ionised to unionised propellant mass flow. For single-charged ions this yields:

$$\eta_m := \frac{\dot{m}_\text{b}}{\dot{m}_\text{p}} = \frac{I_\text{b} m_i}{e} \frac{1}{\dot{m}_\text{p}}, \tag{7.40}$$

$\dot{m}_\text{b}$ ion beam, i.e. ionised propellant flow expelled by the thruster,
$\dot{m}_\text{p}$ injected propellant flow rate into the discharge chamber,
$I_\text{b}$ beam current (A),
$m_i$ mass of a single ion.

The factor is a measure of the chamber's ability to ionise the propellant. Neutral propellant that leaves the engine is lost propellant as it could not be accelerated and thus did not contribute to the thrust. A good discharge chamber design features mass efficiency values in the order of 80 % and laboratory models could demonstrate up to 90 % efficiency [8].

The second ratio describes the *electric efficiency*, it measures the quality of power conversion within the thruster and is defined as beam power divided by the input power:

$$\eta_\text{e} := \frac{P_\text{b}}{P_\text{in}}. \tag{7.41}$$

The beam power can also be written as beam current times beam voltage $I_\text{b} \cdot U_\text{b}$. The input power is defined as the beam power plus all additional power required to generate $P_\text{b}$, e.g. heater power, cathode power, grid current and RF power in case of RITs. This additional power is subsumed to $P_0$ and the input power is then: $P_\text{in} = P_\text{b} + P_0$.

Based on these definitions the total efficiency factor of an electric thruster is:

$$\eta_\text{T} = \gamma^2 \eta_\text{m} \eta_\text{e}. \tag{7.42}$$

### 7.2.4 Specific Impulse and Thrust for Electrostatic Thruster

With the derived equations and relations above we can formulate useful expressions for the specific impulse and thrust for electric thruster. We start with the definition of the specific impulse for electric thruster:

$$I_\text{sp} = \frac{c_e}{g_0} \approx \frac{u_\text{ex}}{g_0}. \tag{5.24}$$

The exhaust velocity, $u_{\text{ex}}$, is related to the ion beam velocity, $v_{\text{b}}$, via the thrust equations for electric thruster, Eqs. 5.4 and 5.12:

$$T = \dot{m}_{\text{p}} u_{\text{ex}} = \dot{m}_{\text{b}} v_{\text{b}}. \tag{7.43}$$

This distinction is important and necessary because as we have seen above that not all atoms that leave the thruster are ionised, $\eta_m < 1$. We use this relation to substitute $u_{\text{ex}}$:

$$I_{\text{sp}} = \frac{v_{\text{b}}}{g_0} \frac{\dot{m}_{\text{b}}}{\dot{m}_{\text{p}}}. \tag{7.44}$$

Inserting the definition of the mass utilisation efficiency, Eqs. 7.40, and 5.10 for $v_{\text{b}}$ yields finally[13]:

$$I_{\text{sp}} = \frac{\eta_m}{g_0} \sqrt{2 \frac{q_i U}{m_i}}. \tag{7.45}$$

This equation is valid for a single-charged ion beam.

We have already derived an expression for the thrust in Sect. 5.3 which is valid for all electrostatic propulsion types:

$$T = I_{\text{b}} \sqrt{2 \frac{m_i}{q_i} U}. \tag{5.14}$$

For Gridded Ion Thruster we can go one step further and consider the limitation in the ion beam current $I_B$ due to space charging as expressed in the Child-Langmuir-Shottky law, Eq. 7.23. Inserting this limit and, $I_{\text{b}} = I_{\text{cls}}$, and considering the thrust density, Eq. 5.34, yields:

$$\begin{aligned} \frac{T}{A} &= j_{\text{cls}} \cdot \sqrt{2 \frac{m_i}{q_i} U}, \\ &= \frac{4}{9} \epsilon_0 \sqrt{2 \frac{q_i}{m_i} \frac{U^{3/2}}{d^2}} \cdot \sqrt{2 \frac{m_i}{q_i} U}, \\ &= \frac{8}{9} \epsilon_0 \left( \frac{U}{d} \right)^2, \\ &\propto \left( \frac{U}{d} \right)^2. \end{aligned} \tag{7.46}$$

[13] Though accurate for ion thruster, this expression for the ion beam velocity serves merely as an approximation for Hall Effect Thruster, as elaborated in Chap. 7 [8]. Despite being an approximation, it adequately serves the purpose of this discussion and does not alter the conclusions.

This equation is the electrical equivalent of Eq. 7.21 for thermal engines. The critical conditions at the throat of the Laval nozzle are just as limiting as the space charging effects at the grids of an ion thruster. In both cases there are two limiting design parameters: one is of physical nature ($U$ and $p_0$ respectively) and one of geometric nature ($d$ and $A_\mathrm{t}$ respectively). Note that even though Eq. 7.46 does not contain the mass flow directly it is indirectly included. Similar to the critical condition for thermal acceleration, space charging necessitates a 'critical' mass flow (here current flow) to hit this limit.

### 7.2.5 Propellant Selection

We already touched the propellant topic at the end of Sect. 5.3 when the principles of electric propulsion systems were introduced. This can be summarised with the relations:

$$(q_i \uparrow, m_i \downarrow) \Rightarrow v_i \uparrow, \tag{5.16}$$

$$(q_i \downarrow, m_i \uparrow) \Rightarrow F \uparrow . \tag{5.17}$$

We stated that the specific impulse of this technology is already very high (several thousand seconds) but the thrust magnitude is extremely low (in the order of hundred mN). This is why propellant that favours thrust is focused on. There are obviously more criteria to be fulfilled, of which the most decisive shall be briefly discussed:

*atomic mass* In order to achieve a meaningful thrust level, higher atomic masses are preferred. From this point of view, such heavy elements as radon, uranium and plutonium would be very good candidates if it were not for their nasty radioactivity,

*ionisation* A low first ionisation threshold is preferred as it defines how much energy must be at least supplied to the thruster. Since multiple ionisation is not desired (Eq. 5.17), the propellant candidate should have high second and third ionisation thresholds,

*molecules versus atoms* Propellant in form of molecules is not desired. They possess a wide spectrum of excitation states that 'eat up collision energy', which is meant for the first ionisation.Furthermore, molecules dissociate before they ionise, which has a double negative impact: energy is lost for dissociation and the ion mass is reduced. In conclusion, a propellant in atomic form is preferred, which is the definition of *noble gases*.

Engineers experimented with a number of substances. Mercury was first used as a propellant. For instance in NASA's experimental mission dubbed SERT-1 (Space Electric Rocket Test I) [17]. The advantages are that it can be stored in liquid form under standard conditions and, as a liquid, it has a high density of 13.5 g/cm$^3$, thereby occupying only a small volume of the spacecraft. However, due to its toxicity and the handling safety requirements, it was discarded.[14] Noble gases instead have become the first choice for electric propulsion. We will see later that using noble gases requires high pressure vessels for storage, see Sect. 9.6.1. Among the noble gases, there is a trade-off to be made: xenon is much more performant than krypton while the latter is much cheaper. The commercial space company SpaceX, for instance, favours krypton as propellant for their Starlink constellation, which is fully in line with their premise of subordinating performance to cost savings. Most other, especially interplanetary missions, favour xenon.[15]

### 7.2.6 *Throttling*

Throttling is the capability to reduce the thrust of an engine in an active way, be it chemical or electric, while trying to maintain the specific impulse. The capability to throttle the engine is vital for deep space mission that apply a solar electric propulsion system (SEP). Table 3.1 demonstrates compellingly the reduction of solar flux with solar distance, which is equivalent to a reduction in input power for the electric thruster. This presents a challenge, and we will discuss the basic principle of electric thruster throttle control below.

Thrust is a linear function of the input power, Eq. 5.29, meanings that we need to actively throttle the input power. The latter is the product of beam voltage and ion beam current. Changing the voltage is not desired, instead the current is varied via the injected propellant. The way to achieve this is straightforward: the less propellant is in the discharge chamber, the higher the effective resistance of the plasma, the less ionisation takes place and finally the lower the beam current. The alternative approach in which the voltage is reduced while maintaining the propellant flow would be a waste of propellant. We will see in Sect. 9.4.2 how the *propellant management system* of electric propulsion systems achieves flow throttling.

Table 7.5 presents predicted flight performance characteristics of NASA's NSTAR ion propulsion system on the Deep Space One mission, that match well with in-flight measurements within 2 % [18]. Since voltage was kept constant over almost the entire throttle spectrum, specific impulse did not change much. A noticeable drop occurs

[14] Spacecraft contamination is also an important parameter, which disqualified mercury in the long-run. Deposits on the surface could form conductive bridges leading to shortcuts.

[15] The price of xenon is actually very volatile and depends strongly on the international market needs.

**Table 7.5** Measured flight performance of NASA's NSTAR ion propulsion system on Deep Space One mission, [18]

| Beam Voltage (V) | Beam Current (A) | Accl. Voltage (V) | Input Power (kW) | Main Flow Rate (sccm) | Thrust (mN) | $I_{sp}$ (s) | $\eta_T$ (-) |
|---|---|---|---|---|---|---|---|
| 1100 | 1.76 | −180 | 2.29 | 23.43 | 92.4 | 3120 | 0.618 |
| 1100 | 1.67 | −180 | 2.17 | 22.19 | 87.6 | 3157 | 0.624 |
| 1100 | 1.58 | −180 | 2.06 | 20.95 | 82.9 | 3185 | 0.630 |
| 1100 | 1.49 | −180 | 1.94 | 19.86 | 78.2 | 3174 | 0.628 |
| 1100 | 1.40 | −180 | 1.82 | 18.51 | 73.4 | 3189 | 0.631 |
| 1100 | 1.30 | −180 | 1.70 | 17.22 | 68.2 | 3177 | 0.626 |
| 1100 | 1.20 | −180 | 1.57 | 15.98 | 63.0 | 3136 | 0.618 |
| 1100 | 1.10 | −180 | 1.44 | 14.41 | 57.8 | 3109 | 0.611 |
| 1100 | 1.00 | −180 | 1.33 | 12.90 | 52.5 | 3067 | 0.596 |
| 1100 | 0.91 | −180 | 1.21 | 11.33 | 47.7 | 3058 | 0.590 |
| 1100 | 0.81 | −180 | 1.09 | 9.82 | 42.5 | 3002 | 0.574 |
| 1100 | 0.71 | −180 | 0.97 | 8.30 | 37.2 | 2935 | 0.554 |
| 1100 | 0.61 | −180 | 0.85 | 6.85 | 32.0 | 2836 | 0.527 |
| 1100 | 0.52 | −180 | 0.74 | 5.77 | 27.4 | 2671 | 0.487 |
| 850 | 0.53 | −180 | 0.60 | 5.82 | 24.5 | 2376 | 0.472 |
| 650 | 0.51 | −180 | 0.47 | 5.98 | 20.6 | 1972 | 0.420 |

for very low mass flow rates, which is due to the drop in propellant efficiency $\eta_m$, i.e. insufficient ionisation and escape of neutral propellant with low velocity. Over the range of constant beam voltage, beam current and thrust level correlate almost perfectly, as expected. The correlation is also there for the mass flow rate, but with a slightly steeper slope.

## References

1. Anderson., J.D. (2020). *Modern compressible flow with historical perspective* (4th ed.). Boston: McGraw-Hill Education.
2. Britannica, T. Editors of Encyclopaedia. "Ernst Mach." Encyclopedia Britannica. Retrieved March 25, 2024, from https://www.britannica.com/biography/Ernst-Mach
3. Oliveira, A. (2019). History of the Bernoulli principle (pp. 1161–1178). https://doi.org/10.1007/978-3-030-20131-9_115
4. Britannica, T. Information Architects of Encyclopaedia. "Leonhard Euler." Encyclopedia Britannica. Retrieved June 6, 2024, from https://www.britannica.com/facts/Leonhard-Euler
5. Loeb, H. W. (1971). Radio frequency ion sources for electrostatic propulsion (hollow cathode design for electron bombardment radio frequency ion thruster source). In *Proceeding of the Symposium on Ion Sources and Formation of Ion Beams* (pp. 77–84).
6. Martinez-Sanchez, M., & Lozan, P. (2015). Lecture Notes in Space Propulsion. Session 10-11. MIT Open Course Ware.

7. Goebel, D., Polk, J., & Sengupta, A. (2004). Discharge chamber performance of the NEXIS ion thruster. https://doi.org/10.2514/6.2004-3813
8. Goebel, D. M., & Katz, I. (2008). Fundamentals of electric propulsion: Ion and hall thrusters. https://api.semanticscholar.org/CorpusID:92746133
9. Simon, J., Volkmar, C., & Probst, U. (2016). High-precision power measurement for accurate characterization of RF ion thrusters.
10. Child, C. D. (1911). Discharge from hot CaO. *Physical Review, 32*, 492.
11. Langmuir, I. (1913). The effect of space charge and residual gases on thermionic currents in high vacuum. *Physical Review, 2*, 450.
12. San Gregorio, L. M., Xie, K., Wang, N., Guo, N., & Zhang, Z. (2018). Ion engine grids: Function, main parameters, issues, configurations, geometries, materials and fabrication methods. *Chinese Journal of Aeronautics, 31*. https://doi.org/10.1016/j.cja.2018.06.005
13. Feynman, R. (1964). The motion of charges in electric and magnetic fields. https://www.feynmanlectures.caltech.edu/II_29.html
14. Hall, S., Jorns, B., Gallimore, A., Kamhawi, H., Haag, T., Mackey, J., Gilland, J., Peterson, P., & Baird, M. (2017). High-power performance of a 100-kW class nested hall thruster.
15. Jackson, J., Allen, M., Myers, R., Hoskins, A., Soendker, E., Welander, B., Tolentino, A., Hablitzel, S., Hall, S. J., Jorns, B. A., Gallimore, A. D., Hofer, R. R., & Pencil, E. (2017). 100-kW nested hall thruster system development, vol. IEPC-2017-219. Atlanta, GA.
16. Hall, S., Jorns, B., Cusson, S., Gallimore, A., Kamhawi, H., Peterson, P., Haag, T., Mackey, J., Baird, M., & Gilland, J. (2021). Performance and high-speed characterization of a 100-kW nested hall thruster. *Journal of Propulsion and Power, 38*, 1–11. https://doi.org/10.2514/1.B38080
17. Sovey, J. S., Rawlin, V. K., & Patterson, M. J. (2001). Ion propulsion development projects in US: Space electric rocket test I to deep space 1. *Journal of Propulsion and Power, 17*, 517–526. https://api.semanticscholar.org/CorpusID:51695395
18. Polk, J., Kakuda, R. Y., Anderson, J., Brophy, J., Rawlin, V. K., Sovey, J., & Hamley, J. (2000). Performance of the NSTAR ion propulsion system on the deep space one mission. In *IEEE Aerospace Conference Proceedings*, vol. 4 (pp. 123–148). https://doi.org/10.1109/AERO.2000.878373

# Chapter 8
# Energy Sources and Power Conversion

**Abstract** The movement of a spaceship requires energy, which makes the energy source a part of the propulsion system. This chapter explores the energy sources and power conversion methods that drive the high velocity mass flow discussed previously. Beginning with the simplest form, thermodynamic energy, used in cold-gas thruster, we then focus on chemical energy and solar-based power generation. The efficient and controlled release of chemical energy at a high rate, known as combustion, occurs in the combustion chamber. Electric propulsion systems, which require substantial electrical power (kW–MW), are also examined. Currently, the most significant method to power electric thrusters is solar energy utilization. The means to harvest solar power, through power generators, is the subject of the second part of this chapter.

## 8.1 Thermodynamic Energy

The arguably simplest form of energy to generate and utilise for accelerating the propellant to notable velocities is *thermodynamic energy*. The thermodynamic state of a medium is characterised by its temperature and pressure. When the surrounding thermodynamic energy level is lower than that of the stored propellant, a gradient in potential thermodynamic energy occurs that can be converted into kinetic energy, i.e. exhaust stream. This is, admittedly, a complex formulation of why an inflated balloon begins to swirl in the room when air is released.

This principle is realised in *cold gas thruster*, they represent the simplest form of a propulsion system. It is basically a Laval nozzle (Sect. 7.1.1) fed by propellant stored at a pressure between 25–30 bar. Since the underlying philosophy of cold gas thruster is to keep things simple, the idea is to use simply the pressure difference to the surrounding like in the balloon example. The temperature of the propellant is not increased by this process, hence the name cold-gas. Propellant storage temperature is usually maintained at about 20° C (293.15 K). Increasing the temperature would increase the exhaust gas velocity (refer to Eq. 7.18) but it would counteract the principle of simplicity. This is also the reason why the inexpensive and easy-to-handle gas *nitrogen* is predominantly used as propellant for cold gas thruster.



F. Gamgami, *Space Propulsion and Spaceship Design*, Springer Praxis Books,
https://doi.org/10.1007/978-3-031-71336-1_8

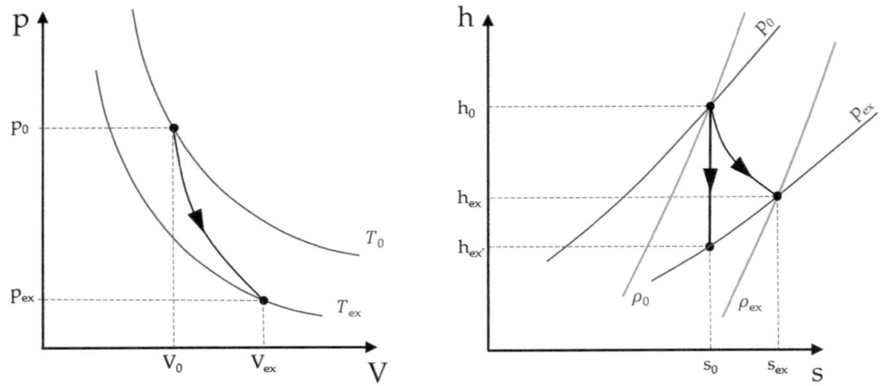


**Fig. 8.1** Expansion process: pressure-volume diagram (left), enthalpy-entropy diagram (right)

Figure 8.1 depicts the basic thermodynamic principle in two useful diagrams: the pressure-volume (p-V) diagram (left) and the enthalpy-entropy (h-s) diagram (right). The propellant is stored in a gaseous state (index 0) and released via a Laval nozzle without further energy input (e.g. combustion) into space (index *ex*).

Both diagrams show the expansion process and the related change in thermodynamic properties from storage condition to nozzle exit. While the p-V diagram is intuitive, it is not as practical as the h-s diagram. This is because a delta in enthalpy corresponds to the change in kinetic energy, refer to Eq. 5.3. Furthermore, the h-s diagram reveals immediately the ideal case of thermodynamic expansion in which entropy $s$ does not change, called *isentropic*,[1] represented by a vertical line from $h_0$ to $h_{ex'}$. The delta in enthalpy is then largest and so is the change in kinetic energy. We have seen in Sect. 5.2 that there is a idealised relation between enthalpy and temperature when the fluid is at rest, indicated by the index 0:

$$h_0 = c_p T_0. \tag{5.5}$$

We can compare the relative performance in terms of exit velocities of a cold gas thruster with one in which the temperature is raised–by combustion or any other mean–using Eq. 5.6. Under the assumption of equal thermodynamic properties and typical temperatures for each system, we receive:

$$\frac{u_{cold}}{u_{hot}} = \sqrt{\frac{T_{cold}}{T_{hot}}} = \sqrt{\frac{293\,K}{2500\,K}} = 0.34. \tag{8.1}$$

[1] In thermodynamics an isentropic process is a process that is fully reversible. Irreversible effects like friction or heat transfer shall not occur. In fact one definition of the physical concept of entropy is closely related to reversibility. If a process cannot be reversed within a *closed* system, the system's entropy has increased. Examples are mixing of different paint, energy dissipation due to friction and ice melting. All processes that cannot be reversed unless energy is introduced from outside the system.

**Table 8.1** Comparison of hydrogen propelled cold gas thruster [1] and Resistojet [2]

| property | Cold gas | Resistojet |
|---|---|---|
| Temperature (K) | 293 | 2500 |
| Specific impulse (s) | 272 | 828 |

The result fits very well with test data as can be seen in Table 8.1. The tests were carried out using hydrogen, which explains the high specific impulse for both types of thruster. Nitrogen is a more common propellant for cold gas thruster. But it yields an $I_{sp}$ of only about 73 s. This clearly demonstrates the limits of this energy source and why it is primarily used for low $\Delta v$ demands and when thrust is more important than specific impulse. Use cases are attitude control or orbit control of pico and micro satellites. Another issue with cold gas propulsion system is closely related to the propulsion management system, Chap. 9, and its scalability.

## 8.2 Chemical Energy

Mastering the utilization of *chemical energy* has fueled the technological evolution of mankind, and similarly, it has been the driving force behind the evolution of rocketry. Considerable amount of research has been spent in understanding and optimizing propellant *injection*, *ignition* and reducing *combustion instabilities*. This section will highlight the basic principles of chemically powered engines, their limitations and the governing properties.

Generally speaking, a chemical reaction is a reformation of the molecular structure of the involved elements. This reformation can either be *exothermic* or *endothermic*. The first group refers to reactions that release energy, which is what we are interested in: to increase the thermal energy state of the propellant. The second group represents chemical reactions that require energy, consequently reducing the thermal energy state, a scenario we aim to avoid. When discussing loss mechanisms, we will see that it is also necessary to address undesirable processes, such as endothermic reactions. The two types are depicted in Fig. 8.2. Note that by definition energy–like any other quantity–that *leaves* the system is negative. Hence burning hydrogen with oxygen is exothermic and the *reaction enthalpy* is therefore negative.

The branch of engineering that deals with chemical reactions is known as *combustion*. We start by considering the simplified chemical reaction of the important propellant combination:

$$2\,H_2 + O_2 \longrightarrow 2\,H_2O + \Delta h_r, \tag{8.2}$$

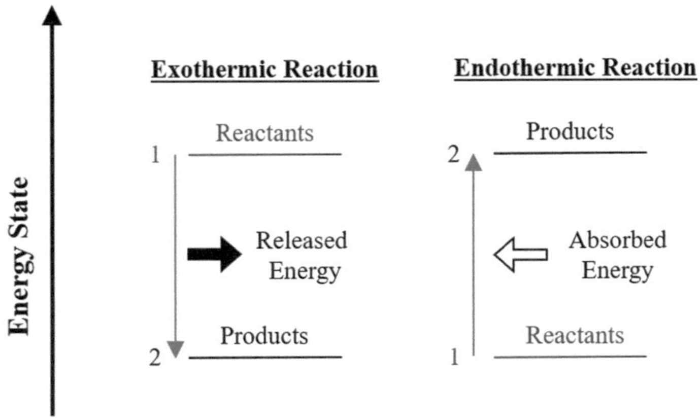


**Fig. 8.2** Energy scheme of exothermic and endothermic reactions

$\Delta h_{\mathrm{r}}$ reaction enthalpy.

Equation 8.2 presents a *stoichiometric* reaction involving two hydrogen, $H_2$, and a single oxygen molecule, $O_2$, at a specific pressure and temperature, yielding two water molecules, $H_2O$. Since single atoms and molecules are not of interest in technical processes, chemists and engineers prefer the term 'mole'. A mole is defined as the amount of substance that contains $6.022 \times 10^{23}$ molecules.[2] It remains to answer the question: how is the reaction enthalpy computed? The easiest way, with a minimum chemistry load is via the *standard formation enthalpy*:

$$\Delta h_{\mathrm{r}}^{\circ} = \underbrace{\sum_{i=1}^{n} n_i h_{p,i}^{\circ}}_{\text{Products}} - \underbrace{\sum_{j=1}^{m} n_j h_{r,j}^{\circ}}_{\text{Reactants}}, \tag{8.3}$$

$\Delta h_{\mathrm{r}}^{\circ}$ reaction enthalpy at standard condition,
$h_{\mathrm{p},i}^{\circ}$ standard formation enthalpy of product i,
$n_i$ mole number of product i,
$h_{\mathrm{r},j}^{\circ}$ standard formation enthalpy of reactant j,
$n_j$ mole number of reactant j.

The standard enthalpy is denoted by a circle (°). It is defined as the reaction enthalpy taking place under standard conditions, that is 25° C reaction temperature and a pressure of 1.013 bar (= 1 atm). The pressure in the combustion chamber of

[2] This number is named after the Italian scientist born in 1776, *Amedeo Avogadro* (1856[†]). He established the law that equal volumes of gases, at the same temperature and pressure, contain the same number of molecules, namely $6.022 \times 10^{23}$. The law and the number have been named in his honour.

**Table 8.2** Comparison of commonly used propellant combinations. Properties provided at storage temperature

| Denotation | Reactants | Main products | $r_{st}$ | $\bar{\rho}$ kg/m$^3$ | $h_r^\circ$ kJ/kg | $h_{r,V}^\circ$ $10^6$ kJ/m$^3$ |
|---|---|---|---|---|---|---|
| Hydrolox* | LH2 + $LO_2$ | $H_2O$ | 7.9 | 425 | 13,423 | 5.7 |
| Methalox* | LCH + 4 + $LO_2$ | $H_2O + CO_2$ | 4.0 | 851 | 10,026 | 8.5 |
| Propalox⋆ L | $C_3H_8$ + $LO_2$ | $H_2O + CO_2$ | 4.4 | 1,055 | 10,014 | 11.0 |
| Kerolox† | $LC_{12}H_{26}$ + $LO_2$ | $H_2O$ + $CO_2$ + CO | 2.3 | 1,022 | 9,655 | 9.9 |
| MMH-MON | $CH_6N_2$ + MON‡ | $H_2O$ + $CO_2$ + N + 2 | 1.0 | 1,084 | 9,424 | 10.0 |
| UDMH-NTO | $C_2H_8N_2$ + $N_2O_4$ | $H_2O$ + $CO_2 + N_2$ | 0.8 | 1,049 | 7,616 | 8.0 |

* 'L' stands for liquid and refers to the storage state
⋆ Not yet flown, but anticipated in Isar Aerospace's Spectrum rocket, a German launcher startup
† RP-1 is highly refined petroleum and specified in MIL-R-25576. Due to its complex chemical structure it is approximated by dodecane, $C_{12}H_{26}$ here
‡ Stands for 'mixed oxides of nitrogen' and refers to solutions of nitric oxide (NO) in dinitrogen tetroxide/nitrogen dioxide ($N_2O_4$ and $NO_2$). Variants are MON1 and MON3, with 1 % and 3 % of NO, respectively

powerful rocket engines is in the range of 100–300 bar, strongly exceeding standard conditions. However, for the purpose of highlighting the different energetic capabilities of the propellant combinations, it is sufficient to consider the standard enthalpy. To compute the reaction enthalpy of hydrogen-oxygen combustion, we need to apply Eqs. 8.3 to 8.2, which leads to:

$$\begin{aligned}
2\,H_2 + O_2 &\longrightarrow 2\,H_2O &+ \Delta h_r^\circ,\\
2h_{H_2}^\circ + h_{O_2}^\circ &\longrightarrow 2h_{H_2O}^\circ &+ \Delta h_r^\circ,\\
2\cdot 0 + 0 &\longrightarrow 2(-241.83\,\text{kJ/mol}) &+ \Delta h_r^\circ,
\end{aligned}$$

$$\Rightarrow \Delta h_r = 483.66\,\text{kJ/mol}.$$

It comes very handy that according to the definition of the standard formation enthalpy, $h_r^\circ$ for both $H_2$ and $O_2$ is zero. Given the molecular mass of 18.02 g/mol for water molecule, we receive in more convenient units: $h_r = 13.42$ MJ/kg. Note that in our example we have received two moles of $H_2O$, consequently we needed to take the molecular mass of two moles into account. In chemistry it is important to maintain correspondence with the written formula turnover. Table 8.2 lists the reaction enthalpy of a selection of propellant combinations assuming stoichiometric and complete combustion.

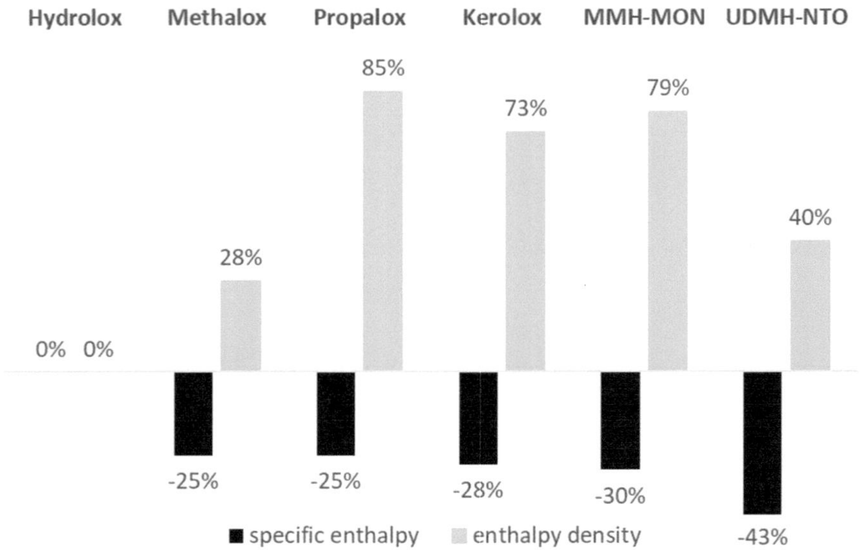


**Fig. 8.3** Comparison of bi-propellant performances relative to hydrolox

The list could be continued almost endlessly [3]. Fluorine in combination with hydrogen exhibits an excellent performance but is not further investigated due to its extreme toxicity. Other promising satellite propellants, so called 'green propellants', are still in development and have been omitted. They will be briefly discussed below. Figure 8.3 visualizes the reaction enthalpy of Table 8.2 relative to hydrolox.

The energetic superiority of hydrolox becomes apparent as it leads the ranking in specific enthalpy (J/kg) by 25 to 40 %. The weakness, though becomes clear too, namely it's poor enthalpy density (J/m$^3$), which is the reason why launcher manufacturer tend to denser propellant combinations for the first stage: the lower density of hydrolox results in larger storage volumes and consequentially larger structural mass.

Further, hydrolox engines tend to be heavier [13]. In the case of launcher stages, larger tank volumes lead to higher aerodynamic drag losses due to the increased frontal and/or skin area, see Eq. 6.22. For orbital systems, this is an issue because the volume inside the fairing is limited. We will encounter these constraints in Sect. 10.4. The higher specific impulse, however, could overcome these disadvantage and a careful trade-off[3] is needed.

Table 8.3 translates the reaction enthalpy of Table 8.2 to specific impulse with help of Eq. 5.4. Although these values are theoretical and cannot be achieved due to inevitable losses, they clearly show the performance ranking.

---

[3] Technical trade-offs are never free from programmatic constraints. If the desired technology is too expensive or not available to the country the company is located, due to export restrictions, then the optimum is often what you have at hand.

**Table 8.3** Theoretical $I_{sp}$ of selected propellant combinations

| | |
|---|---|
| Hydrolox | 528 s |
| Methalox | 457 s |
| Propalox | 456 s |
| Kerolox | 448 s |
| MON-MMH | 443 s |
| NTO-UDMH- | 398 s |

The propellant mixture ratio $r$ is defined as:

$$r = \frac{\text{mass of oxidiser}}{\text{mass of fuel}} = \frac{O}{F}. \tag{8.4}$$

The analyses above have assumed *stoichiometric* mixing ratios. This is plausible, since stoichiometric combustion maximizes the energy output and consequently the temperature. However, we have discovered in Sect. 5.2 that besides temperature there is a second factor relevant for a high exhaust velocity, i.e. specific impulse, the molecular mass:

$$u_{\max} \propto \sqrt{\frac{T_0}{\mathcal{M}}}. \tag{5.7}$$

This fact prompted rocket engineers of the early days to experiment with relative fuel rich mixture ratios in cases where this leads to combustion products with lower molecular mass.[4] This approach seems counter-intuitive as the excess fuel remains unburned and therefore cannot release any energy. The idea was to use the heat of combustion of the stoichiometric part of the propellant combination in the flow to heat the adjacent unburned excess fuel, as shown in Fig. 8.4 for hydrolox.

The idea was quickly confirmed. Figure 8.5 shows the dependency of the relevant factors ($\mathcal{M}$, $T_0$, $I_{sp}$) on the mixture ratio, computed with NASA's Chemical Equilibrium with Applications (short CEA) [7] for hydrolox combustion.

The left graph confirms that the combustion temperature is maximal for a stoichiometric mixture ratio ($r = 7.93$), whereas the molecular mass is a monotonic function. Considering these relations in view of relation 5.7, it becomes obvious that we are dealing with *competing requirements*: to achieve a high temperature, $r$ should be stoichiometric, whereas lower $r$ (i.e. relatively fuel rich) reduces the molecular mass. The right graph confirms this qualitative reasoning: the maximum specific impulse is not congruent with maximum combustion temperature. The graph suggests that to maximise the specific impulse the ideal[5] $r$ should be around 4.5 and thus considerably *lower* than the stoichiometric mixture ratio as falsely suggested

[4] Fuel rich relative to the stoichiometric ratio. As long as the ratio is $> 1$ the mixture ratio is absolutely oxidiser rich.

[5] Note that this value differs slightly with combustion pressure and expansion ratio, without changing the essence of the message.

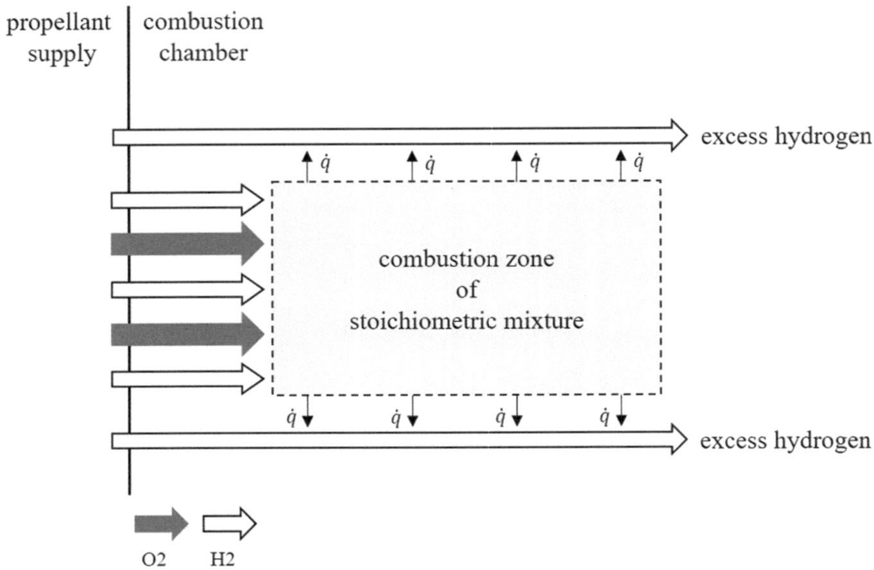


**Fig. 8.4** Sketch of heat transfer in fuel rich hydrolox combustion towards unburned excess hydrogen

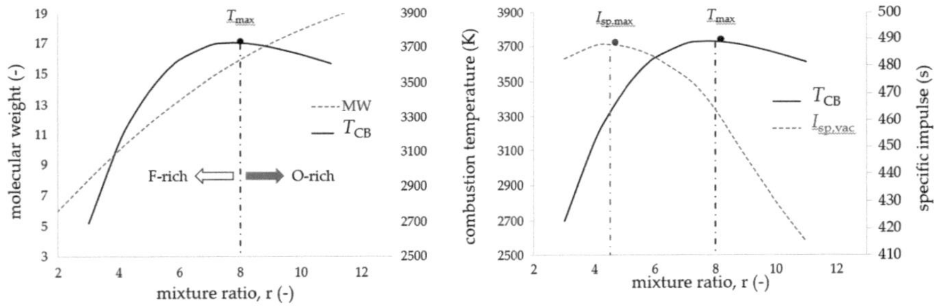


**Fig. 8.5** Dependency of molecular mass, combustion temperature and specific impulse on mixture ratio, computed with NASA's CEA [7] for 100 bar combustion chamber pressure and $\epsilon = 100$

by a pure thermal consideration.

It is therefore worthwhile to analyse the finding from a system perspective. A mixture ratio below the stoichiometric value might suggests to load *more* fuel than needed for the sole purpose to reduce the molecular mass. Although remaining unburned, this additional fuel still contributes to the specific impulse, as it has been heated by adjacent burnt fuel. The following questions arise in this context:

(a) Is the totally loaded propellant mass higher in case of a relatively fuel rich mixture ratio compared to a stoichiometric ratio for the same $\Delta v$ requirement?
(b) What is the system level impact when deviating from the stoichiometric (i.e. thermal idea) mixture ratio?

**Table 8.4** Trade-off matrix for the optimum mixture ratio. Not all impact factors included

| Mixture ratio | Positive | | | | Negative | | |
|---|---|---|---|---|---|---|---|
| $r \rightarrow r_{\mathrm{st}}$ | $\bar{\rho} \uparrow$ | | | | $I_{\mathrm{sp}} \downarrow$ | $T_{\mathrm{Load}} \uparrow$ | $m_{\mathrm{p}} \uparrow$ |
| $r \rightarrow r_{I_{\mathrm{sp}}}$ | $I_{\mathrm{sp}} \uparrow$ | $T_{\mathrm{Load}} \downarrow$ | $m_{\mathrm{p}} \downarrow$ | | $\bar{\rho} \downarrow$ | | |

These questions can be answered with help of the Tsiolkovsky equation

$$\Delta m_{\mathrm{p}} = m_0 \left[ 1 - e^{-\frac{\Delta v}{I_{\mathrm{sp}} g_0}} \right]. \tag{6.10}$$

Assuming a constant total wet mass of $m_0$, a mixture ratio that maximizes the specific impulse will also minimize the absolute amount of propellant needed. Hence, fuel rich in this context is equivalent to less propellant.

As so often, if things look too good and beneficial there is at least one hitch to be identified and evaluated. In case of hydrolox, hydrogen has a significantly lower density than oxygen even if both are stored in liquid form, LH2-LOX, namely 71 kg/m$^3$ and 1142 kg/m$^3$. This leads to an overall lower average propellant density and consequently to larger tank volumes and more structural mass. Therefore, depending on absolute figures, propellant savings may be counteracted by additional structural mass and in case of launch systems this also means increased drag losses during the atmospheric flight. These arguments necessitate a careful system-level trade-off.

The trade-off matrix in Table 8.4 provides at a glance the impact of the mixture ratio on component (i.e. engine) and system level (e.g. stage design). A characteristic of the matrix is its skew-symmetry, $\mathbf{A}^T = -\mathbf{A}$. The matrix is not complete and there are several key parameters that need to be considered in a full analysis. In fact, a detailed system level analysis reveals that an $r$ of 4.5 does not lead to an optimum from a *system perspective*: current hydrolox launcher have engines that operate with a mixture ratio varied in the range of 5.5–6. Still far from stoichiometry but not as low as suggested by a pure engine performance analysis.

The following statement is in general correct irrespective of the propellant combination:

$$r_{\mathrm{opt},I_{\mathrm{sp}}} < r_{\mathrm{opt,sys}} < r_{\mathrm{st}}. \tag{8.5}$$

This is a classical case where an inconspicuous parameter on component level (i.e. engine) has a profound impact on the system level layout. Mathematically speaking: the local parametric optimum is superseded by the global optimum. It is, therefore, the task of systems engineers to evaluate the extent of the influence of these parameters on the system and mission. In some cases, it is sufficient to involve a few further disciplines (e.g. thermo-mechanical architecture and aerothermodynamic analyses for an atmospheric flight) in other cases it is required to consider for instance system reliability, vehicle assembly and/or transportation to finally conclude on the *optimum* design parameter.[6]

Burning Atomic Hydrogen

An admittedly exotic but instructive case is the idea of burning atomic hydrogen. To understand the concept, it is worth visualising the underlying concept of chemical energy release, which is closely related to molecular *bond energy*. This is basically the energy required to pull a molecule apart into its atomic constituents. Obviously pulling atoms apart that otherwise would like to stay together is not an exothermic process. But the concept can be reversed and the bond energy becomes the attainable energy, if bonding is enabled. Hydrogen for instance has a bond energy of 436 kJ/mol which is almost twice the energy of hydrolox but at an improved molecular mass of 2.016 g/mol instead of 14.11 g/mol for a mixture ratio of 6. Applying Eq. 5.4, an atomic hydrogen rocket would achieve in a loss free combustion an exhaust velocity of:

$$u_{\max} = \sqrt{\frac{2 \times 436 \cdot 10^3}{2.016 \cdot 10^{-3}} \frac{\text{J mol}}{\text{kg mol}}} = 20,798 \frac{\text{m}}{\text{s}},$$

corresponding to a specific impulse of 2, 120 s. This is the highest achievable performance based on pure chemical energy,[7] [8]. Future fundamental research will reveal whether the exotic state of atomic hydrogen remains within the realm of science fiction or becomes achievable through technical advancements.

### *8.2.1 Engine Cooling and Energy Losses of Thermo-Chemical Engines*

The phenomena discussed in this section are partially applicable to all engines that utilize the thermal acceleration principle, as detailed in Sect. 5.2. However, for precision and to provide depth into the discussion, we will primarily focus on combustion processes.

Propellant injection into the combustion chamber is a challenge for any combustion motor, be it in a car, plane or rocket. The difference is that high performance rocket engines have an operating pressure of up to 300 bar and high combustion temperatures of more than 3,500 K, Fig. 8.5. The engine experiences simultaneously high thermal as well as high mechanical loads. The mechanical load is typically managed

---

[6] This domain is known as Multi-Disciplinary Optimisation MDO. It is wide spread in aeronautics and launch vehicle design [20] but to a lesser extent for satellites and space probes.

[7] Astrophysicists speculate that the interior of Jupiter is made of *metallic hydrogen* due to its strong magnetic field, its primary composition (i.e. hydrogen) and the extraordinary high pressure of approximately 450 GPa. The attribute *metallic* stems from its lattice structure akin to metals. The fact that researcher on solid state matter seriously speculate on the existence of metallic hydrogen under ambient conditions in a *meta-stable* state [9], gives hope that this energy form might be technically accessible in the future.

through clever construction methods or, less ingeniously, by simply adding mass. Thermal loads, on the other hand, pose a greater challenge. Modern rocket engine materials, such as rhenium/iridium, have an impressive operating temperature of 2,400 K [3]. However, this is still 1,000 K short of what is required for the combustion chamber and nozzle to withstand the combustion temperatures. Wall cooling techniques are inevitable for all engine classes. We will discuss cooling measures further below in the context engine thermal losses. Other performance losses are incomplete combustion and internal molecular excitation. All factors combined lead to undesirable deviations from the theoretical performances provided previously in Table 8.2 and are characterized by a single efficiency coefficient for clarity:

$$h_{\mathrm{r}}^{\mathrm{real}} = \prod_{i}^{N} \eta_i \cdot \Delta\, h_{\mathrm{r}}^{\mathrm{id}}, \tag{8.6}$$

$$h_{\mathrm{r}}^{\mathrm{real}} = \eta_{\mathrm{T}} \cdot \Delta\, h_{\mathrm{r}}^{\mathrm{id}}, \tag{8.7}$$

$h_{\mathrm{r}}^{\mathrm{real}}$ real reaction enthalpy with losses, at standard condition,
$h_{\mathrm{r}}^{\mathrm{id}}$ ideal reaction enthalpy without losses, at standard condition,
$\eta_i$ single efficiencies for each loss mechanism related to the reaction enthalpy,
$\eta_{\mathrm{T}}$ total engine efficiency

As cooling mechanisms indirectly linked to efficiency losses, we will discuss the effects in the following together.

### Incomplete Mixing–Intentionally and Unintentionally

Good propellant injection quality is characterised by a number of factors, of which the two most important are: minimum pressure drop and atomisation. The importance of limiting the pressure drop becomes obvious when visualising again that thrust is a function of chamber pressure, Eq. 7.21. We will focus in the following on *atomisation*. The design principle of combustion engineers is "what's mixed is burned" and atomisation is key to ensure a high degree of mixing. The term describes the complex and highly dynamic process of propellant vaporisation and droplet formation on a very small scale, before the picture is further complicated by chemical reactions and the formation of combustion products that begin to interact with the unburned propellant.

In this convoluted situation, lumps of propellant could slip away and stay unburned. This is the unintentional part of incomplete mixing. The intentional part is needed to maintain structural integrity of the combustion chamber walls and large part of the expansion nozzle: one mean to protect the walls from the very hot combustion gas ($\approx$3, 000 K) is by injecting colder fuel close to the walls. The idea is to form a protective *cooling film* layer between the hot inner part, where the vigorous combustion process takes place and the wall. The fuel is injected into the chamber near the wall and from outside at several locations, as depicted in Fig. 8.6. It is relevant to note that the film does not only protect the wall from hot gas but also from the detrimental chemical effect of atomic oxygen and hydroxyl (OH) in case of hydrolox

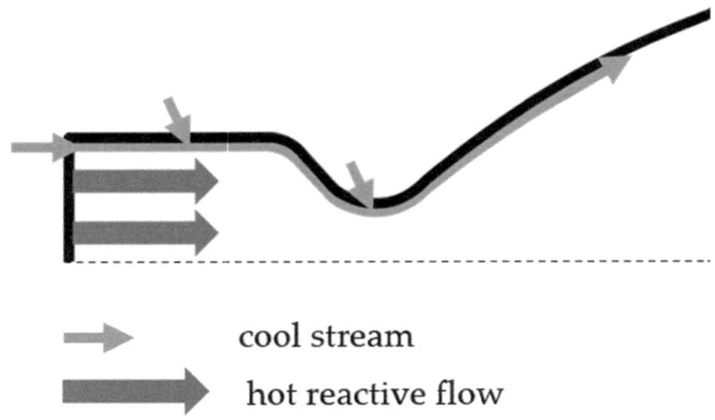


**Fig. 8.6** Sketch of film cooling and injection location

as propellant combination.
The throat of the nozzle is subject to high heat fluxes for two main reasons. Firstly, as discussed in Sect. 7.1.1, despite gas expansion, the temperature at the throat remains significantly high, dropping to only 88 % of the value in the combustion chamber. Secondly, the geometry of the throat exacerbates the heat transfer. Its surface area is much smaller than that of the combustion chamber, but it contains about 88 % of the energy density of the chamber. This combination makes the throat highly susceptible to thermo-mechanical failure. As we will see in the next section, there is a further method to cool the nozzle wall temperature, regenerative cooling.

Regenerative Cooling

The above introduced film-cooling concept is a form of *active cooling* and despite its merits it is often not sufficient to maintain wall temperature within its integrity limits. This was already discovered in the late 30 s of the past century by the German rocket engineer Walter Thiel, the chief designer of the A4 (Aggregate 4) rocket motor.[8] The ground-breaking counter measure to prevent the walls from thermo-mechanical stress and the engine from bursting is known as *regenerative cooling*. Walter Thiel and his team designed a double walled thrust chamber through which the fuel was pumped before entering the combustion chamber, Fig. 8.7. From a heat transfer point of view, regenerative cooling is equivalent to convective cooling, where a fluid (e.g. cool air or water) flows past an object (e.g. pipe or hot cup of coffee) and removes heat. The term 'regenerative' is derived from the fact that in a rocket, a portion of the propellant (in general, fuel) is utilised for cooling prior to combustion.
Modern rocket engines apply jackets, channels and passages within the walls of the nozzle. *Pressure losses* in these passages are significant and need to be considered when designing the propellant feed system, as will be detailed in Sect. 9.3. Hence, it is important to understand that cooling of the combustion chamber and expansion

[8] Walter Thiel (1943†) was a key figure in the German A4 rocket program at the Army Research Center in Peenemünde. He led the rocket motor group and made significant achievements in controlling combustion instabilities and excessive thermo-mechanical loads in the chamber. His work was crucial in enabling the development of modern high-thrust, high-performance space propulsion engines. He died with his family in an Allied bombing raid on Peenemünde [10].

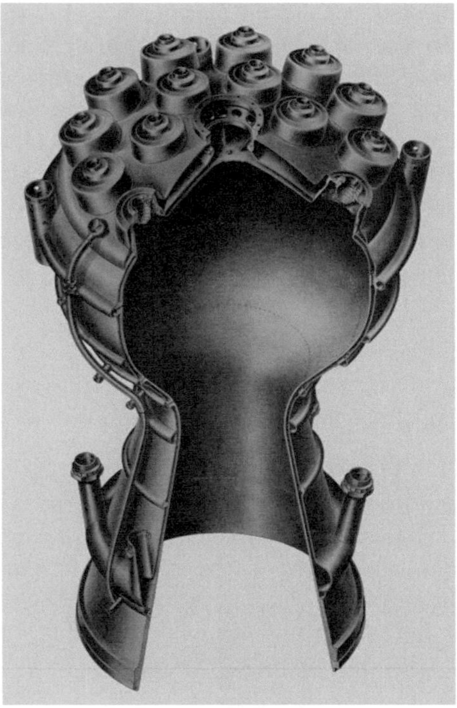

**Fig. 8.7** Regenerative cooling of German A4 rocket motor [11]. Credit: NASA. Courtesy: Dave Christensen

nozzle is a necessary evil: the heat transfer into the walls reduces the average temperature of combustion and thus the thermal state of the fluid, extra fluid is required to create a thermal insulation film, the pressure drop in the regenerative cooling system drives the propellant feed system and above all the design and manufacturing process is much more complex. Therefore, in an engineer's ideal world, the nozzle is made of a lightweight heat resistant material that does not require cooling means.

The alternative to active cooling, is passive-cooling. Two methods are known: *ablative cooling* and radiative cooling. The core idea of ablative cooling is to sacrifice on purpose a layer of wall that faces the hot combustion process and flow. That particular layer is in general made of silica-phenolic composite material, [12]. As it burns, it consumes heat for the phase change and forms a protective film on the wall as it is dragged away by the hot stream. This is obviously simpler and thus cheaper than regenerative-cooling but also less efficient.[9] Solid rocket motors apply this method due to the lack of a fluid. The simplest cooling method, however, is not to cool at all. In this case the thrust chamber and expansion nozzle form a *heat sink* and the walls are made of thick heat resistant material with a high heat capacity as well as a high conductivity to prevent structural disintegration. This is clearly the simplest

[9] The first generation of SpaceX's Merlin engine used this method for the Falcon 1 to limit development cost and speed-up development time. Current versions of the engine apply regenerative cooling as all modern engines.

method, but also the most inefficient in terms of maximum tolerable temperatures, dry mass and most importantly firing time.
Smaller thruster ($< 1,000\,\mathrm{N}$) that operate in space and do not suffer from excessive heat loads do rely solely on *radiative cooling* even for steady-state operation. We have seen in Sect. 3.1 that the heat ejection capability per square meter of a black body can be calculated with help of the Stefan-Boltzmann law:

$$q_{\mathrm{ej}} = \epsilon \sigma T^4, \tag{8.8}$$

| | |
|---|---|
| $q_{\mathrm{ej}}$ | heat flux ($\mathrm{W/m^2}$), |
| $\epsilon$ | body's emissivity, |
| $\sigma$ | Stefan-Boltzmann constant $5.67037 \cdot 10^{-8}\mathrm{W/m^{-2}K^{-4}}$, |
| $T$ | black body temperature (K). |

The thermo-mechanical architect of the spacecraft has then to make sure that the thrusters are placed such as to allow for a larger field of view into space with little to no obstacles in their vicinity. This is not only required for safe thruster operation but also as a protective measure for the spacecraft itself as this could lead to local overheating. Further, illumination by the Sun and albedo from Earth or Moon need to be considered as well and might restrict the optimum operation point of the engines or require dedicated thermal control manoeuvre, e.g. pointing to deep space for nozzle cooling before operation. As a rule of thumb engines with thrust levels below about 1 kN thrust do not require regenerative cooling but can rely on radiative cooling in combination with film cooling.

The general issue with thermal loads is strongly related to the size of the combustion chamber, as can be easily understood in case of a spherical chamber. The ratio of heat rejection capability to the rate of energy release is directly connected to the chamber's geometry:

$$\frac{\text{heat ejection capability}}{\text{rate of energy release}} \propto \frac{\text{surface}}{\text{volume}} \propto \frac{r^2}{r^3} = \frac{1}{r}. \tag{8.9}$$

This equation says that the larger combustion chambers, have an unfavourable surface-to-volume ratio and active cooling measures become inevitable, which drive the design and complicate the manufacturing process.

In case of engines with storable propellant within an intermediate thrust range of 2–7 kN there is the issue that the current storable fuel, MMH or hydrazine is limited in the amount of thermal heat that it can take, thus limiting the effectiveness of regenerative cooling. The limit is related to its hypergolic nature, the propellant decomposes uncontrolled if a temperature limit is exceeded causing and undesired exothermic reaction. Engineers study therefore a mixed approach in which part of the engine is cooled with fuel and the other with the oxidiser, NTO or MON, such that the burden is shared among the two propellants [14].
In case of purely film-cooled engines, the relation suggests that the conditions improve for smaller chambers, which is the case, but smaller chambers suffer from another challenge. The reason is that smaller thruster with a smaller chamber geometries have less fuel mass flow available for film-cooling since a minimum mass flow is needed to achieve the required energy release.
These considerations demonstrate that engine cooling is a challenge for both high-thrust and small-thrust thermal engines.

Frozen Flow versus Chemical Equilibrium

It is often falsely assumed that all chemical reactions take place in the combustion chamber and the expansion in the subsequent nozzle is a pure gas dynamic process without further chemistry. This is clearly not the case, which prompted engineers to distinguish between a *frozen flow* and flow in *chemical equilibrium*. We shall discuss this phenomenon by studying the example of hydrolox combustion in more detail. Equation 8.2 is actually a simplification of the real chemical reaction as it ignores all by-products besides the main product, water. Equation 8.10 is a more complete formula of the reaction, though still omitting small contributions of ionisation products for the sake of clarity.

$$\underbrace{H_2 + O_2}_{\text{reactants}} \longrightarrow \underbrace{H_2O}_{\text{main product}} + \underbrace{H + H_2 + H_2O_2 + OH + O + O_2}_{\text{by-products}} + \Delta h_r, \tag{8.10}$$

| | |
|---|---|
| OH | hydroxyl radical, |
| $H_2O_2$ | hydrogen peroxide. |

The by-products emerge due to unburned propellant and dissociation processes that break the molecules–an endothermic process. Both factors reduce the maximum available heat that can be obtained from the combustion process, which is expressed in a lower combustion temperature. In a flow, that is said to be in equilibrium, this missing energy can be regained during the expansion process via recombination.

Table 8.5 lists the mole fraction[10] of all combustion products along the nozzle for an equilibrium flow with snapshots at the combustion chamber, the throat and the nozzle exit. Furthermore, three expansion ratios have been analysed of which the

[10] The mole fraction is defined as the ratio of the number of particles (molecules or atoms) of constituent $n_i$ divided by the total number of particles in the mole: $x_i = n_i/N_{\text{tot}}$. The sum of mole fractions of all components in a mixture is always 1.

**Table 8.5** Mole fraction evolution for an equilibrium flow within a nozzle for LOX-LH2 combustion computed with NASA's CEA program, assuming $p_c$ = 100 bar and O/F = 8

| Element | Comb. chamber | Throat | $\epsilon_{50}$ | $\epsilon_{100}$ | $\epsilon_{1000}$ |
|---|---|---|---|---|---|
| H | 0.03301 | 0.02852 | 0.00086 | 0.00019 | 0.00000 |
| $HO_2$ | 0.00028 | 0.00019 | 0.00000 | 0.00000 | 0.00000 |
| $H_2$ | 0.11777 | 0.10873 | 0.01406 | 0.00546 | 0.00000 |
| $H_2O_2$ | 0.00006 | 0.00004 | 0.00000 | 0.00000 | 0.00000 |
| O | 0.01600 | 0.01351 | 0.00031 | 0.00007 | 0.00000 |
| OH | 0.10685 | 0.09538 | 0.00802 | 0.00323 | 0.00001 |
| $O_2$ | 0.03569 | 0.03429 | 0.00902 | 0.00589 | 0.00397 |
| Sum of by-products | 0.30966 | 0.28066 | 0.03227 | 0.01484 | 0.00398 |
| Main product $H_2O$ | 0.69034 | 0.71934 | 0.96774 | 0.98515 | 0.99602 |

**Table 8.6** LOX-LH2n: $p_c$ = 100 bar, $r$ = 8, with NASA's CEA

| Assumption | $I_{sp,vac}$ |
|---|---|
| Frozen (comb. chamber) | 424 s |
| Frozen (throat) | 432 s |
| Equilibrium | 464 s |

highest one, $\epsilon_{1000}$, is rather hypothetical and shall serve as an instructive asymptotic case. The underlying decisive physical properties are chamber resident time $t_{ch}$ and chamber length, $L_{ch}$. In the current analysis, the expansion ratio is a measure for both properties: the larger $\epsilon$ the larger $t_{ch}$ and $L_{ch}$.

It can be observed that the sum of all by-products decreases along the nozzle. The by-product OH, for instance, amounts to 10 % of the gas mixture in the combustion chamber and reduces to merely 0.8 % towards the nozzle exit for an expansion ratio of 50. The trend is obvious: the larger the expansion ratio the lower the amount of unburned by-products. This is because the longer the nozzle the more time is given to the molecules to finish the chemical reaction process as well as dissociated and ionised components to recombine thereby releasing energy.

The antagonist of equilibrium flow is *frozen flow*. As the name suggests, in a frozen flow model, it is assumed that chemical reactions stop at a certain point along the nozzle. As a result, untapped chemical energy is carried out of the nozzle thereby reducing the achievable specific impulse. The impact is considerable as listed in Table 8.6, where two hard freezing points are assumed: after the combustion chamber and at the nozzle throat, respectively. The difference in specific impulse is between 30 to 40 s, which amounts to 10 %.

**Table 8.7** Comparison of selected high thrust engine efficiencies based on vacuum specific impulse and benchmarked with NASA's CEA program [13]

| Engine | Prop. | $\eta_T$ | $I_{sp}^{real}$ | $I_{sp}^{id}$ | $\epsilon$ | $r$ | $p_c$ |
|---|---|---|---|---|---|---|---|
| RS-25 | Hydrolox | 0.973 | 452 s | 465 s | 78 | 6.03 | 206 bar |
| RL-10B | Hydrolox | 0.968 | 466 s | 481 s | 280 | 5.88 | 44 bar |
| YF-75D | Hydrolox | 0.954 | 443 s | 464 s | 80 | 6.00 | 41 bar |
| J-2 | Hydrolox | 0.945 | 421 s | 446 s | 27 | 5.50 | 53 bar |
| Raptor (vac.) | Methalox | 0.987 | 363 s | 368 s | 34 | 3.60 | 300 bar |
| RD180 | Kerolox | 0.943 | 338 s | 359 s | 37 | 2.72 | 257 bar |
| Merlin | Kerolox | 0.935 | 348 s | 372 s | 165 | 2.34 | 67 bar |
| F-1 | Kerolox | 0.909 | 304 s | 334 s | 16 | 2.27 | 70 bar |
| Aestus | MON-MMH | 0.938 | 324 s | 346 s | 84 | 1.90 | 11 bar |
| HiPAT™ | MON-Hydrazine | 0.937 | 329 s | 351 s | 300 | 1.00 | 9.4 bar |

In reality, there is no instant freezing of chemistry for all chemical reactions at a specific location in the thrust chamber or expansion nozzle. Instead, the speed of chemical reactions including combustion and recombination processes decrease gradually between the throat and the nozzle outlet until it ceases and the state freezes–hence the name 'frozen flow'. Numerical analysis of the governing equations is needed to determine the actual energy release and achievable specific impulse.[11]

Molecular Excitation

The vigorous combustion processes occurring inside the thrust chamber cause molecules to vibrate and rotate around their main axes. In addition, also the internal molecular and atomic state is affected: particle collisions cause electrons to level-up and attain higher energetic states or are stripped away, leaving the particle (atom or molecule) ionised, thus changing the particle's electron configuration. All these phenomena are endothermic and referred to as *internal molecular excitation modes*. They are obviously not desired in thermal engines as they reduce the transverse kinetic energy of the fluid, Fig. 5.3. Although inevitable they need to be quantified by numerical modelling and tests in order to assess the quality of the engine's design.

Real Engine Efficiency

To conclude the discussion on combustion losses, efficiencies of a selection of rocket motors will be compared. Table 8.7 groups engines according to their propellant combination and lists them in order of decreasing efficiency. The benchmark, the ideal case, was computed by NASA's CEA program assuming equilibrium flow and vacuum conditions ($p_{amb} = 0$).
The results show, that kerolox engines do not exceed 94 %. Saturn's mighty first stage engine, the F-1, has only an efficiency of 91 %. The group of hydrolox engines is led

[11] This discipline is called computational fluid dynamics (CFD). The governing equations are the Navier-Stokes equations coupled with a chemical reaction model.

by the upper stage engine RL-10B and the RS-25 with 97 % efficiency.[12] The most remarkable performance is achieved by SpaceX's Raptor engine, reaching 98.7 %. This is a result of the engine's highly sophisticated *full-flown combustion cycle*. An in-depth analysis reveals that the dominating factor of the engine's efficiency is in fact the combustion cycle process [13], subject of Sect. 9.5.

### 8.2.2 The Choice of Propellant

A key feature of chemical propulsion systems is that the energy is stored within the propellant and no further mass to supply energy is required, in contrast to electric propulsion. However, this advantage comes with constraints. We have seen that once the propellant has been selected, the theoretical specific impulse is determined by its chemistry. The choice must therefore be made wisely and depends very much on mission objectives, environmental requirements, availability and operating costs, to name but a few. In this section we will highlight the chemical and physical properties that are of technological and system relevance:

| | |
|---|---|
| *chemical*: | reaction enthalpy, flash point, toxicity, chemical stability, material compatibility, |
| *physical*: | average density, heat capacity, enthalpy of evaporation, vapor pressure, boiling & freezing point, |
| *operational*: | storability, safety, ground handling complexity, |
| *programmatic*: | availability, experience, cost, business model and long-term strategy, |

Some of these properties are related to each other, for instance, the storability of a propellant depends on its boiling temperature as well as chemical stability. So is the safety aspect of a propellant closely linked to its reactiveness, flash point and toxicity.

New space companies with a strong commercial mindset, such as SpaceX, have pivoted away from the classical approach in which the focus lied purely on physical and chemical performance indicators. A sustainable and successful business model must give emphasis to 'programmatic' aspects. The company SpaceX, for instance, has selected kerolox as propellant combination due to the *availability* of this technology through technology transfer programmes with NASA. The company developed and improved the Merlin engine, which powers the company's highly reliable Falcon rocket family. Despite the company's extensive knowledge of kerolox, SpaceX has switched to the fully cryogenic propellant combination methalox for its new launcher,

[12] The RS-25 is also known as Space Shuttle Main Engine (SSME) of the Space Transportation System (STS). Three engines propelled the STS and four engines propel NASA's new super heavy lift launch vehicle Space Launch System (SLS).

Starship.[13] What appears to be an odd move is actually a *strategic* one. The company plans to establish a permanent human presence on Mars, and intends to mine the propellant required to commute between Mars and Earth on Mars itself, Sect. 4.5. Since current satellite based data has proven the existence of carbon-dioxide on Mars, Sect. 4.5, the choice becomes comprehensible. The company plans to extract methane and oxygen by means of In-Situ Resource Utilization (ISRU). This is a remarkable example of rigorous and far-sighted decision-making based on a *strategic objective*.[14]

Although a single factor could be decisive in the propellant trade-off, the other properties still go into the equation and need to fulfill threshold requirements. After all they pose important design parameter to be accounted for–on propulsion system level as well as on spaceship architect level. The above listed properties shall be briefly detailed:

*Reaction enthalpy* determines the maximum theoretical specific impulse that can be attained from a chemical reaction: in case of a single propellant (mono-prop) by decomposition or in case of a propellant combination (bi- or tri-prop) by chemical combination (i.e. redox reaction).

*Ignitability* refers to the level of difficulty to ignite a propellant. The easiest form is given by *hypergolic* propellant and propellant combinations, like Hydrazine, MMH, UDMH, that ignite with NTO or MON variants by contact. This has several operational advantages: restart capability is a natural feature as only valves have to be opened to bring the propellants in contact in the thrust chamber, without the need of a special ignition device. On the other hand, the safety measures for ground handling are complicated, as unintentional combustion must be prevented.[15] The situation is different with cryogenic and partially cryogenic propellants, like hydrolox, methalox and kerolox. These combinations require active support to start the ignition process. Rocketdyne J-2 engine, that propelled Saturn V's second and third stages, used a reliable pyrotechnic ignition system in which an electric current ignited a solid propellant charge. The disadvantage of this system is obviously the lack of re-startability, which led to the introduction of lase-based ignition systems [19].

*Flash point* means the lowest temperature, $T_{\text{flp}}$ at which the propellant vapour (above the liquid) ignites in presence of an ignition source. The resulting flame is not necessarily sustainable and the ignition could lead to a darting flame only–hence *flash*. The strength of the flash depends on the fluid's vapour pressure at storage temperature. It indicates the *fire hazard* of a fuel and *safety measures* need to account for it.

---

[13] Starship is the world's largest ever built rocket. It is a great leap in space access and rocket design for several reason: it is the first fully reusable launcher, it four folds current capacity from 25 t to 100 t and it could reduce the kg price down to USD 200 per kg [15].

[14] This is of course a condensed line of argumentation. For instance, hydrogen can be gained on Mars as well but a re-usable hydrolox launcher, which was a key requirement for SpaceX, would be even taller than the current Starship, 121 m and would come with many new challenges.

[15] A tragic example is the 'Nedelin catastrophe' in 1960, when an R-16 intercontinental ballistic missile exploded on the launch pad during launch preparation because the second stage accidentally ignited due to the hypergolic nature of the propellant, UDMH-$N_2O_4$. At least 100 ground personnel lost their lives [18].

*Toxicity* of a propellant implies complex safety procedures on ground. Not only with regard to ground operations in preparation for flight (i.e. the exploitation phase), but also during the development phase. Furthermore, the exhaust products of these propellants are also toxic and require elaborate cleaning procedures, which primarily pose a health risk to those involved and delay engine development time and increase costs. This is the reason why fluorine, $F_2$, and related chemicals ($OF_2$, $ClF_2$, $ClF_3$ and $ClF_5$) have been avoided despite their high-performance figures as oxidiser.

*Chemical stability* of a propellant must be guaranteed under all operating conditions. Degradation is in particular an issue for solid propellant and some mono-propellants due to the radiation exposure.

*Material compatibility* of the tank and the propellant must be guaranteed. Hydrogen molecules are capable to leak out of the tank due to their tiny size (smallest molecule!). Oxygen, on the other hand, is very reactive, making the development of linerless CFRP tanks for LOX very challenging, see Sect. 9.6.4.1.

*Average density* of the propellant combination, $\bar{\rho}_{\text{prop}}$, is a measure for the compactness of the engine, the storage system and for large amounts of propellant for the spacecraft's overall dimension. We have stated above that the mixture ratio, $r$, is key a parameter as it is directly related to the average propellant density. The corresponding formula is:

$$\bar{\rho}_{\text{prop}} = \frac{(1+r)\cdot\rho_{\text{o}} + \rho_{\text{f}}}{\rho_{\text{o}} + r\cdot\rho_{\text{f}}}. \tag{8.11}$$

The wish to reduce the overall system mass–or to load more propellant–some launch provider load *sub-cooled* propellant to increase the average density. Among them SpaceX for Falcon 9 and Glavcosmos for Soyuz. The specific design of the Russian rocket allows to achieve overall propellant *densification* since the RG-1 (Soviet version of RP-1) tank of the main booster stage is partially surrounded by much cooler LOX tanks ($\approx$90 K) [16]. Active sub-cooling has also been studied extensively for hydrogen and is regarded an enabling technology for long duration in-space storage [17], as it reduces *boil-off losses*. The technology is not only beneficial for chemical propulsion systems but also for hydrogen propelled thermo-nuclear propulsion as we will see in Chap. 11.

*Heat capacity* of a fluid is a measure of how much energy needs to be provided–or can be absorbed in case of wall material–to increase its temperature by 1 K. For the engineer, choosing the right one depends on how it is used: if the fluid needs to be heated, such as in a resistojet, then a low heat capacity is desirable as this reduces the heating energy required. If the fluid is used for regenerative cooling then a high heat capacity is desired to maximise heat absorption. Cryogenic propellant suffers from boil-off due to parasitic heat fluxes entering the storage tank. In this case a high heat capacity is a blessing as it reduces the boil-off rate and thus propellant *storage losses*. This case plays a crucial role for cryogenic long-term storage in space [17]. Heat capacities are either measured under constant pressure, $c_p$, or constant volume, $c_V$,. The first is of interest for space propulsion processes, since combustion or heat transfer is well approximated under constant pressure.

*Enthalpy of evaporation* measures the enthalpy required to induce a phase change of the medium, that is either boiling or sublimation. Similar to the heat capacity, it forms a crucial factor for *cryogenic* propellant. The larger the enthalpy of evaporation the better storage of cryogenic propellant becomes due to reduced *boil-off losses*:

$$\dot{m}_{\text{boil}} = \frac{\dot{Q}_{\text{tot}}}{h_{\text{evp}}}. \tag{8.12}$$

*Vapour pressure* of a fluid characterizes the storage pressure under which an equilibrium[16] of evaporation and condensation is attained given a constant storage temperature, $T_{st}$. If the propellant is stored below its vapour pressure, $p_{vp}$, it will continue to evaporate and increase the pressure in the gas phase until $p_{vp}$ is reached. A low $p_{vp}$ combined with a low *flash point*, $T_{flp}$ poses a safety challenge. Another technical aspect is also related to the vapour pressure, *cavitation* in the propellant feed-system and turbomachinery, see Sect. 9.3. In addition low vapour pressure propellant have unfavourable ignition and combustion properties. As a rule of thumb, *storage pressure* shall be at least twice the propellant's vapour pressure.

*Boiling & freezing point* temperatures are important for propellant storage and operation. Hydrazine for instance, is a popular mono-propellant for satellites. It is used for orbit maintenance, reaction wheel desaturation, attitude control and other small $\Delta v$ tasks. Its boiling point, $T_{bp}$, of 386 K (113° C) is less of a challenge to maintain,[17] but its freezing point, $T_{fp}$, of 275 K (1.85° C) needs to be taken into account during spacecraft operation. Propellant tank temperature is therefore continuously monitored and in general actively controlled. A list of temperature limits of the above discussed propellant is provided in Table 8.8.

*Storability* of a propellant is measured for *standard conditions*. A propellant is said to be storable, if it can remain liquid at these conditions (1 atm and 25° C). An arguably anthropocentric view, but given the fact that this temperature can be achieved for a vessel in Earth's vicinity (i.e. 1 AU) without great effort, Sect. 3.1, it is reasonable. High performance propellant combinations like hydrolox and methalox are gaseous at ambient conditions due to their very low boiling temperatures, hence

[16] The water in an open glass at warm summer day with an ambient pressure of slightly above $p_{vp}$ has no chance to reach equilibrium with its environment (Earth's atmosphere) and will eventually completely evaporate.

[17] Before the propellant boils due to temperature increase, the tank will explode due to the pressure increase, Sect. 9.6.1.

**Table 8.8** Comparison of chemical and physical properties of liquid rocket propellant. Source NIST WebBook, SRD69

| Propellant | Toxicity | Storable | $\rho$ kg/m$^3$ | $T_{fp}$ K | $T_{bp}$ K | $c_p$ kJ/kg·K | $h_{evp}$ kJ/kg |
|---|---|---|---|---|---|---|---|
| LOX (C) | No | No | 1,142 | 54 | 90 | 1.6978 | 213.1 |
| LH2 (C) | No | No | 71 | 14 | 20 | 9.4762 | 446.4 |
| LCH4 (C) | No | No | 422 | 90 | 111 | 3.4669 | 509.6 |
| RP-1 | No | Yes | 820 | 233 | 573 | 2.2117 | 362.8 |
| NTO | Yes | Yes | 1,450 | 264 | 294 | 1.5487 | 430.4 |
| UDMH | Yes | Yes | 793 | 216 | 336 | 2.7296 | 586.0 |
| MMH | Yes | Yes | 866 | 220 | 364 | 2.9288 | 876.3 |
| Hydrazine | Yes | Yes | 1,004 | 275 | 386 | 3.0835 | 388.7 |

(C) cryogenic
$\rho$, $c_p$ and $h_{evp}$ at typical storage conditions: $p_t = 4$ bar and $T_t \leq T_{bp}$, for cryogenic propellant, else $p_t = 4$ bar and $T_t = 25°$ C

the name *cryogenic*. Propellant that cannot exist in liquid form for standard conditions is also referred to as volatile and passive or even active means are needed to maintain the liquid state. On the contrary hypergolic propellant combinations currently in use are storable and no sophisticated means are needed to maintain them in their operational temperature range under standard conditions. Naturally, this made hypergolic propellant the first choice for in-space propulsion systems of satellites and spacecrafts.

*Safety* aspects in view of propellant selection are strongly related to toxicity, corrosion and explosion or fire hazard.

*Ground handling* subsumes all effort required to store, transport and fuelling. RP-1 is an example of a benign propellant that is storable: it has a high flash point and is not toxic. The related effort is therefore low. Liquid hydrogen, due to its cryogenic nature, and the hypergolic MMH, due to its toxicity, are more demanding examples regarding the required procedures, facilities and devices for ground handling.

*Cost* of a propellant might seem less crucial given the high operational cost of a launcher, a spacecraft or any other space vessel. However, in view of launcher re-usability and given the increasingly competitive space market propellant cost does play nowadays an increasing role in the economic balance sheet. This is especially true for electric propulsion in the context of mega constellations of up to 40,000 satellites, Sect. 7.2.5.

*Long-term strategy* is best exemplified by SpaceX's choice of methalox for the Starship propulsion system—see discussion above.

In conclusion, the eventually selected propellant combination has significant design impacts on the spaceship's system architecture and overall performance. This impact increases the larger the $\Delta v$ requirement becomes. The selection process is in general driven by only a handful key criteria from system perspective but the implications drive the overall system design and operation requirements.

## 8.3 Solar Power Generators

Solar-powered generators[18] (SPG) have been used in space since the earliest days. With a few exceptions, they are the first choice for satellites, space probes and space stations. The International Space Station (ISS) for instance has the largest solar power generator so far installed in space, at around 215 kW. When it comes to electric propulsion systems, solar power is currently the only practical choice. Due to their power-hungry nature, all electric propulsion systems are currently powered by solar generators. This will not change until sufficient progress has been made in space qualified nuclear electric generators, see Chap. 11.
In particular, geostationary telecommunication satellites profit from the efficiency of electric propulsion systems. Their advantage at system level is twofold, as they power both, the payload and the propulsion system.[19] Spacecrafts powered by solar electric propulsion (SEP) are strongly driven in their design by the solar generator which in turn are driven by the propulsion requirements. This relation will be explained in more detail later in the chapter. To understand this mutual dependence, it is inevitable to establish a sound understanding of the physics and technology of photo voltaic arrays first.

The section starts with a concise overview of the physics and technology of solar cells, followed by an explanation of the solar wing architecture. After discussing failure modes and degradation effects, the section concludes with an examination of contemporary solar generator designs and their associated performance figures.

### *8.3.1 Solar Cell Physics and Modeling*

The lowest level building block of an SPG is the *solar cell*, a marvellous semi conductor device that, with correct extra-treatment, is capable of delivering the required electric current to power all consumers in the spacecraft. It is therefore called, current source. This is the main difference to batteries, which are a voltage sources and

[18] Other terms are solar array (SA), photo voltaic array (PVA). There is no stringent unified definition and it is necessary to check what exactly is included and excluded or whether it is synonymously used.

[19] This advantage in case of geostationary telecommunications satellites is due to the fact that their radio frequency (RF) payload is as power hungry as the electric propulsion system (EPPS). This means that both subsystems (payload and EPPS) share the SPG, i.e. they share one resource.

deplete, like a fuel tank, if not re-charged. In contrast, solar cells will provide electricity for as long as there is a light source, or more precisely, a source of electromagnetic radiation–in theory, indefinitely.[20]

Figure 8.8 shows a basic physical schematic of a single junction solar cell. It consists of a *p-doped* silicon layer at the bottom and an *n-doped* silicon layer on top. The interface between the two is called a pn-junction, or short *junction*. Silicon is a semiconductor and not a complete conductor like metals. A semiconductor must first be motivated to become a conductor. For instance, by heat or irradiation. The physics behind this phenomenon is based on quantum mechanics, which, although not overly complex, is beyond the scope of this book. Instead, we will focus on the pre-treatment of the semiconductor material–specifically silicon–that is crucial for establishing the aforementioned layers. The n-doped side consists of silicon mixed with a few phosphorus atoms, with a relatively low but sufficient doping level of $10^{-6}$. Phosphorus is in the 5th column of the periodic table and has five valence electrons, which means it needs three more to reach the *noble gas* configuration.

The basic material silicon has four valence electrons and is missing four other electrons. It shares the remaining four electrons with neighbouring atoms. If the neighbour is a phosphorus atom, five valence electrons are available, but as only four are needed, one remains loosely bound to its parent atom, phosphorus, left in Fig. 8.9. By pumping energy into the material by means of electromagnetic radiation, this electron can be released much more easily than the double-bound silicon valence electrons. The p-doped side on the other hand consists of silicon mixed with a few atoms of the third group in the period system, like boron. These atoms feature only three valence electrons and lack the five more electrons needed to reach noble gas configuration. This lack is manifested by a hole, as shown in the right-side depiction

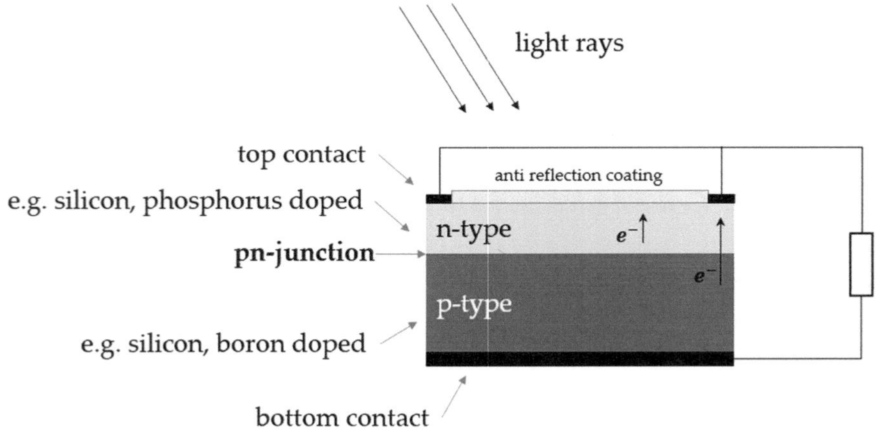


**Fig. 8.8** Basic physical schematic of a single junction solar cell

---

[20] In most satellites, the solar array works hand in hand with the batteries. The latter are charged during sunlight and depleted, if the spacecraft enters the planetary shadow.

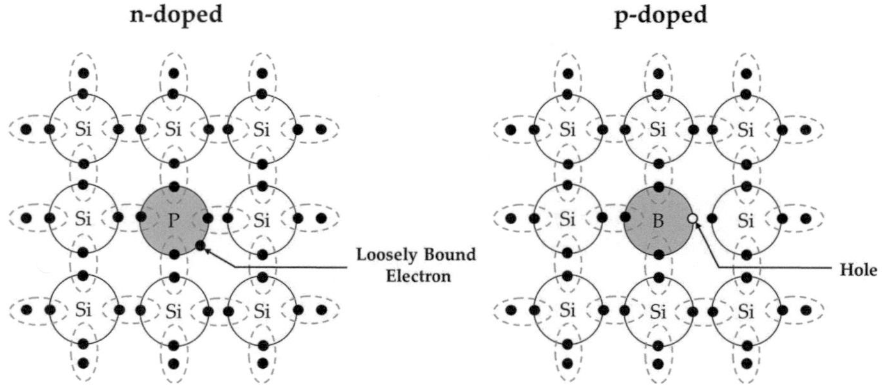


**Fig. 8.9** Silicon lattice doped with phosphorous (left) and boron (right)

in Fig. 8.9. It is important to note that both n- and p-doped materials are neutral in themselves and not charged. However, the concentration of electrons and holes in these materials differs. When brought into contact, this difference triggers a *diffusion-driven* balancing process; electrons diffuse from the n-doped side to the p-doped side.

This diffusion process would cease after a short time due to resulting electric fields that build up inside the material. But if electrons are given an alternative path, like shown in Fig. 8.8, a continuously working *electric current machine* can be engineered.

The above discussed silicon based single junction solar cell can reach a maximum efficiency of 20 %, under Airmass Zero (AM0) and 25° C standard conditions, and is the standard solar cell type for terrestrial use. Since mass and volume are limited resources in space applications, there is an acute need in the space market for higher efficiency solar cells. The next step of improvement is to use gallium-arsenide (GaAs) solar cells instead of silicon. The MIR space station, launched in 1986 by the Soviet Union, was powered by 10 kW worth of GaAs solar cells, which provided a power density of 180 W/m$^2$ [21]. Gallium-Arsenide enables a higher electron mobility, thus increasing the total current compared to silicon, and it exhibits a significantly reduced thickness due to its direct electronic bandgap. However, the cost of materials and production of GaAs cells is much greater than that of Si cells. Besides changing the semi conductor material, there is a more sophisticated way of further increasing the cell efficiency; namely by using *multi-junction* (MJ) solar cells. This method combines different basic materials in a stacked configuration. The reason for this lies in the so-called *gap energy* which is the energy required to make an electron break its bond and move freely in the lattice. Figure 8.11 depicts the energy thresholds in the

Fig. 8.10 Schematic of triple-junction (TJ) solar cell

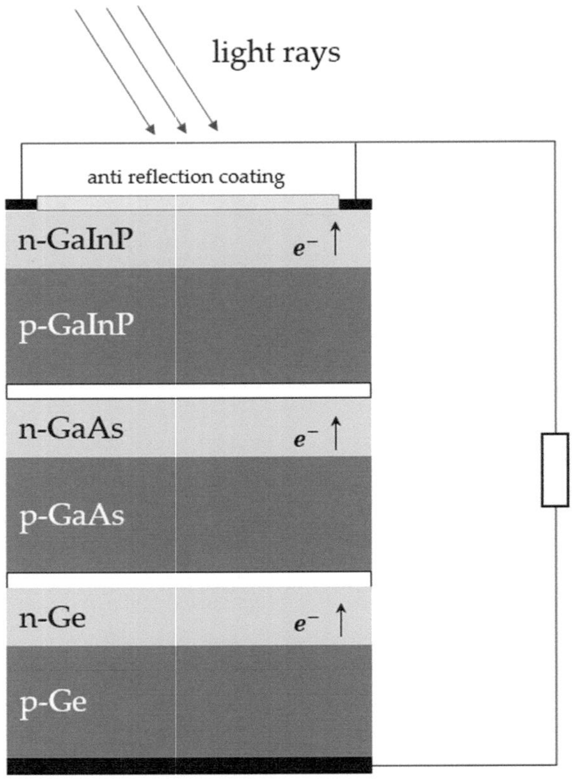


context of the solar spectrum, approximated as a black body radiator of 5800 K. A single-junction solar cell absorbs only photons left of the marked limiting wavelength and misses all photons right of this limit as they are not energetic enough. On the other hand, photons with very short wavelengths are too energetic. They not only knock off an electron but also transfer kinetic energy to it. This excess energy heats the material, which negatively impacts cell efficiency and other performance factors. The core idea of multi-junction solar cells, therefore, is to engineer a solar cell *compound* consisting of different semiconductor materials with different energy thresholds in order to harvest as much of the solar spectrum as possible. The stacking order is of utmost importance to make this 'filtering' system work as desired.

A prominent example is the triple-junction (TJ) InGaP/GaAs/Ge solar cell with an efficiency over 30 %, which has been the new standard for space-based solar power generators for many years. Figure 8.10 shows the schematic configuration of this cell type. Obviously, this increase in efficiency has its price in terms of production cost.

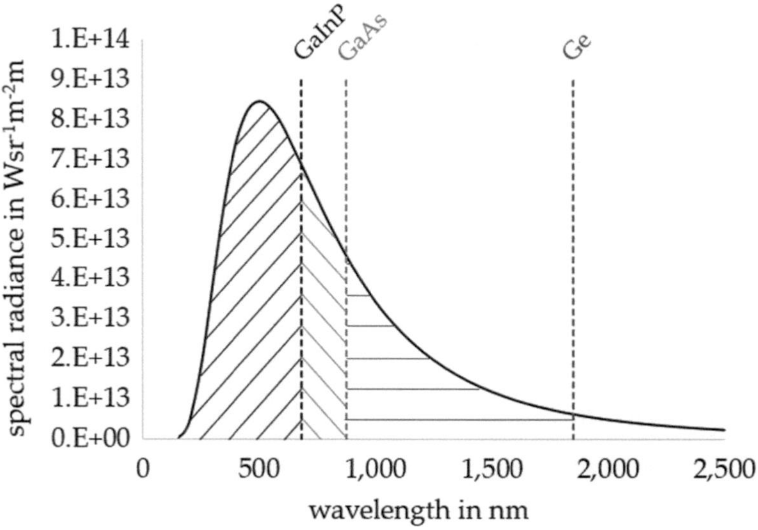


**Fig. 8.11** Solar spectrum and energy bandwidth usage for a triple-junction solar cell made of InGaP/GaAs/Ge

## Equivalent Circuit

As highlighted above, a solar cell is a current source and not a voltage source, unlike batteries. Figure 8.12 depicts the electric model of a solar cell. It consists of a:

*current source* $I_i$: represents an ideal solar cell without losses. It is also called light-generated current and is directly proportional to the irradiance,

*diode*: represents the electrons that receive enough energy to break loose but do not make it out of the crystal, for instance by falling into a hole. The related current is a loss,

*shunt resistance* $R_{sh}$: represents an alternative current path for the electrons and is due to the non-ideal nature of the semiconductor crystals, i.e. manufacturing defects. The related current and the voltage drop are losses,

*series resistance* $R_s$: represents all resistance contributions that electrons experience on their way to the metal contact points for extraction.

Applying Kirchhoff's laws and subsequent re-arrangement leads to the solar cell current equation [22]:

$$I = I_i - I_s \cdot \left( e^{\frac{V + I R_s}{a_d V_T}} - 1 \right) - \frac{V + I R_s}{R_{sh}}. \tag{8.13}$$

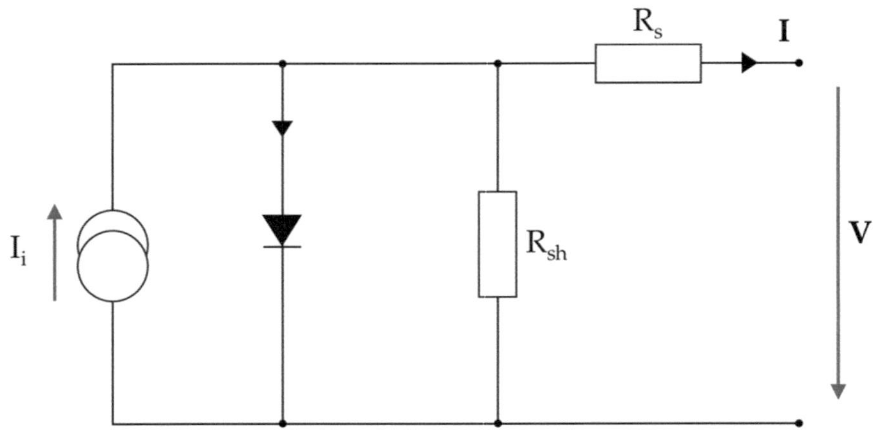


**Fig. 8.12** Circuit model of a solar cell

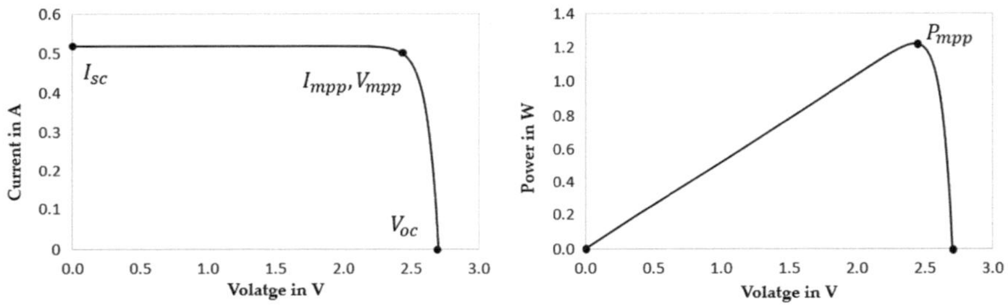


**Fig. 8.13** Solar cell characteristic curves: I-V curve (left), P-V curve (right)

The parameter $V_T$ is called thermal voltage defined as $k \cdot T/q$, with $k$ is the Boltzmann constant, $T$ the temperature, and $q$ the electron charge. The parameter $a_d$ is the diode ideality factor. Solving this equation leads to the famous I-V curve that describes the performance of solar cells and entire arrays alike, left-side depiction of Fig. 8.13. This diagram has three characteristic points, quantified by four parameters specific to each solar cell:

$I_{sc}$ short circuit current,
$V_{oc}$ open circuit voltage,
$I_{mpp}$ current at maximum power point,
$V_{mpp}$ voltage at maximum power point.

The right figure shows the P-V curve, characterised by the maximum power point $P_{mpp}$.

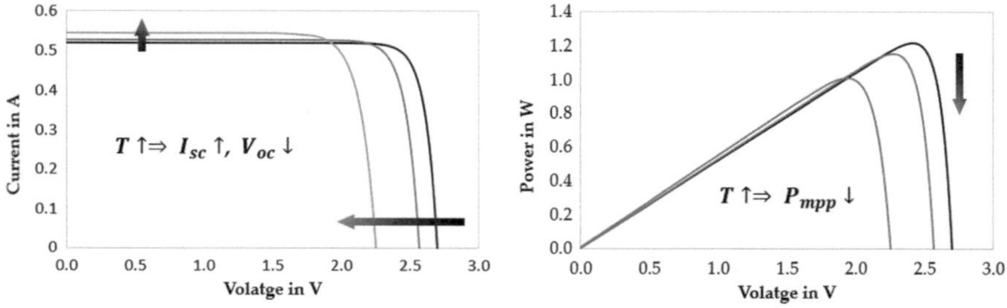


**Fig. 8.14** Temperature dependence of solar cell characteristic curves. Left $I - V$ curve and right $P_{\mathrm{mpp}} - V$

## 8.3.2 Environmental Impact on SPG Performance

The efficiency of a solar cell is significantly affected by its environmental conditions, particularly temperature and radiation–both, electromagnetic and corpuscular, see Sect. 3.2.

Temperature

Figure 8.14 depicts the impact of cell temperature on the I-V curve and the power output. The drop in $P_{\mathrm{mpp}}$ is a clear indication of the solar cell performance decrease with increased temperature. The related temperature gradients are quantified in Table 8.9. Congruent to the figure the temperature impact on the open circuit voltage is stronger than on the short circuit current. This is especially relevant for body-mounted solar panels, as they are unable to efficiently dissipate heat generated from the array–caused by $R_{sh}$ and $R_s$–through their backside. Missions to the inner solar system are negatively affected in a similar way: the higher solar flux causes the solar array to settle at a high equilibrium temperature, see Sect. 3.1, causing a decrease in the electrical output of the array.

Prominent examples are the two ESA led missions BepiColombo to Mercury and Solar Orbiter. The latter approached the Sun as close as 0.28 AU which corresponds to 12.8 times the intensity experienced at 1 AU, 1350 W/m$^2$. The most extreme example, however, is NASA's Parker Solar Probe launched in 2018 with a Delta IV

**Table 8.9** Temperature coefficients of TJ Solar Cell 3G30C from Azur Space GmbH at Begin-of-Life (BOL) from data sheet

| | | |
|---|---|---|
| $\Delta I_{\mathrm{sc}}/\Delta T$ | mA/°C | 0.36 |
| $\Delta V_{\mathrm{oc}}/\Delta T$ | mV/°C | –6.20 |
| $\Delta I_{\mathrm{mpp}}/\Delta T$ | mA/°C | 0.24 |
| $\Delta V_{\mathrm{mpp}}/\Delta T$ | mV/°C | –6.70 |

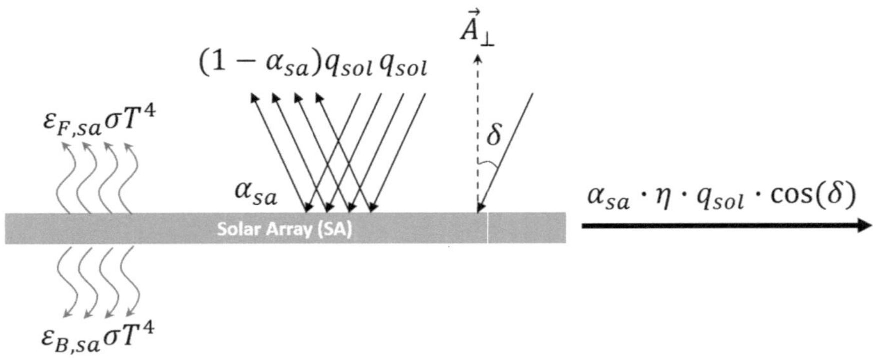


**Fig. 8.15** Simple solar array thermal model

heavy.[21] Its perihelion is at 0.046 AU, a distance where the solar flux is staggering 577 times higher than at 1 AU. Protective structural measures needed to be taken into account including a sophisticated heat shield that reached up to 450° C locally [23]. The solar array is tilted towards the Sun to minimise irradiation and thus the equilibrium temperature. On the other hand, deep space missions to the outer solar system benefit from a colder equilibrium temperature, but suffer from a lower absolute irradiance from the Sun, which corresponds to a lower current source generation, $I_i$. The combined effects can be assessed with a simplified thermal model, as depicted in Fig. 8.15.

This model neglects the irradiance from the spacecraft itself, and assumes that there is no further planetary heat source in the vicinity like Earth's Albedo or the Moon's thermal radiation. In particular, the latter is a challenge for any spacecraft from a thermal management point of view. The model further assumes that no photon is capable of transmitting through the solar panel, i.e. pass through unhindered. Establishing a heat balance equation for an incident angle $\delta = 0°$:

$$(\epsilon_{\text{F,sa}} + \epsilon_{\text{B,sa}})\sigma T_{\text{sa}}^4 A_{\text{sa}} = \alpha_{\text{sa}} q_{\odot} - \alpha_{\text{sa}} \eta_{\text{cell}} q_{\text{sol}}, \tag{8.14}$$

and solving for the solar array temperature, yields:

$$T_{\text{sa}} = \left( \frac{\alpha_{\text{sa}} q_{\odot} - \alpha_{\text{sa}} \eta_{\text{cell}} q_{\odot}}{(\epsilon_{\text{F,sa}} + \epsilon_{\text{B,sa}})\sigma} \right)^{0.25}. \tag{8.15}$$

With help from the calculated equilibrium temperature, it is possible to study the change of the I-V curve as function of distance, Fig. 8.16.

[21] Missions to the inner Solar System are very energy-intensive. The Parker Solar Probe, with a launch mass of just 685 kg, not only required one of the most powerful launchers of its time, it needed in addition a solid state kick-stage, Star 48, like. It is the same rocket motor used by NASA's New Horizons (Pluto) and Dawn (Vesta & Ceres) missions.

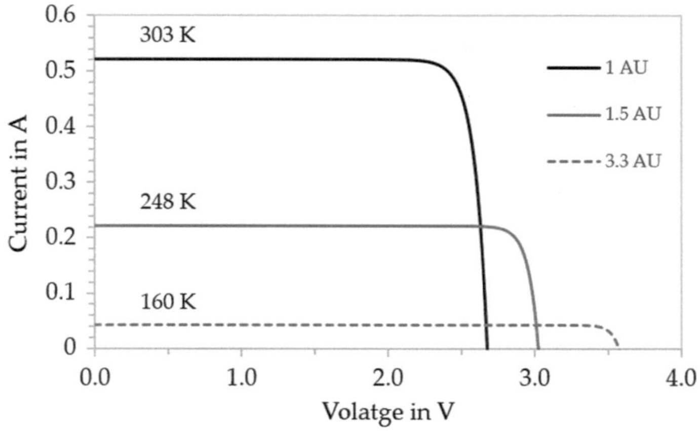


**Fig. 8.16** Variation of solar cell I-V curve with distance from the Sun

The considerable decrease of the SPG's power generation capacity with solar distance and the cold conditions gave rise to the development of dedicated solar cells optimised for low intensity, low temperature conditions (LILT) [26]. In the case of NASA's Dawn mission to Ceres, for example, the solar flux drops from an average of 1,350 W/m$^2$ at 1 AU to 176 W/m$^2$ at Ceres,[22] which corresponds to a received power decrease by 87 %. This natural condition has two implications:

(a) the solar power generator needs to be oversised at BOL to meet minimum power generation conditions at EOL,
(b) the electric propulsion system must be throttleable.

A detailed look of previous deep space missions offer valuable insights into the design of the two systems. The initial array output of the Dawn spacecraft was 10.5 kW at 1 AU but reduced to 1.4 kW at 3 AU [27]. The total spacecraft power consumption was 3.1 kW of which the Gridded Ion Engine (NSTAR) consumed most, 2.5 kW at full throttle (80 $V_{dc}$). Thus, the array generated 3 times more power than actually needed and only at a distance of 1.85 AU was it necessary to start throttling down the engine to a final input power level of 524 W (at 160 $V_{dc}$) [29]. The SEP design of NASA's Psyche mission followed the same approach. At 1 AU (71° C) the 62.6 m$^2$ array produced 21 kW, which decreased to 2.3 kW at 3.3 AU due to the decrease of equilibrium temperature to −105° C and degradation effects. Psyche used HETs (SPT-140), of which a single thruster consumed a maximum of 4.5 kW and could be throttled down to 0.9 kW. The spacecraft operated two of the four thrusters simultaneously [28]. The array capability was again oversized by a factor of more than two. This oversizing allowed engine operation for a long part of the journey without the need of throttling and losing performance.

[22] Due to the eccentricity of Ceres' orbit, the flux varies between 152 and 207 W m$^{-2}$. The value above assumes a distance of 2.77 AU.

**Table 8.10** The illuminated I-V parameter of a typical InGaP/GaAs/Ge triple-junction solar cell on Ge substrate* at BOL and three electron equivalent fluences

| | | | BOL | $2.5 \cdot 10^{14}$ | $5 \cdot 10^{14}$ | $1 \cdot 10^{15}$ |
|---|---|---|---|---|---|---|
| Open circuit voltage | $V_{oc}$ | [mV] | 2,700 | 2,616 | 2,564 | 2,522 |
| Short circuit current | $I_{sc}$ | [mA] | 520 | 519 | 514 | 502 |
| Voltage at max. power | $V_{mpp}$ | [mV] | 2,411 | 2,345 | 2,290 | 2,246 |
| Current at max. power | $I_{mpp}$ | [mA] | 504 | 503 | 501 | 487 |
| Efficiency | $\eta$ | [–] | 29.5 | 28.6 | 27.8 | 26.5 |

* From the TJ Solar Cell datasheet (3G30C from Azur Space GmbH)

Radiation

While temperature affects cell efficiency, it is reversible in nature. Another effect, however, is irreversible and detrimental to solar cells, *corpuscular radiation*. This type of radiation consists of highly energetic charged particles, electrons, protons and heavy ions as discussed in Sect. 3.2. The high energetic particles damage the crystal structure (e.g. lattice) in side the solar cell by *atom displacement*. In other words, the semiconductor atoms are kicked off from their places, causing defects in their crystalline lattice [25]. These defects affect the electron-hole generation by photons and more importantly their diffusion through the semiconductor towards the metal contact at the surface. In fact, damage by charged particles is the primary cause of performance degradation in space solar cells. Table 8.10 lists the main cell characteristics for a triple-junction solar cell. The radiation impact is measured in 1 MeV *equivalent electron fluence*. Fluence is the *total* number of particles that pass through a surface of 1 $cm^2$, refer to the definition in Sect. 3.2, Eq. 3.6. The energy equivalence of 1 MeV means that each of these electrons has an energy of 1 MeV. The fluence is therefore *monoenergetic*. The difference of the radiation level listed in Table 8.10 is due to the magnitude of the fluence, i.e. cumulated total number of electrons, and not the energy level of the single particle.[23] A fluence level of $1 \times 10^{14}\,e/cm^2$, corresponds to a 10 year mission lifetime in a typical LEO environment, while the $1 \times 10^{15}\,e/cm^2$ corresponds to a typical 15 years mission in GEO. The load is highest in MEO with an equivalent fluence of approximately $5 \times 10^{15}\,e/cm^2$. The detrimental radiation impact can be significantly diminished by applying a *cover-glass* to the top

[23] The attentive reader may have noticed that the space radiation environment does not consist of monoenergetic electrons, as presented in Sect. 3.2. The reason for this approach lies in the fact that the realistic spectrum of space radiation is very difficult to simulate on Earth. Radiation tests are carried out in particle accelerators that can only produce mono-energetic particles. The challenge is to determine the terrestrial equivalent of space radiation in terms of generated damage within the constraint of monoenergetic particles. Time-consuming and costly correlation measurements are required, which is the main reason for the high cost of space-qualified (also known as space-hardened) electronics.

**Table 8.11** Typical solar cell efficiency factors for LEO

| | | |
|---|---|---|
| Cover glass | 0.99 | BOL |
| Cell mismatch | 0.99 | BOL |
| Calibration error | 0.995 | BOL |
| Miscellaneous (Micrometeorites) | 0.97 | EOL |
| Random loss | 0.97 | EOL |
| Plume impingement | 0.95 | EOL |
| Total | 0.872 | |

surface of the solar cell. The cover glass absorbs low energy protons and reduces the kinetic energy of electrons. The cover glass also has the additional advantage that it improves the infrared emissivity of the solar cell assembly and thus lowers its temperature. This protective measure, however, comes at the expense of mass.

Solar Cell Efficiency

Table 8.11 provides typical solar cell efficiency factors for a LEO mission. Note that the causing effects are either from the beginning there (i.e. BOL) or occur or accumulated in the course of the mission (i.e. EOL). A detailed analysis is needed to confirm the efficiency losses as this depends on the type of mission, e.g. LEO, MEO, GEO, Cis-Lunar or interplanetary and mission duration. Each environment has its own challenges.

### *8.3.3 Architecture of Solar Power Generator*

We have discussed the physics and technology of solar cells in the previous section. In this section we will see how to arrange solar cells to form a *solar power generator* (SPG) that meets the requirements of the spaceship and its main power consumers– in our case the electric propulsion system. Figure 8.17 shows the principle of how solar cells are combined to generate the desired system voltage and current: cells connected in series cumulate the voltage, while cells connected in parallel cumulate the current. The SPG is comprised of the following elements:

*solar cell*: fundamental element of solar power generator, it determines the generator's main characteristics in terms of size, lifetime and resilience,

*solar string*: comprised of solar cells connected in series, it sets the maximum voltage level,

*solar segment*: comprised of solar strings connected in parallel,

*solar panel*: comprised of solar segments connected in parallel,

*solar wing*: comprised of panels that are connected in parallel,

*solar array*: comprised of two or more wings.

The schematic of a representative example of a solar power generator is depicted in Fig. 8.18. Five cells are connected in series to reach the required array voltage. Together they form a string. Three strings connected in parallel form a segment and a panel is comprised of two parallel segments. The wing consists of four panels.

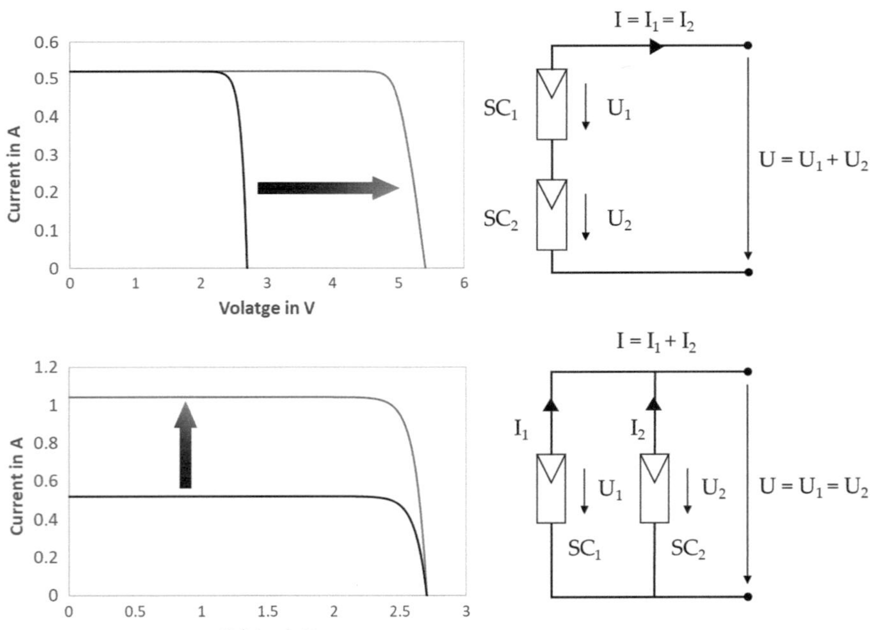


**Fig. 8.17** Cumulation of current (up) and voltage (down) through solar cells in series connection and parallel connection respectively

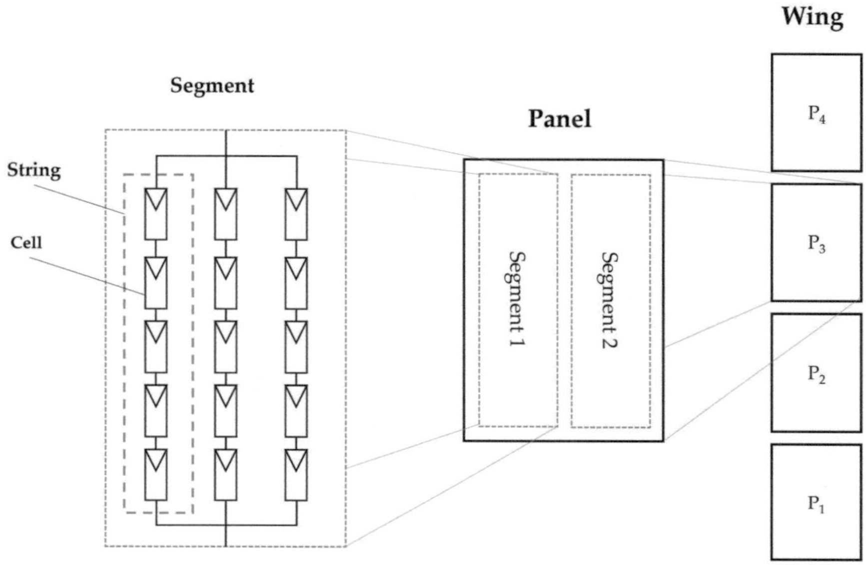


**Fig. 8.18** Schematic of a classical solar power generator with rigid panels

Solar Generator Protection

As shown above, the solar cell is the fundamental unit of the solar power generator and its malfunctioning can endanger the entire array. Such malfunction can occur due to

- temporary shadowing of the panel because of spacecraft appendages, like booms and antenna, or
- damage due to space debris and micro-meteorites.

The malfunctioning cells stop producing a current and can become a full or partial resistor. This leads to local hot spots and current backflow from adjacent solar cells and strings into the defect cell–an effect that reduces the overall system performance and could even destroy the entire panel. This problem is solved by diodes placed in parallel to the string (called bypass diodes) and in series to the string (called blocking diodes), Fig. 8.19. As the name suggests the first prevents the current to flow through the malfunctioning cell, while the latter prevents backflow of the current from strings connected in parallel. In the nominal case, it forms an additional resistance for the current that passes through it, which is the price to pay for this protective measure. Solar arrays are typically designed with one additional string to maintain redundancy in case of an irreversible failure.

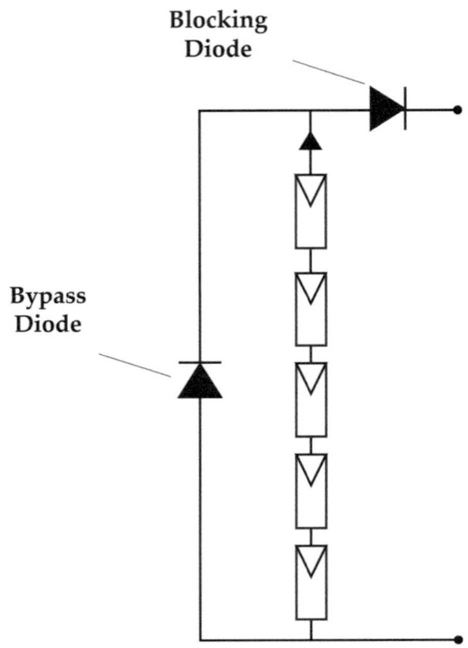


**Fig. 8.19** Placement of safety diodes around solar cell string

## 8.3.4 Design Approach for Solar Power Generator

There are several design approaches for a solar power generator. We show one that is instructive and straight forward, following a step-by-step approach.

*Step 1:* Define BOL and EOL power demand.

*Step 2* Select solar cell type and define operating environment as well as related efficiency factors.

*Step 3*: Define the BOL output voltage of the array, consider a provision of 10% for voltage losses due to the PVA diodes.

*Step 4*: Compute the number of cells in row, $N_{cr}$ with EOL cell performance, $V_{mp}$.

*Step 5*: Compute the required total PVA current, $I_{pva}$ with the EOL power demand

*Step 6*: Define number of wings, $N_w$, number of booms per wing, $N_b$ and number of blankets per boom, $N_{bk}$.

*Step 7*: Compute the current per blanket, $I_{bk}$

*Step 8*: Compute number of parallel strings per blanket, $N_{s,bk}$:

*Step 9*: Compute the total number of parallel strings and add 2 strings, $N_{s,tot}$:

This algorithm is applied to the reference mission C-One and detailed in annex D.

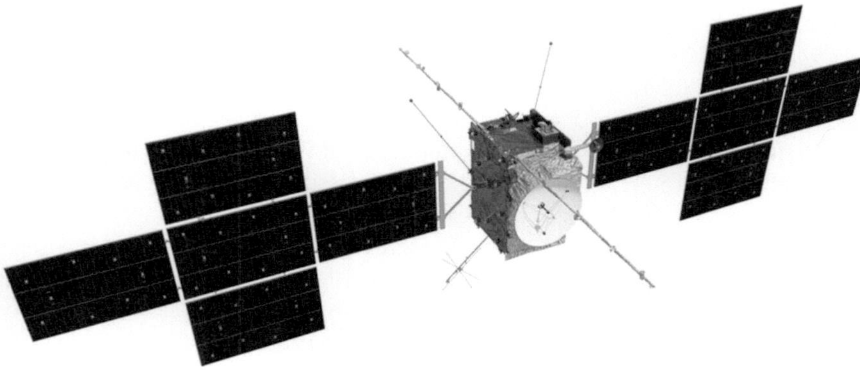

**Fig. 8.20** JUICE with deployed solar arrays, launched in April 2023. Credit: European Space Agency–ESA/ATG medialab

### 8.3.5 Solar Generator Technologies

The largest solar generator used on a spacecraft to date was for ESA's Jupiter Icy Moons Explorer (JUICE) which was launched April 2023 and is on a 7.5 year journey to Jupiter. The spacecraft has a total solar panel area of approximately $85\,\text{m}^2$, with each panel measuring $3.47\,\text{m} \times 2.48\,\text{m}$. The panels are arranged in two cross-shaped arrays, with five rigid panels on each side see Fig. 8.20. The panels are comprised of 124 parallel strings, each with 19 solar cells. The strings on a panel are grouped in 10 segments (also called sections). A remarkable feature is the cover-glass thickness of $300\,\mu\text{m}$ which is three times thicker than the cell itself and required to limit cell performance degradation due to the intense radiation environment at Jupiter [30]. The solar power generator produces more than 20 kW (BOL) at 1 AU, a capacity that drops to about 800 W when the spacecraft reaches its destination. Due to the mission profile of JUICE which encompassed a swing-by at Venus and that brought the spacecraft to a distance of 0.64 AU to the Sun, it experienced a solar flux variation ranging from $3300\,\text{W/m}^2$ down to $46\,\text{W/m}^2$ at Jupiter (5.4 AU). The solar array was tilted to reduce the solar flux and thus high temperatures in its closest approach to the Sun. With a nominal operating temperature of -130° C (143 K) the design focus was on the cold environment[24] at Jupiter which necessitates the use of LILT optimised cells [30], Sect. 8.3.2.

Flexible Solar Cells

Rigid solar arrays, like for JUICE, face challenges in terms of mass and packing density, if the spacecraft demands very high power levels. Flexible solar arrays are a viable solution to break this limitation. This is because it is possible to attach the solar cells to a flexible substrate, thus avoiding rigid material. This offers unrivaled stowage

[24] Measurements have shown an increased cell efficiency with lower temperature: from 26.8 % for 298 K to 34.1 % for 143 K.

options. The result is a *flexible blanket* that can be stowed in a canister (called a *mandrel*) and rolled out in space like a Persian carpet. The concept is known as ROSA, short for Rollout Solar Arrays. ROSA can reduce the mass of a 15 kW wing by 33 % and its stowed volume by 75 % compared to rigid panel arrays [32]. NASA's Dawn spacecraft for instance was capable of producing 80 W/kg at 1 AU, whereas ROSA for ISS produces 150 W/kg. This can be made possible by enabling technologies, like STEM booms, which stands for storable tubular extendable members. They are made of composite material, these booms store *strain energy* and release it without the use of an electric motor and other mechanisms, thus reducing weight and sources of error. The result is a lightweight photovoltaic solar power generator that can be stored in a canister,and which enables unprecedented packaging efficiency [33]. The absence of a truss structure, however, leads to a less stiff structural assembly. While this is not a limiting factor per se, it is a point that requires attention, particularly for large assemblies and when considering interactions with solar wind and attitude control frequencies.

In 2021 NASA installed two of six planned ROSA on the international space, each capable of producing about 14 kW BOL. The Chinese space station Tiangong is also powered by flexible solar blankets with a capacity of 110 kW. In addition to space stations, ROSA technology is being currently used to power deep-space exploration and scientific missions, including the Double Asteroid Redirection Test (DART mission) and Gateway's Power and Propulsion Element (in development). Furthermore, the design of ROSA has been incorporated into commercial satellites. The next generation telecommunication satellite, OneSat, of Airbus is another prominent example.

Flexible solar cells, in particular IMMs (Inverted Metamorphic Multijunction) types, promise further advancement. Invented at the National Renewable Energy Laboratory (NREL) of the U.S. department of energy, IMMs are also multijunction (MJ) solar cells without the disadvantage of classical MJ cells. As discussed earlier, MJ cells take advantage of different bandgap energies which allows them to exploit larger portions of the energy spectrum.

However, each semiconductor layer features different lattice spacing, thus forming a kind of natural defect, which consequently increases the internal resistance. Table 8.12 compares solar cell performances of different types and manufactures. IMM solar cells reduce layer transition issues by incorporating a metamorphic layer between the various doped materials. This layer gradually changes in composition, resulting in a smoother transition between the layers and fewer defects, which naturally occur due to the lattice spacing mismatch between the two semiconductor layers. The manufacturing procedure enables the removal of the parent substrate, which is the material on which the epitaxial cell layers are grown. The cell can be grown directly on the handling substance of choice, e.g. Ge or GaAs wafer, resulting in an ultra-light and ultra-thin device that is highly flexible after the removal of the growth substrate.[25] IMM's production process allows for the addition of a reflector on the cell's rear surface. This reflector redirects any lost light energy back to the

[25] https://www.nrel.gov/docs/fy11osti/49151.pdf.

**Table 8.12** Performance figures of InGaP/GaAs/Ge triple-junction solar cell. Upper table Begin-of-Life (BOL) figures, lower table normalised $P_{mpp}$ for three fluences$^{\dagger}$. Sources: SolAero and AZUR SPACE

| $m_{cell}$ | [g] | IMM-$\alpha$ 1.48 | ZTJ 2.54 | 3G30C 2.60 | XTJ Prime 2.54 | Z4 2.54 |
|---|---|---|---|---|---|---|
| $\eta_{\text{BOL}}$ | [%] | 32 | 29.5 | 29.8 | 30.7 | 30.0 |
| $V_{\text{oc}}$ | [V] | 4.780 | 2.726 | 2.700 | 2.72 | 3.95 |
| $V_{\text{mpp}}$ | [V] | 4.280 | 2.410 | 2.411 | 2.406 | 3.540 |
| $I_{\text{sc}}$ | [A] | 0.322 | 0.525 | 0.520 | 0.543 | 0.362 |
| $I\text{mpp}$ | [A] | 0.305 | 0.498 | 0.504 | 0.528 | 0.347 |
| $5 \cdot 10^{14}$ | [–] | 0.91 | 0.90 | 0.94 | 0.90 | 0.92 |
| $1 \cdot 10^{15}$ | [–] | 0.87 | 0.85 | 0.90 | 0.87 | 0.90 |
| $5 \cdot 10^{15}(\ddagger)$ | [–] | 0.71 | 0.74 | – | 0.79 | 0.78 |

All values @ AM0 (1353 W/m$^2$), 28° C, † 1 MeV e/cm$^2$, ‡ interpolated for ZTJ and XTJ Prime

active PV layers, effectively boosting the cell's power efficiency. Furthermore, these reflectors are designed to discard much of the infrared radiation, which keeps the PV cells cooler and operating more efficiently.

### *8.3.6 Power Management and Thermal Radiator*

The power generated by the SPG (or a nuclear reactor) needs to be converted into the correct voltage and then distributed to the onboard consumer. This task is performed by the Power Management and Distribution system (PMAD). It consists mainly of power processing units (PPUs) and harness. Both dissipate power, while performing their task, that needs to be ejected into space. Heat ejection is achieved by *thermal radiator* or simply radiator. It is basically a plate with a special surface finish to enable good heat ejection capabilities while minimising heat absorption. Heat ejection is achieved via thermal radiation according to Stefan Boltzmann law, Sect. 3.1–hence the name *thermal radiator*. It is essential that the heat is transported from the source to the radiator and then distributed efficiently across the radiator's surface to avoid temperature peaks above the specified level. The simplest approach, adopted by most spacecrafts, is to mount the heat dissipating source directly on the radiator: the equipment is still within the spacecraft but the space facing side of the panel is effectively a radiator. Thermal control engineers speak then of passive control. It is therefore desired to select a surface on the spacecraft that 'sees' the cold deep space and which experiences little to no external radiation flux, e.g. from the Sun or the spacecraft itself as this would hamper the heat ejection capability. Deposits from the plumes of thrusters causes the radiator to degrade and lose performance.

If the amount of dissipated heat is large, passive means are not sufficient anymore and thermal engineers employ *heat pipes*. These devices consist of capillary tubes in

**Table 8.13** Radiator area ($m^2$) as function of solar incident angle, $\delta$, surface temperature, $T_{rad}$ and $\dot{Q}_{diss} = 1,000$ W. Surface properties correspond to white paint, $\alpha = 0.2$ and $\alpha = 0.8$

| $T_{rad}$ | 90° | 80° | 70° | 60° | 50° | 40° | 30° | 20° | 10° | 0° |
|---|---|---|---|---|---|---|---|---|---|---|
| 30° C | 1.31 | 1.39 | 1.49 | 1.59 | 1.69 | 1.79 | 1.88 | 1.96 | 2.00 | 2.02 |
| 60° C | 0.90 | 0.94 | 0.98 | 1.02 | 1.06 | 1.10 | 1.13 | 1.16 | 1.18 | 1.18 |
| 100° C | 0.57 | 0.59 | 0.60 | 0.62 | 0.63 | 0.65 | 0.66 | 0.67 | 0.67 | 0.67 |

which a fluid circulates thereby transporting very efficiently heat across the radiator. If the outer surface of the spacecraft is not sufficient, a deployable radiator is commonly used. The radiator needs to be designed for the hot case and checked for the cold case. For an interplanetary mission, the hot case is at BOL, since both the dissipation heat of the propulsion system and the solar flux are highest. As the spaceship ventures further into space, away from the Sun, these factors decrease, which could lead to excessive cooling by the thermal radiator–a scenario that needs to be prevented.
The starting point for the radiator sizing is a similar heat balance equation like for the solar array, Eq. 8.14:

$$2\epsilon_{rad}\sigma T_{rad}^4 A_{rad} = \alpha_{rad} q_{\odot} \cos(\delta) A_{rad} + \dot{Q}_{diss}, \tag{8.16}$$

and solving for the radiator area, yields:

$$A_{rad} = \frac{\dot{Q}_{diss}}{2\epsilon_{rad}\sigma T_{rad}^4 - \alpha_{rad} q_{\odot} \cos(\delta)}. \tag{8.17}$$

While the SPG should always be pointed to the Sun ($\delta = 0°$), the thermal radiator should not ($\delta = 90°$). In case of a fixed radiator, this is not always achievable and a rest solar irradiance angle remains.
Table 8.13 shows the required radiator area as function of the radiator's temperature and the solar irradiance angle for $\dot{Q}_{diss} = 1,000$ W at 1 AU.

The results in the table consider white paint, which has an absorptivity ($\alpha$) in the range of 0.1–0.3 and an emissivity ($\epsilon$) of 0.7–0.95 [31]. The impact of $T_{rad}$ is significant due to the fourth potency in Eq. 8.17. If equipment is directly mounted on the radiator, a temperature drop of 15 K can be assumed as a rule of thumb from the heat dissipating unit to the radiators surface. As example, if the operating temperature of the equipment is 45° C then a $T_{rad}$ of 30° C can be assumed for a preliminary sizing estimate.

## References


1. Ranjan, R., Chou, S. K., Riaz, F., & Karthikeyan, K. (2017). Cold gas micro propulsion development for satellite application. Energy Procedia, 143, 754–761. https://doi.org/10.1016/j.egypro.2017.12.758
2. Page, R. J., Stoner, W. A., & Barker, L. (1988). Preliminary design study of hydrogen and ammonia resistojets for prime and auxiliary thrusters. https://api.semanticscholar.org/CorpusID:92925432
3. Thunnissen, D., Guernsey, C., Baker, R., & Miyake, R.: Advanced Space Storable Propellants for Outer Planet Exploration. 40th AIAA/ASME/SAE/ASEE Joint Propulsion Conference and Exhibit (2004). https://doi.org/10.2514/6.2004-3488
4. Zurawski, R. L. (1986). Current evaluation of the tripropellant concept. https://api.semanticscholar.org/CorpusID:137491333
5. Mungas, G., Vozoff, M., & Rishikof, B. (2012, October). NOFBXTM: A New Non-Toxic, " Green " Propulsion Technology with High Performance and Low Cost. AIAA SPACE Conference and Exposition 2012. https://doi.org/10.2514/6.2012-5235
6. Fukuchi, A., Nagase, S., Maruizumi, H., & Ayabe, M. (2010). HAN/HN-Based Monopropellant Thrusters. https://api.semanticscholar.org/CorpusID:103227084
7. NASA Glenn Research Center, (2022) Chemical Equilibrium with Applications (CEA) https://www1.grc.nasa.gov/research-and-engineering/ceaweb/
8. Silvera, I., & Cole, J. (2010). Metallic hydrogen: The most powerful rocket fuel yet to exist. *Journal of Physics: Conference Series, 215*, 012194. https://doi.org/10.1088/1742-6596/215/1/012194
9. Tenney, C., Sharkey, K., & McMahon, J. (2017). On the possibility of metastable metallic hydrogen. *Physical Review B, 102*,. https://doi.org/10.1103/PhysRevB.102.224108
10. Dornberger, W. (1 Dec. 1989). Peenemünde: die Geschichte der V-Waffen. Ullstein; 14. edition
11. A4/V-2 private collection of Christensen D., http://heroicrelics.org/info/v-2/combustion-chamber-cut-away.html
12. Richter, G. P., & Smith, T. D. (1995). Ablative material testing for low-pressure, low-cost rocket engines. https://api.semanticscholar.org/CorpusID:136541306
13. Gamgami, F., Sippel, M., & Dumont, E. (2010). Statistical Analysis and Classification of rocket motor efficiency, thrust to mass ratio and structural index [Techreport]. DLR. https://elib.dlr.de/75246/
14. Arnold, R., Goetz, A., Philipp, P., Maeding, C., Obermaier, G., Caruana, J-N.,(2011). Technology Development Program for a New Storable Propellant Low-Thrust Engine Demonstrator.
15. Payload Space, The Starship Report. (2024). https://payloadspace.com/starship-report/
16. Nguyen, K., & Knowles, T. (2002). *July)*. Propellant Densification for Launch Vehicles: Simulation and Testing. https://doi.org/10.2514/6.2002-4293
17. Mustafi, S., Johnson, W., Kashani, A., Jurns, J., Kutter, B., Kirk, D., & Shull, J. (2010). August). *Subcooling for Long Duration In-Space Cryogenic Propellant Storage*. https://doi.org/10.2514/6.2010-8869
18. Gingerich, D. (2015). THE RUSSIAN R-16 NEDELIN DISASTER: AN HISTORICAL ANALYSIS OF FAILED SAFETY MANAGEMENT. *Journal of Space Safety Engineering, 2*, 65.
19. Soller, S., Rackemann, N., Preuss, A., & Kroupa, G. (2016, May). APPLICATION OF LASER-IGNITION SYSTEMS IN LIQUID ROCKET ENGINES.
20. Wiegand, A., Schmierer, C., & Weikert, S. (2019). MULTI-DISCIPLINARY VEHICLE DESIGN BASED ON THE TOOLS ASTOS AND ESPSS. https://api.semanticscholar.org/CorpusID:221170010
21. Bouisset, P., Nguyen, V. D., Akatov, Y. A., Siegrist, M., Parmentier, N., Archangelsky, V. V., et al. (1992). Quality factor and does equivalent investigations aboard the soviet space station MIR. *Adv Space Res, 12*(2–3), 363–7. https://doi.org/10.1016/0273-1177(92)90130-P

22. Pindado, S., Cubas, J., & Manuel, C. (2014). Explicit Expressions for Solar Panel Equivalent Circuit Parameters Based on Analytical Formulation and the Lambert W-Function. *Energies, 7*, 4098–4115. https://doi.org/10.3390/en7074098
23. Lindner, A., Oberhuttinzer, C., Paarmann, C., Muller, J., Strandmoe, S., & Costello, I. (2019). Solar Orbiter Solar Array - Exceptional Design for a Hot Mission. 1–7. https://doi.org/10.1109/ESPC.2019.8932039
24. Kennel, C., & Petschek, H. (1969). Van Allen belt plasma physics.
25. Wildfeuer, M. (Nov 2022). Radiation degradation of III-V multijunction space solar cells. https://nbn-resolving.de/urn/resolver.pl?urn:nbn:de:bvb:91-diss-20221202-1596596-1-7
26. Appelbaum, J. Low-Intensity Low-Temperature (LILT) Measurements and Coefficients on New Photovoltaic Structures. Progress in Photovoltaics: Research and Applications, 1996.
27. Stella, P., DiStefano, S., Rayman, M., & Ulloa-Severino, A. (2009). Early mission power assessment of the Dawn solar array. 001617–001621. https://doi.org/10.1109/PVSC.2009.5411392
28. Snyder, J. S., Goebel, D. M., Chaplin, V. H., Ortega, A. L., Mikellides, I. G., Aghazadeh, F., k. Johnson, I., & Kerl, T. (2019). Electric Propulsion for the Psyche Mission. https://api.semanticscholar.org/CorpusID:220493686
29. Brophy, J., Marcucci, M., Ganapathi, G., Garner, C., Henry, M., Nakazono, B., & Noon, D. (2003). The Ion Propulsion System for Dawn. https://doi.org/10.2514/6.2003-4542
30. Kroon, M., Bongers, E., Cavel, C., Baur, C., Faleg, F., & Riva, S. (2019). Low-Intensity Low-Temperature (LILT) Power prediction of JUICE solar array. *European Space Power Conference (ESPC), 2019*, 1–5. https://doi.org/10.1109/ESPC.2019.8932020
31. Tachikawa, S., Nagano, H., Ohnishi, A., & Nagasaka, Y. (2022). Advanced Passive Thermal Control Materials and Devices for Spacecraft: A Review. *International Journal of Thermophysics, 43*,. https://doi.org/10.1007/s10765-022-03010-3
32. Banik, J. A., Kiefer, S. H., LaPointe, M., & LaCorte, P. (2018). On-orbit validation of the roll-out solar array. 2018 IEEE Aerospace Conference, 1–9. https://api.semanticscholar.org/CorpusID:49539277
33. Chamberlain, M., Kiefer, S., LaPointe, M., & LaCorte, P. (2021). On-orbit flight testing of the Roll-Out Solar Array. *Acta Astronautica, 179*, 407–414. https://doi.org/10.1016/j.actaastro.2020.10.024
34. Pavon, S., Tregubow, V., Peukert, M., & Lescouzères, R. (2012). Engineering validation model for the Exomars bipropellant propulsion subsystem. https://api.semanticscholar.org/CorpusID:198186179
35. Henry, C. (2002). An Introduction to the Design of the Cassini Spacecraft. *Space Science Reviews, 104*, 129–153. https://doi.org/10.1023/A:1023696808894
36. Gamgami, F., Rohrbeck, M., Wollenhaupt, B., & Andersson, B. (2016). *A Dual-Mode Propulsion System with Arcjets as an Alternative Propulsion System for the SGEO Platform, 3125075, Space Propulsion Conference 2016*. Italien: Rom.
37. Wiley, S., Dommer, K., & Mosher, L. E. (2003). Design and Development of the MESSENGER Propulsion System. https://api.semanticscholar.org/CorpusID:113052626
38. Tam, W., Dommer, K., Wiley, S., Mosher, L., & Persons, D. (2002). July). *Design and Manufacture of the Messenger Propellant Tank Assembly*. https://doi.org/10.2514/6.2002-4139
39. Intuitive Machines ODAR – Version 1.0, Attachment D, Intuitive Machines-1 Orbital Debris Assessment Report (ODAR) IM-1-ODAR-1.0 https://apps.fcc.gov/els/GetAtt.html?id=265886&x=
40. NSSDCA ID: IM-1-NOVA https://nssdc.gsfc.nasa.gov/nmc/spacecraft/display.action?id=IM-1-NOVA

# Chapter 9
# Propellant Management System

**Abstract** The propellant management system plays a crucial role in the propulsion system and is a critical factor in large space transportation systems across all three domains: to space, in space, and from space. This chapter systematically discusses the two key elements of the propulsion subsystem: the propellant feed system and the storage system. It covers both chemical and electrical propulsion systems, spanning a thrust range of ten orders of magnitude.

## 9.1 PMS Architecture Overview

The primary task of the *propellant management system* (PMS) is to *store* and *supply* propellant in a controlled way to the engines at specified operating conditions, most importantly, flow rates, temperature and pressure. These conditions are the engine's propellant inlet requirements. Though not sharp requirements, they form a relatively narrow tolerance range that needs to be met, in order to ensure nominal engine operation. In particular, the transient start-up and shut-down sequences pose a risk due to dynamic flow phenomena in the piping, piping accessories and, in the case of high-thrust systems, also in the turbomachinery.[1] In essence, the PMS ensures the engine operates as intended.

Figure 9.1 depicts a graphical overview of the PMS and its main interfaces to other spaceship subsystems. The propellant equipment (e.g. valves) is generally controlled either directly by the Energy Power Subsystem (EPS), a dedicated Propulsion Drive Electronic (PDE), or a Remote Interface Unit (RIU), depending on the avionic architecture. This control is based on telemetry gathered by sensors and evaluated by the Data Handling System (DHS). However, in the functional chain, it is the attitude control software, referred to as flight software, that commands the respective units to power on or off the PMS equipment. The *thermal* and *mechanical* design aspects related to PMS have in general strong visible impact on the entire spacecraft design

[1] Different but largely synonymous terms are used for this subsystem and its elements. The feed lines are also referred to as 'tubing' or 'piping'. Correspondingly all involved control elements like valves, orifices etc. are called 'control equipment'.

 
F. Gamgami, *Space Propulsion and Spaceship Design*, Springer Praxis Books,
https://doi.org/10.1007/978-3-031-71336-1_9

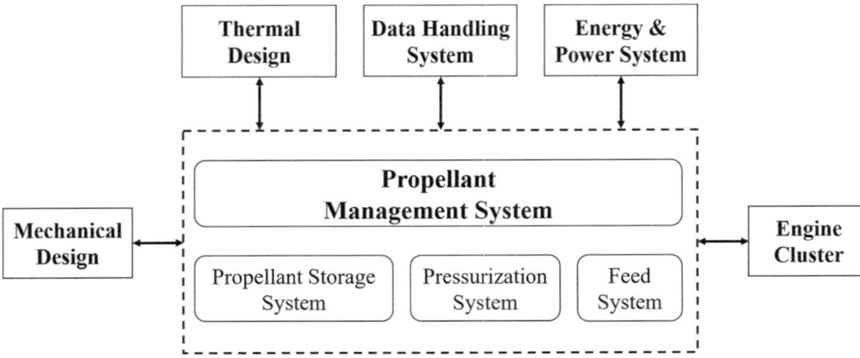


**Fig. 9.1** Interface overview of the propellant management system with other spacecraft subsystems

the larger the propellant mass fraction $\mu_p$. Both are particularly relevant for cryogenic propellant storage: this is partly because of the low temperatures that must be maintained, and partly because of the large quantities of propellant that determine the spacecraft's design. The EPS is crucial for electric propulsion systems.

The PMS consists of three subsystems that are closely related to each other:

*propellant storage system*: tanks and related support structure required to store the propellant for a specified duration, temperature and pressure with minimum losses,

*pressurization system*: all flow control equipment, gas provision and sensors required to prevent under-pressure within the draining propellant tanks,

*feed system*: all equipment required to supply and gauge the propellant from the tanks to the engine interfaces.

The nature of the three depends strongly on the engine thrust level, more precisely on the chamber pressure $P_{ch}$, see Eq. 7.21. In short, the higher the required thrust, the higher the chamber pressure and consequently the required inlet pressure. This correlation is not visible for small thrusters and engines (1 N–30 kN) but fully unfolds for very high thrust levels (>100 kN). For instance, the inlet pressure requirement for a 10 N (bi-prop) thruster is similar to that of a 30 kN (bi-prop) engine. The latter has its nominal operation point around 15 bar while the former is capable to deal with a range from 5 to 22 bar. This is not a coincidence but has a very practical reason: the integration of all thrusters into a unified feed system would result in a notable reduction in the complexity of the design, an improvement in the overall robustness of the system, and a reduction in the associated costs. This system aspect justifies a performance compromise for high thrust engines.

There are three PMS types that can be grouped in two categories:

*pressure-fed:*
*blowdown*: transient pressure drop
*pressure-regulated*: controlled pressure drop with steady state operation

*pump-fed*: actively controlled pressure increase with steady state operation by means of a *turbopump*

The first two types of the *pressure-fed* category feed the thrusters from a *pressure reservoir*. These systems are *passive* and the main task of the respective PMS is to manage the pressure drop from a pressure reservoir within the tanks to the lower thruster inlet pressure. The picture changes for *pump-fed* systems. Here the propellant is stored at relatively low pressure of 3–6 bar and needs to be raised by turbopumps to a high engine inlet pressure. In essence, PMS is primarily concerned with the management of pressure; either from high pressure to low pressure or vice versa. Therefore, a well-designed propellant feed system is characterised by low pressure losses.

Table 9.1 provides a list of engines and the applied PMS type in increasing thrust order. Rocket Lab's Rutherford engine is the smallest engine in terms of thrust that is pump-fed while Rocketdyne's AJ10-137 (main engine of Apollo's Service module) was the largest pressure-fed engine in service so far. A threshold exists in relation to

**Table 9.1** Selection of chemical engines and applied feed system

| Engine | Application | $T_{vac}$ | $P_{ch}$ (bar) | Propellant | Feed Type |
|---|---|---|---|---|---|
| S400 | GEO satellites | 420 N | 15 | MON-MMH | pressure-fed |
| RD-843 | VEGA | 2.5 kN | 20 | $N_2O_4$-UDMH | pressure-fed |
| Rutherford | Electron⋆ | 25 kN | – | LOX-RP1 | pump-fed |
| Aestus | Ariane 5 | 30 kN | 11 | $N_2O_4$-MMH | pressure-fed |
| AJ10-190 | Space Shuttle OMS* | 47 kN | 8 | $N_2O_4$-MMH | pressure-fed |
| CE-7.5 | GSLV | 74 kN | 58 | LOX-LH2 | turbopump-fed |
| AJ10-137 | Apollo Service Module | 91 kN | 7 | $N_2O_4$-A50$^+$ | pressure-fed |
| RL-10B-2 | Atlas V & Delta IV | 110 kN | 44 | LOX-LH2 | pump-fed |
| LE-5B | H-IIA & H-IIB | 137 kN | 36 | LOX-LH2 | pump-fed |
| Vinci | Ariane 6 | 180 kN | 61 | LOX-LH2 | pump-fed |

⋆ By means of an electropump
*Orbital Maneuvering System
$^+$ Aerozine 50 is 1:1 mixture of Hydrazine and UDMH, a high energetic propellant used in ICBMs

inlet pressure, above which pressurised feed systems become too heavy and inefficient in comparison to pump-fed systems. This is evident, given that high-pressure tanks are typically heavy in order to sustain the pressure, and the greater the propellant that needs to be stored, the heavier the tanks. At a certain point it is inevitable to shift to a pump-fed PMS so save mass. The threshold is not a harsh line but rather a region in which a detailed and non-biased trade-off needs to be performed to identify the optimal choice—mission-wise and programmatic-wise. Moreover, the required $\Delta v$ is in principle part of the equation, as it determines the amount of propellant to be stored.

We will discuss in the following sections these PMS types and the related system architectures. In order to stay close to the functional application and to keep the subject matter tangible, we will discuss the piping control equipment in the order in which they appear. The same is applied to propellant storage and pressurisation, with the exception of integral tanks, which are dealt with separately in Sect. 9.6.

## 9.2 Blowdown PMS

We start the discussion with the simplest PMS, the pressure-fed system operated in blowdown. The idea is straightforward: propellant is stored under pressure in a tank, and can flow to the thrusters by simply opening an isolation valve is opened. Blowdown systems are generally used for attitude and orbit control of satellites but also as vernier thrusters for roll-control in launch vehicle during ascent. For instance, the Italian launcher VEGA-C uses hydrazine mono-prop system in blowdown. The $\Delta v$ capacity is rather small: in the order of some hundred m/s, if combined with catalytic thrusters and merely about 50 m/s, if combined with cold gas thrusters, see Sect. 8.1.

This uncomplicated propulsion system will be employed here to elucidate the function of the key propellant piping equipment, as well as the related operational procedures, e.g. re-pressurisation and water-hammer. it shall be stressed that the equipment discussed here is also relevant for pressure-regulated systems presented in the subsequent sections, such as the classical bi-propellant system.

Figure 9.2 shows the schematic of a typical Blowdown system as applied for Globalstar, a cellular communication satellite and part of a constellation of 48 satellites. It consists of:

- 1 tank (surface tension)
- 1 Filter (F)
- 1 Propellant Isolation Latch Valve (PIV-L)
- 1 Flow Restrictor, e.g. Orifice (FR)
- 2 Propellant Fill & Drain Valve (FDV1, FDV2)
- 1 Pressure Transducer (PT)
- 1 Temperature Transducer (T)
- 5x1 Newton Thruster, controlled by Flow Control Valves (FCV)

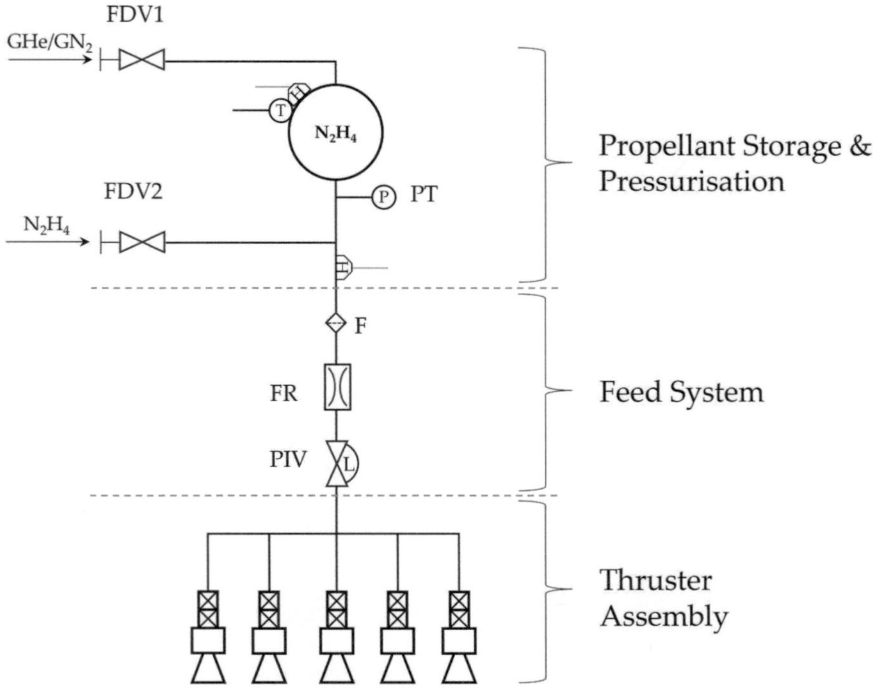


**Fig. 9.2** Mono-propellant propulsion system of Globalstar [1]

The propellant is stored under medium pressure between 20 and 30 bar. Opening the pressure isolation valve (PIV) and the thruster flow control valve (FCV) will cause the propellant to drain out of the tank due to the pressure gradient. As a direct consequence the tank pressure will gradually decrease. Consequently, the thrust magnitude also decreases as the tank depletes. This is an important characteristic of blowdown systems: the thrust is not constant. Mission planning must carefully consider the decrease in thrust, as there might not be sufficient thrust for specific manoeuvres at the end of life (EOL) if miscalculated. For instance, de-orbiting imposes an acceleration requirement, which will be discussed later using the example of the satellite Themis, see Sect. 9.2.2.

This pressure decrease can be modeled as an adiabatic process.[2] The adiabatic relation states that:

$$\frac{P_2}{P_1} = \left(\frac{V_1}{V_2}\right)^{\gamma}, \tag{9.1}$$

[2] An adiabatic process assumes that no heat transfer occurs between the system of interest and the environment, which is valid if the technical or natural process happens rapidly. However, in reality, heat transfer can often not be ignored, making it more practical to consider a *polytropic* expansion. Additionally, note that while an isentropic process is both adiabatic and frictionless, adiabatic processes are not necessarily frictionless and can involve irreversibilities.

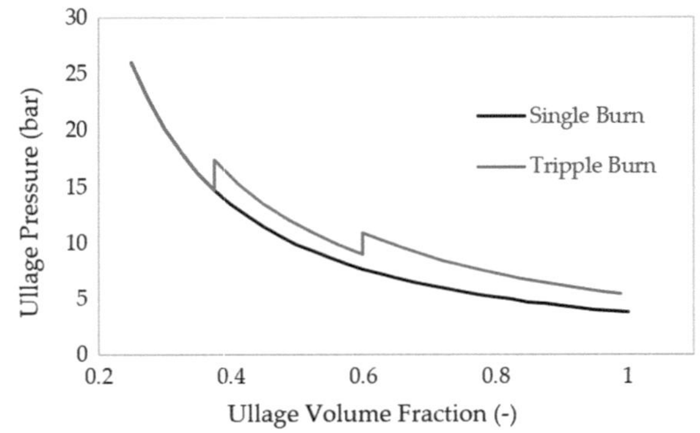


**Fig. 9.3** Pressure drop modelling for a single and a triple burn

| | |
|---|---|
| $P_1, P_2$ | pressure at state 1 before expansion and state 2 after expansion, |
| $V_1, V_2$ | volume at state 1 before expansion and state 2 after expansion, |
| $\gamma$ | adiabatic exponent. |

The volume after expansion is related to the volume before expansion by the ejected propellant volume, $V_p$.

To illustrate the physical effect, we will model the depletion of a tank in two ways a) complete depletion in a single burn b) depletion in three consecutive steps. In the latter case sufficient time in between the burns should pass to allow the propellant and the gas in the ullage to reach thermal equilibrium with its surrounding, here 20 °C, which is also the starting temperature. Figure 9.3 shows the ullage pressure over ullage volume relative to the total available volume $V_0$. The term *ullage* denotes the gas part in the tank, which is responsible for the tank pressure—irrespective of the nature of the feed system type. In the triple burn case, thermal equilibrium after a burn allows the ullage pressure to recover slightly and to start the next expansion phase from a higher level. Since pressure is equivalent to energy density, thermal equilibrium means that heat flows into the tank, either by the surrounding heat dissipating equipment or by dedicated tank heater devices. This physical model is qualitatively correct, however, the process is in reality not quite adiabatic but somewhere between adiabatic and isothermal. This is expressed in thermodynamics by the polytropic coefficient $\kappa$, which is not a fundamental quantity of the material alone, like $\gamma$, but a process quantity.[3] Simply said, the better the tank insulation the closer the process to the adiabatic case.

[3] The issue with the polytropic coefficient, is that knowledge about the process is required, which is a priori not available. It must be deduced from either detailed modelling, measurements or analogy considerations.

### 9.2.1 Feed System

The genuine task of the feed system is to provide propellant to the engines at the right conditions. The two dominant control parameters are pressure and mass-flow. The design of the feed system follows key design principles:

- minimise and control *internal* and *external leakage*
- comply to maximum unit *cycling limit*
- comply to maximum unit *propellant throughput capacity*
- comply to minimum *reliability* requirement
- take provisions for *failure cases* in terms of *redundancy*
- comply with *safety* regulations for on-ground operation

In particular, *internal leakage* poses a threat to the structural integrity of the tubing due to pressure build-up. The design rule is to establish an order ranging from low leakage rates (high tightness) down to larger leakage rates (lower tightness). Globalstar, Fig. 9.2, is an instructive example to discuss the functionality of a feed system. It consists merely of the basic required equipment fulfil the main requirements: a filter (F), a flow restrictor (FR) and a single isolation *propellant latch valve* (PIV) and the tubing itself—the latter is often referred to as *feed lines*. The filter prevents particles that have entered the tank during the filling process or have come loose from the piping equipment from flowing to the thruster and damaging them. The flow restrictor is a plain orifice that restricts the flow by abruptly reducing the cross-sectional area thereby causing on purpose a pressure drop. We will see below, Sect. 9.2.3, why this is necessary to prevent an infamous failure mode called *adiabatic decomposition*. The primary task of the latch valve is to isolate the tank section from the thruster assembly. The PIV is nominally closed and opened after launch to enable thruster operation. Latch valves are usually *bi-stable valves* that open or close upon a power pulse-command, but power is not required to maintain either state. The thrusters used in Globalstar are flight-proven *catalytic* monopropellant hydrazine ($N_2H_4$) thrusters of 1 Newton ($\pm 0.1$ N) at 22 bar (BOL). Each thruster has two valves within that are called flow control valves (FCV). These are dual seat *solenoid* valves and contrary to latch valves, solenoids are *mono-stable*. They are in general nominally closed (NC) whereby remaining closed until activated by powering the valve, which causes a current to flow through a coil inside the valve. This current induces a temporary magnetic field in the valve armature, typically a ferromagnetic material, that forms one part of the seat/seal assembly. The valve closes as soon as the power is switched off. In case of Globalstar, the flow control valves consumed 10.2 W at 28 VDC each [1].

Blowdown monopropellant systems are used for a wide range of space applications. They are the first choice for almost all LEO satellites and are also popular for roll control in launchers, e.g. VEGA and Ariane 5. The general rule says that this simple and robust propulsion system is to be preferred, if the $\Delta v$ demand is small ($<100$ m/s).

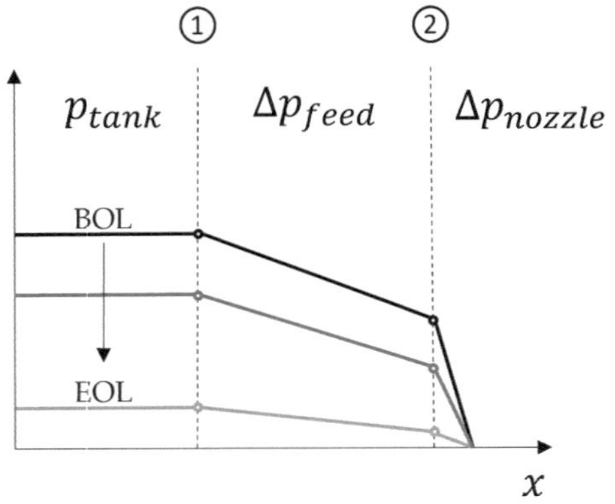


**Fig. 9.4** Pressure drop within blowdown system, tank outlet ①, thruster inlet ②

## Pressure Drop

The flow in a blowdown system experiences a superimposed pressure drop, consisting of a transient component due to tank depletion and a pressure drop within the feed system due to friction in the tubing and due to obstacles formed by the piping equipment. The pressure drop along the flow path $x$ and time (BOL→EOL) is depicted in Fig. 9.4. Equations 9.2 and 9.3 quantify the pressure decrease. The parameter in question must be obtained through testing. Each equipment needs to be characterised in all relevant dimensions, e.g. temperature and mass flow rate. The pressure drop caused by the tubing itself depends on many factors, surface roughness, length, bending angle and number of bendings.

A pressure drop analysis is in general mandatory for all PMS types (not only blowdown) to ensure a sufficient inlet pressure level in all operational modes until spacecraft decommissioning. Nevertheless, in practice, this is often omitted for low thrust levels in the order of 1 N due to the large amount of experience data and the robustness of the system. If conducted, the analysis needs to be performed for maximum operational flow rate, meaning all engines that are required according to the nominal—and off-nominal—manoeuvre plan need to be reflected in the *design cases*.[4]

$$\Delta p_{tube} = \zeta \cdot \rho \frac{v^2}{2}, \quad \zeta = \frac{C_1}{Re} + C_2, \tag{9.2}$$

$$\Delta p_i = K_i \cdot \rho \frac{v^2}{2}, \tag{9.3}$$

$$\Delta p_{feed} = \Delta p_{tube} + \sum_1^N \Delta p_i. \tag{9.4}$$

[4] Note that for any subsystem there are several design cases under which the set of requirements needs to be fulfilled, similarly to a spider diagram. In particular EPS and TCS designers are notoriously known for large design case matrices.

### The Three Barrier-Rule—A Matter of Safety

In the now retired Atlas V user guide there is the following safety requirement for the propulsion system [2]:

'Range Safety requirements require *three inhibits* (dual-fault tolerance) if system failures could result in catastrophic events and *two inhibits* (single-fault tolerance) if failures could result in critical events.'

This requirement is not specific to the now retired Atlas V, all launch vehicles impose safety requirements on their payload to protect the ground personal and the launcher. Above all, propellants are flammable, in certain conditions even explosive and some are toxic and therefore require special consideration. The origin of these requirements is rooted in the respective national space transportation law. In case of the US it is the FAA, stating[5]:

'A propulsion system must be dual fault tolerant to inadvertently becoming propulsive. Propulsion systems must be single fault tolerant to inadvertent mixing of fuel and oxidizer.'

This became the famous *three-barrier rule* in the jargon of propulsion engineers. According to general interpretation, a barrier is provided by a seat within an isolation equipment. The above discussed flow control valves (FCV) within the thruster provides two such barriers. it is important that the two seats are operated independently of each other to count as two barriers. Such seat types are called dual seat solenoid valve, see Fig. 9.5. The fill and drain valve may seem an exception but in fact it is a three seat device. We will encounter this rule in all feed systems.

### Propellant Thermal Control

Thermal control is relevant for all spacecraft subsystems. In case of the propulsion system, its the propellant that requires attention and imposes requirements.[6] Hydrazine, for instance, freezes below 1.7 °C at 1 bar, Table 8.8 The spacecraft's TCS needs to prevent that. It has two means available, thermal *sensor-heater* combination or simple mechanical thermostats. The first requires software and therefore the DHS to analyse the sensor readings. It is common practice to use three sensors for redundancy and to perform *majority voting* in case one of the three diverges from the other two. Heater output is actively modulated by on/off time, triggered by pre-set temperature thresholds which may be adjusted in flight. The duty cycle is commonly about 70% leaving a margin of 30%, in case heat losses have been underestimated. This approach gives maximum flexibility but is also the more complex

[5] FAA 14 CFR Part 400 requirements, Chap. 3–Commercial Space Transportation, Federal Aviation Administration, Department of Transportation

[6] Requirements that a subsystem, here propulsion, imposes on another subsystem, here TCS, are called *interfacing requirement* as opposed to system requirements that are broken down to the respective subsystem. This interface is not necessarily tangible but could be functional.

to implement, test and validate. The second and less sophisticated mean to control the temperature is by *mechanical thermostats*. The heater has an integrated thermo-mechanical sensor and is adjusted on-ground without any regularpossibility to adjust its cycle behaviour.[7] They switch on automatically when the local temperature falls below a fix lower threshold and switch off when the temperature rises above a fix upper threshold. Thermostats are limited in the number of cycles they can perform, a fact that must be considered when setting device thresholds and conducting thermal analysis—narrow operating bandwidth means many heater cycles. Mission duration and the thermal environment are directly impacting factors.

### 9.2.2 Propellant Storage and Pressurisation

The tank for Globalstar,[8] in Fig. 9.2 is loaded first with Hydrazine via FDV2 and then pressurised via FDV1 with nitrogen to a pressure of 22 bar [1]. The typical ratio is 25% ullage volume to 75% propellant volume. On ground, there is a clear stratification due to gravity but in microgravity pressurant and propellant start to move and fragment within the tank. Tank design and propulsion system operation need to prevent pressurant gas from entering into the feed lines since bubbles passing through a thruster will alter the thrust generated and consequently the result of the manoeuvre —in the worst case even be detrimental to the thruster itself. Three methods to prevent this are discussed in Sect. 9.6.

Thruster operation in orbit will deplete the propellant mass and, consequently, the pressure level within the tank. A simplified pressure drop is depicted in Fig. 9.4. The blowdown ratio for this system is 4:1 meaning that the final tank pressure will have dropped by a factor of 4 down to about 5.5 bar at end of life (EOL).

This leads to a corresponding thrust level drop from 1 N (BOL) to 0.25 N (EOL). The attitude and orbit control subsystem must consider this and the EPS needs to command longer thrusting times in order to achieve the required $\Delta v$ or angular momentum.

In general LEO mission requirements for orbit control cope well with this thrust decrease. However, new policies require active de-orbiting of the spacecraft to reduce the risk of more space debris [5]. This, as well as other mission-specific reasons, mean that *re-pressurization* is required to boost thrust level for this critical final manoeuvre. Blowdown propulsion systems that feature re-pressurization are also called *rechargeable*. Figure 9.5 shows an augmented architecture with an additional pressurisation tank as used in the mission Themis [3]. The propellant tank is initially pressurised to 32 bar prior launch and continuous propellant consumption leads to a pressure drop down to 6 bar. The propellant tank can be recharged by help of the pressure reservoir

[7] Operators have their hacks to circumvent the spacecraft manual. One way to shut them temporarily off is to cut their power supply, if the power distribution unit (PDU) offers this option. However, it is by design not possible to deliberately switch thermostats on.

[8] Called OST 31/0 manufactured by Astrium AG, now Airbus Defence and Space.

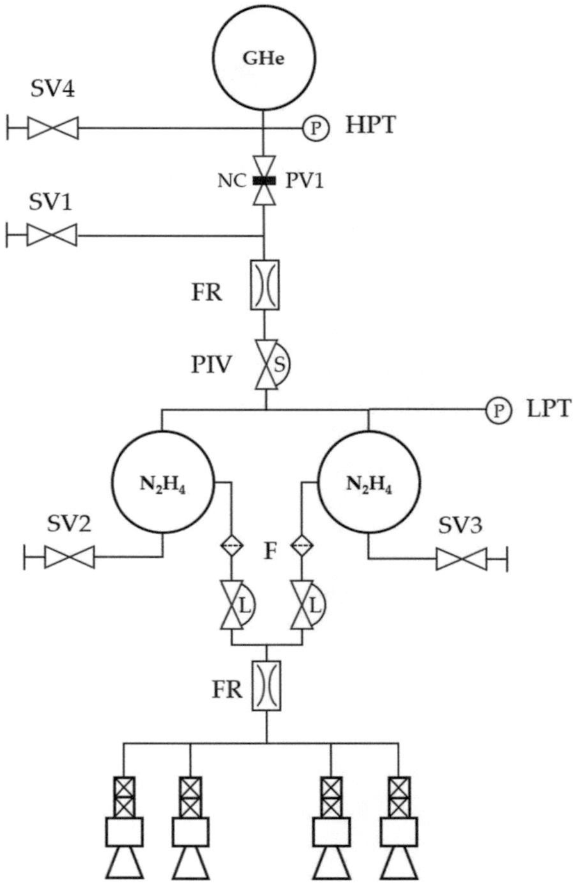


**Fig. 9.5** Flow schematics of Themis' Reaction Control System (RCS) with re-pressurisation according to [4]

in the helium pressurant tank. The two tanks, propellant and pressurant, however need to be *hermetically* isolated from each other until re-pressurisation is needed. This is achieved by means of a nominally closed pyrotechnic valve (NCPV).

Pyrotechnic valves, or pyrovalves (PV) for short, are the third important flow control equipment for isolating or opening tubing segments. As the name suggests they work with a small explosives. Those that open are nominally closed (NC) prior to activation those that shall isolate are nominally open (NO). Isolation is needed, for instance, after placing the satellite in a graveyard orbit to prevent inadvertent propulsion activation through post-mortal discharging effects or propellant mixing. This is part of the satellite's End-of-Life passivation procedure.[9] We will encounter more cases below where pyrovalves are required.

[9] Another example of a necessary passivation action in a spacecraft is the complete discharge of the battery.

Since a pyrovalve cuts through a virgin tubing segment, they are inherently hermetic. It is evident that in instances where explosives are required to initiate the opening or closure of a tubing passage, it signifies that the tubing is hermetically sealed and leakage is by design not possible.

### 9.2.3 Operation in Space

The commissioning of the propulsion system must follow a clear order to prevent hazardous incidents. Two steps are required, *venting* and *priming*. Venting is the process of releasing trapped gas between the propellant isolation valve(s) and the thrusters by opening all of flow control valves (FCV) in the thrusters. This process may take some time until the pressure is sufficiently low and is driven by the thruster injector diameter. The TCVs are then closed hereafter and the priming process starts. The propellant isolation valve, located between tank and thrusters, is opened and the propellant is allowed to enter the lower feed system.

Priming as well as other transient changes in propellant flow, like throttling or cut-off of high thrust engines cause a transient fluid dynamic phenomenon known as *water hammer* and is related to locally steep pressure gradients and high amplitudes. This is not restricted to blowdown systems but shall be discussed here briefly on this occasion. The propellant entering the empty lines accelerates to a limit speed before it hits the FCVs, which causes a (longitudinal) pressure wave, that propagates in the feed lines. While pressure waves are typical for hydraulic systems, it could be harmful due to its strength. The pressure amplitude could lead to over-pressuring equipment beyond the qualified proof pressure thereby causing mechanical damage to the flow control equipment [6]. This can be avoided by an *orifice*. The role of this passive flow device is to reduce the velocity of the flow which is equivalent to dissipating the flow's kinetic energy. In consequence, the pressure amplitude of the wave is reduced.[10]

The effect is not relevant for all propellant types. Liquid hydrogen, for instance, is a benign fluid in that respect causing harmless pressure amplitudes. Hydrazine on the other hand is a critical propellant and sensitive to large pressure amplitudes, which bears the risk of *adiabatic decomposition*. This is because the local rise in pressure is accompanied by a rise in temperature—heating due to gasdynamic compression.[11] If a limit is exceeded, the hydrazine in the feed lines begins to decompose thereby producing gas, just as it is supposed to do in the thrust chamber. The feed lines have a small volume, and the released gas causes the pressure to increase further. In the worst case, this leads to an over pressurisation, that the equipment cannot withstand. It is again the orifice that needs to be designed to prevent this hazardous case.

[10] Pressure is an energy density, therefore, physically speaking water hammer is the conversion of kinetic energy into pressure energy.

[11] Gasdynamic compression is the prime cause of vehicle heating during atmospheric entry and not friction.

Finally, it should be mentioned that from an operational point of view, it is important to be aware of the need of thruster *pre-heating*. This is specific to mono-propellant thrusters, that require a catalyst-bed to decompose the propellant to release its energy, Sect. 5.2.1. To reach operational temperatures (typically $> 160\,°C$) can take up to 30 min and has to be considered in the power budget as well as the mission planning [11]. Cold-starts lead to thruster degradation and their number should be kept limited. In the best case nominal cold-starts are completely avoided and only reserved for contingency cases to prevent more damage, e.g. unscheduled collision avoidance. Some new green propellants, called High Performance Green Propulsion (HPGP), require pre-heating above 350 °C and do not allow for cold starts [7].

### *9.2.4 SLIM—Smart Lander for Investigating Moon*

An intriguing example of a blowdown system is the propulsion system of JAXA's Smart Lander for Investigating Moon (SLIM). It is unique in many ways. The spacecraft has a launch mass 590 kg of which 470 kg was propellant, corresponding to a propellant mass fraction, $\mu_p$, of 80%. Its flow schematic is depicted in Fig. 9.6. It consists of two orbital main engines (OMEs) of 500 N each for lunar orbit insertion, descent and landing. In addition, there are $12 \times 22$ N thrusters for attitude control. The propulsion system is a bi-propellant, that runs with Hydrazine-MON3 [8]. The PMS of bi-prop systems is commonly pressure regulated. The peculiar combination of both blowdown and bi-prop, in case of SLIM, means that the propulsion engineers were less interested in thrust but rather in the superior specific impulse of this propellant combination. The principal innovation, however, resides in the propellant tank. Due to the limitations of the domestic H-IIA launcher in terms of trans-lunar injec-

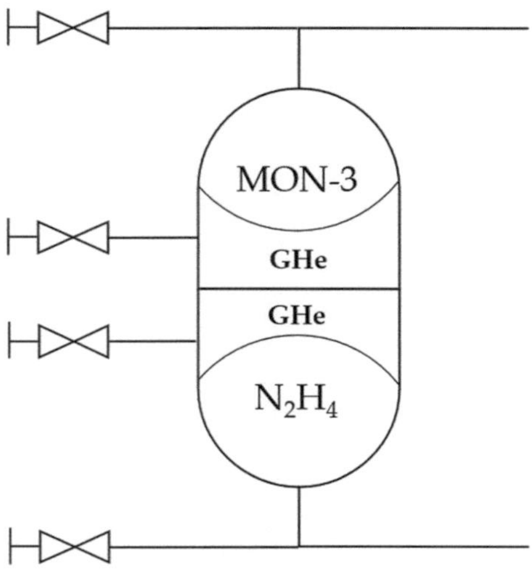


**Fig. 9.6** Combined tank design of SLIM [9]

tion performance, engineers were compelled to pursue strategies for the reduction of dry mass wherever possible. This was achieved by a combination of innovative design choices. Firstly, instead of separated tanks, fuel, oxidiser and even pressurant were stored in a single tank. Secondly, this single tank served as the structural backbone of the spacecraft meaning that it served as the load bearing structure, which is a noteworthy novelty for this small class of spacecraft. Load carrying tanks are common for large launcher stages, and are known as integral tanks, see Sect. 9.6.4.

Furthermore, the unorthodox propulsion system had a built-in advantage specific to planetary landing. To understand these circumstances, it is important to visualise that landing on a planetary body, like the Moon, requires continuous thrust reduction to enable a smooth touch-down. This interesting topic will be detailed in Sect. 10.2. For now it is relevant to note that there are two ways of thrust reduction: either by *throttling*, i.e. mass flow reduction, or by operating the engines in *pulse-mode*. I.e. on-off operation, whereby the *effective* thrust level can be adjusted by modulating the on-off time of the thruster. Orchestrating all relevant engines leads to a quasi-smooth thrust profile. This approach was used for SLIM [9]. Hence, t fact that SLIM uses a blowdown system fits very well in into the overall mission: the reduction in tank pressure results in a corresponding decline in thrust level, which is beneficial for achieving the desired landing thrust profile.

## 9.3 Pressure-Regulated PMS—Chemical Propulsion

Pressure regulated propulsion management systems are applied in chemical propulsion systems as well as in electric propulsion systems that run with gaseous propellant. For the sake of clarity we will discuss the latter in a dedicated section below. Figure 9.7 shows the schematic of a classical bipropellant system.

This type of bi-prop system was used in the satellite platform Spacebus of the Italo-French company Thales Alenia Space with varying level of redundancy. This platform became a successful workhorse in the GEO satellite communication business. It is propelled by the hypergolic propellant combination MON-1—MMH, which self-ignites when it comes in contact and thus renders an ignition system obsolete, Sect. 8.2.2. There is a single main engine, designated LAE (liquid apogee engine) and 16 smaller thrusters, referred to as RCTs (reaction control thruster). The system fulfills four *main* system requirements:

| | | |
|---|---|---|
| (1) | perform orbit insertion from GTO to GEO | → LAE |
| (2) | perform East-West station keeping | → RCTs |
| (3) | perform North-South station keeping | → RCTs |
| (4) | de-saturate reaction wheels | → RCTs |

The flow schematic is undoubtedly more intricate than that of a mono-propellant, but it is not inherently more complex once the underlying logic is deciphered. This will be demonstrated in a step-by-step manner.

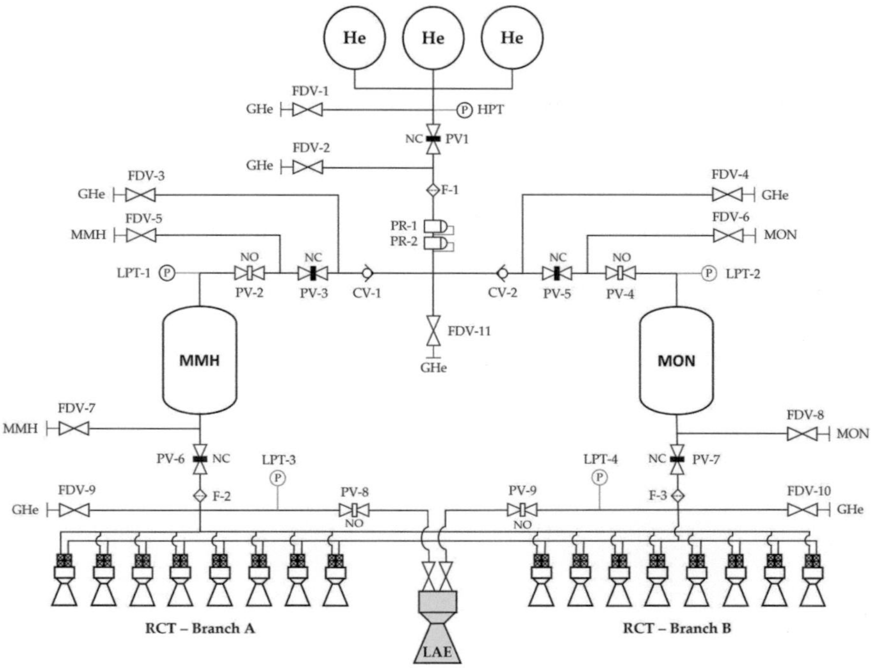


**Fig. 9.7** Schematic of a bi-propellant propulsion based on Spacebus, a GEO satellite plattform of Thales Alenia [12]

### *9.3.1 Pressurant Control Assembly—PCA*

The section between the pressurant (Helium) tanks and the propellant tanks is known as the pressurant control assembly (PCA), Fig. 9.8. Its purpose is to supply the propellant tanks with pressurant gas in a controlled way. The pressurant is stored under high pressure, typically 310 bar, in composite overwrapped pressure vessels (COPV), see Sect. 9.6. The pressurant tanks are filled on ground via a fill and drain valve (FDV-1). The gas is prevented to distribute towards the propellant tanks via the nominally closed pyrovalve (NCPV-1). It is opened during the in-orbit commissioning and check-out phase after separation from the launcher's upper stage.

Air is not allowed in the system nor vacuum. Therefore, helium is filled into the tubes below PV-1 via FDV-2 and into the lower compartment vie FDV-3. A filter is needed to stop particles above 15 μm to propagate through the tubing caused by the pyrovalve activation. Particles that could enter the system during the propellant and pressurant loading are filtered to 2 μm. Thereafter follows the core component of the PCA, the mechanical *pressure regulator* (PR-1). It regulates the gas pressure from storage pressure (310 bar) down to about 18.5 bar, the exact value depends mainly on the inlet pressure requirement of the supplied thruster and the pressure drop in

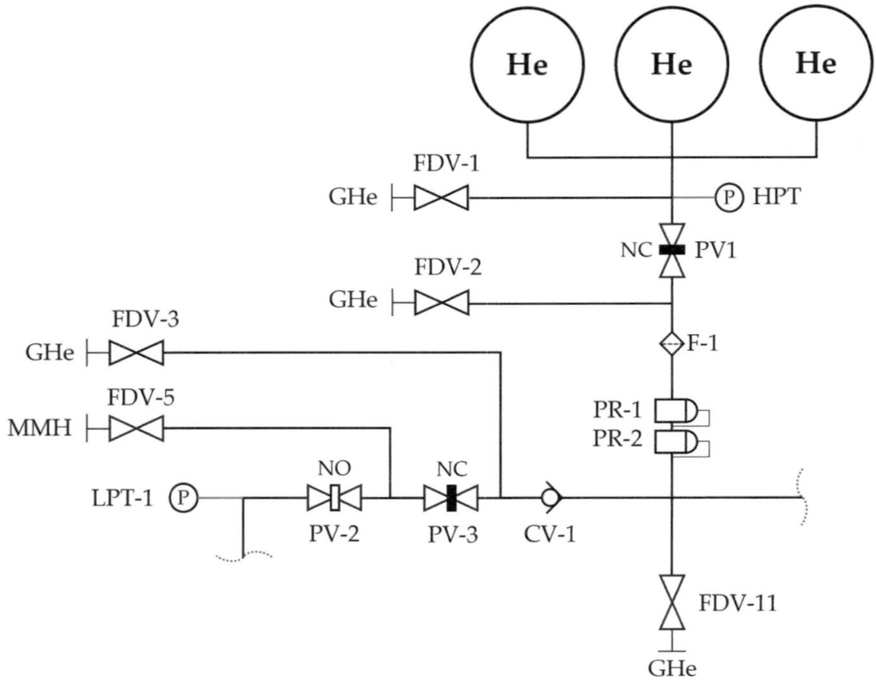


**Fig. 9.8** Pressurant Control Assembly of Spacebus type CPPS [12]

the feed system. Due to its importance, there are two pressure regulators in series in case one fails with the redundant stage being set at 0.5 bar higher than the primary stage. Almost all current bi-prop systems are pressure-regulated with SLIM being one of the rare exceptions, Sect. 9.2.4. They utilse in general a mechanical pressure regulator with a predetermined pressure set-point, like in the Spacebus case. Downstream of the two pressure regulators there is a junction, the adjoining branches are symmetrical, which is why we discuss only the left-hand branch, i.e. in direction of the MMH tank. The first element is a check valve (CV-1) also called 'non-return valve'. It acts like a diode in electronics and does not allow propellant vapour to pass in the reverse flow direction into the upstream lines. The rationale is evident: if fuel and oxidizer could migrate upstream, i.e. towards the junction, they could potentially mix and react chemically due to their hypergolic nature. This failure mode is especially of concern during launch due to the heavy vibration loads and long mission durations. A nominally closed pyrovalve (PV-3) is therefore implemented to prevent this failure mode. It will be opened as part of spacecraft in-orbit commissioning. The tubing between check valve and pyrovalve is filled with helium on ground via FDV-3 and FDV-4 respectively. Propellant is filled into the system via FDV-5 and FDV-6 respectively. A nominally open pyrovalve (NOPV-2) is placed before the propellant tank. For GEO satellites like Spacebus, this valve is typically fired after the transfer from GTO to GEO is completed thus hermetically isolating the two branches. The remaining propellant in the tanks, approximately 1/3 of the initial amount, is then used in blowdown mode.

### 9.3.2 *Propellant Isolation Assembly—PIA*

The feed system downstream of the propellant tank is called propellant isolation assembly (PIA). The equipments here control the propellant flow to the thruster (Fig. 9.9).

Due to different mass flow requirements, tubing for the RCT's is smaller in diameter than for the LAE. The first elements are a nominally closed pyrovalve (PV-6) followed by a filter to capture any dirt resulting from pyrovalve activation. Together with the two series, solenoid flow control valves in the thruster, the design is compliant with the three barrier rule. FDV-7 is required for loading and offloading of propellant simulant after ground testing. Note that the propellant tanks are loaded 'bottom-up'. Downstream is the last pair of fill and drain valves FDV-9 and FDV-10, respectively. They are required for ground testing of the thrusters and also to fill the tubing between the closed pyrovalves and thruster valves with helium prior to launch. Contrary to most blowdown systems, venting is always performed in bi-prop systems with MON-MMH. PV-8 and PV-9 respectively are required for passivation.

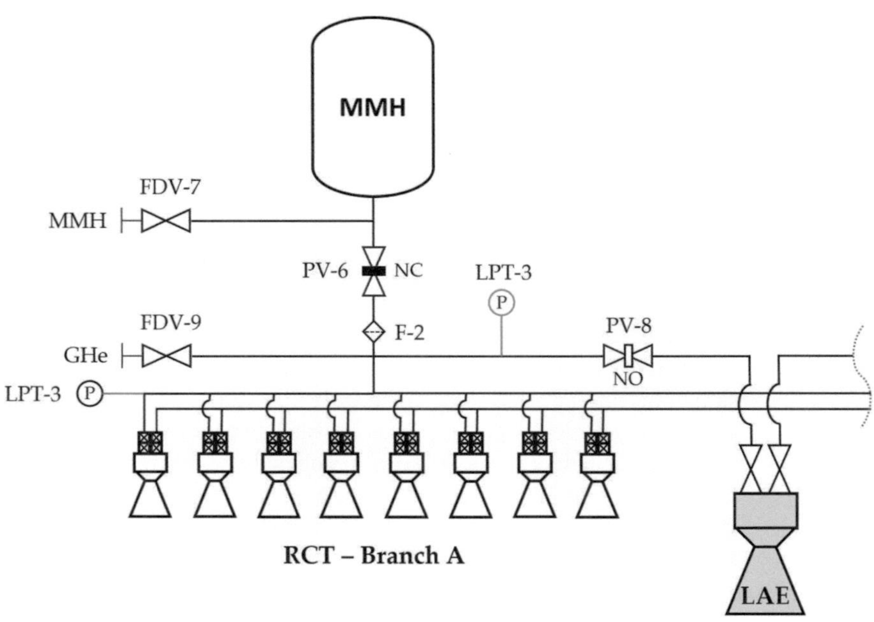


**Fig. 9.9** PIA fuel side of Spacebus type CPPS [12]

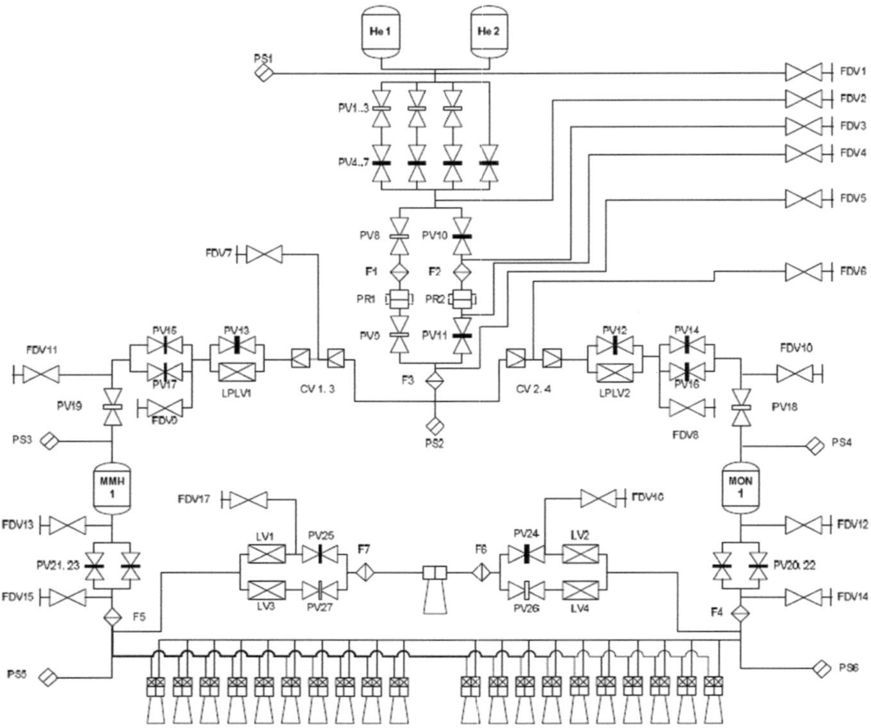


**Fig. 9.10** Propulsion flow schematic of ESA's ExoMars Trace Gas Orbiter [56]

They prevent the residual propellant to leak to the thruster and cause inadvertent reactions after decommissioning of the spacecraft.

The bi-prop system discussed above is straight forward. It does not feature redundancy except of the pressure regulator. All this changes in case of scientific missions. Firstly, because they are institutional missions and executing agencies rarely take risks. Secondly, because there is a lot at stake: the instruments are cutting edge technology and require many years of development. The work should not have been in vain because of pressure isolation valve.

Figure 9.10 shows the flow schematic of ESA's ExoMars Trace Gas Orbiter, a scientific mission with a high degree of redundancy.[12] The pyrovalve ladder of three nominally open and four nominally closed PVs is activated in between large manoeuvres en-route to Mars. The ladder is opened before each main engine manoeuvre and closed hereafter. This is necessary to prevent leakage through the high pressure gas regulation system into the propellant tanks. Control equipment in parallel indicate redundancy in case the nominal equipment does not open (i.e. fail to open). The PV pair PV15 and PV17 is such an example. The feed system is said to be *fault tolerant*,

[12] ExoMars TGO was launched 2016 by a Russian Proton-M rocket.

specifically single fault tolerant, meaning that it can absorb a failure without affecting its functionality and performance.

The striking feature of the PIA are the latch valve pyrovalve combinations, for instance, LV3-PV27 in parallel to LV1-PV25. LV3 is nominally used to isolate the propellant from the thruster. In case it fails to open, the mission would be lost. The idea is then to isolate the segment via PV27, to open PV25 and to use the redundant latch valve LV1.[13] Comparing the PIA of ExoMars with that of Spacebus reveals that the latter entirely omitted the use of a latch valve between tank and thruster. The reason is because the main engine is only required for GEO insertion, a manoeuvre split usually in three burns and which takes only a couple of days, before the entire branch is sealed by a pyrovalve. The root problem of inadvertent propellant mixing because of leakage through the FCVs into the main engine's combustion chamber is hence excluded. The situation is obviously different in case of an interplanetary mission like ExoMars where the main engine is needed up until arrival for orbit insertion—a 9 months cruise phase. The latch valves enable *short-term* isolation of the main engine. The spacecraft operators then vent the tubing downstream of the latch valve after each manoeuvre. This reduces the contact time between the propellant and the engine's FCVs, which could lead to swelling of the Teflon parts in these valves [56].

### 9.3.3 Examples of Pressure-Regulated Systems

Several examples of bi-propellant systems shall be highlighted to understand the variety of pressure-regulated propellant management systems. We first discuss MON-MMH based systems that have been combined with mono-prop systems. The second variant is the Dual Mode System, which is closely related to the classical MON-MMH combination. It is in general used for the same applications but offers additional advantages. The third is about the upper stage of the retired Ariane 5G, EPS (Étage á Propergols Stockables). The example will illustrate that pressure regulated PMS can support bi-prop systems with a thrust magnitude of up to 30 kN which is almost 2 orders of magnitude above the classical thrust range of a pressure-regulated bi-prop system, 400–600 N. The record is held by the AJ10-137 from Aerojet Rocketdyne, though. It propelled the Apollo Service Module with a staggering thrust magnitude of 97 kN, a record for a pressure-fed system. A variant of this engine, down-scaled to 27 kN of thrust, is still operational on the European Service Module (ESM) in support of the NASA lunar spaceship Orion. This vehicle is notable for its unconventional tank design. The last example highlights a small cryogenic propulsion system of the commercial US company, Intuitive Machines. It holds the record for the smallest cryogenic propulsion system used in space.

[13] Note that isolation of the faulty branch is commonly not performed unless nominal operation of the parallel branch is proven, if the satellite operators do have this option.

### Separated and Unified Propulsion Systems

The above discussed propulsion systems of Spacebus and ExoMars Trace Gas Orbiter belong to the class of *unified propulsion system* (UPS). The first UPS designed in Europe was for the geostationary communication satellite Olympus-1, developed by BAE Space in Stevenage England and launched 1989 [10]. Before the introduction of UPS, large satellites with a high $\Delta v$ demand had two independent propulsion systems on-board: a pressure regulated bi-prop for the large apogee manoeuvre with 400–460 N thrust and a mono-prop system operated in blowdown for the remaining tasks, i.e. reaction control and orbit maintenance.[14] Another famous representative of this separated architecture is the Cassini space probe to Saturn. Like GEO satellites, it used a helium pressurised bi-prop system with MON-MMH to fuel the two main engines, 445 N each, and a blowdown mono-prop system with hydrazine for attitude control referred to as reaction control system (RCS). The bi-prop system had a propellant mass of 3000 kg of which 1,132 kg was oxidiser and 1,868 kg was fuel. The mono-prop system was rechargeable, it had a dedicated small helium pressurant vessel to re-pressurise the propellant tank, similar to Themis, see Sect. 9.2.2. It required 132 kg of hydrazine. With a launch mass of 5,655 kg and a total propellant mass fraction $\mu_p$, of 52%, it becomes clear that the propulsion system was a major design driver of Cassini's overall architecture[15] [57]. The then newly introduced unification brought system advantages in terms of cost, engineering and AIT (Assembly, Integration and Test) effort.

### Dual Mode System

Some companies pursued a different path towards unification. The result was a rivalling UPS to MON-MMH, which became known under the term *dual mode*

---

[14] The very first propulsion technology used by satellites for GEO insertion was based on solid-fuel. The satellite was spin-stabilized to compensate for the inherent thrust vector misalignment of the small solid-fuel motor and integration misalignment. The satellite needed to be de-spun after manoeuvre completion. These small solid rocket motors added to the space debris problematic since their exhaust gas contained molten solid particles.

[15] Although not directly related to PMS it is worth to highlight more design features of Cassini's propulsion systems. Firstly, Cassini had for redundancy reasons two main engines. In addition, the designers were afraid that micrometeoroids could destroy the thin coating inside the nozzle and decided to protect each engine by a retractable cover. The main engine was used for all manoeuvres that required a 'linear' $\Delta v$ above 5 m/s. Each engine had a gimbal mechanism in order to aligned the engine's thrust vector with the current CoM position, thus reducing the parasitic torques and consequentially propellant mass of the RCS. The RCS consisted of $24 \times 1$ N thruster arranged in four branches of eight thruster. Among its tasks were course attitude control—fine attitude control was performed by the reactions wheels (RW)—course correction manoeuvre that required less than 5 m/s and RW de-saturation. The thruster cluster was attached to four thruster booms to increase the lever arm.

*system.*[16] It features hydrazine as fuel instead of MMH and MON-3 as oxidiser. The PCA of both can be identical but the use of Hydrazine for attitude control allows for a less complicated tubing manifold as depicted in Fig. 9.11.

The advantage of Hydrazine is, that it can be used in two modes—hence the name—as fuel in a bi-prop system or as a single propellant in a mono-prop system. In bi-prop mode, it achieves an even higher specific impulse of at least 5 s compared to MON-MMH [58]. Besides these two advantages, it is argued that hydrazine thrusters have better pulse control properties when operated in pulse mode. This, in turn, improves attitude control performance due to the more precise thrust profile.

NASA'S famous deep space mission Messenger, for instance, utilised a Dual Mode system. Launched 2004, its sophisticated mission profile to the inner planet Mercury is strongly reflected in the complexity of the space probe's propulsion system. It consists of three main propellant tanks, one for the oxidizer and two for the fuel. In addition, it features a refillable auxiliary fuel tank. The flow schemati depicted in Fig. 9.12. All thruster except of the main engine, Leros-1b, were mono-propellant thruster.

In order to prevent an off-axis movement of the centre of mass (CoM), the main propellant tanks of the spacecraft are positioned symmetrically around the centre-line. In the actual configuration (different to the shown schematics), the central oxidiser tank is flanked by two propellant tanks on each side [59]. Each main fuel tank carried 178 kg of hydrazine and 231 kg of MON3. The auxiliary tank carried 9.34 kg of fuel. Only 2.3 kg of helium stored in a 67 litre tank under 230 bar was sufficient to pressurise the system [59]. Messenger's propulsion system required a total of 17 thrusters of three types:

1 x 667 N bi-propellant main engine, Leros 1b, a proven apogee engine for GEO satellites with a specific impulse of 316 s.

4 x22 N mono-propellant thrusters, MR-106, with a specific impulse of 230 s required to stabilise the spacecraft during main engine burns similar to vernier engines for some launcher. This is required due to thrust misalignment of the main engine, which is the reason why the Messenger team called them thrust vector control thruster [11]

12 x4.4 N mono-propellant thrusters, MR-111, with a specific impulse of 220 s for fine attitude control burns and momentum management.

The probably most intriguing part of this propulsion system is the fact that the 6Al-4V titanium auxiliary tank can be *refilled* during flight. This is achieved by a common manifold, grey bar to the left in Fig. 9.12. Its function is to create a capacity of equal pressure. Once the pressure of the auxiliary tank has dropped below the main tank pressure of the fuel, it can be refilled by opening AFTLV2. It remains the question, however, why is an additional tank needed at all? As we will see below in Sect. 9.6, a diaphragm tank offers the advantage to reliably perform a manoeuvre in any attitude.

[16] An interesting historical fact is that in Europe, MON-MMH-based UPS thruster technology was pursued, whereas in the US and Japan, both MON-hydrazine and MON-MMH-based technologies were developed.

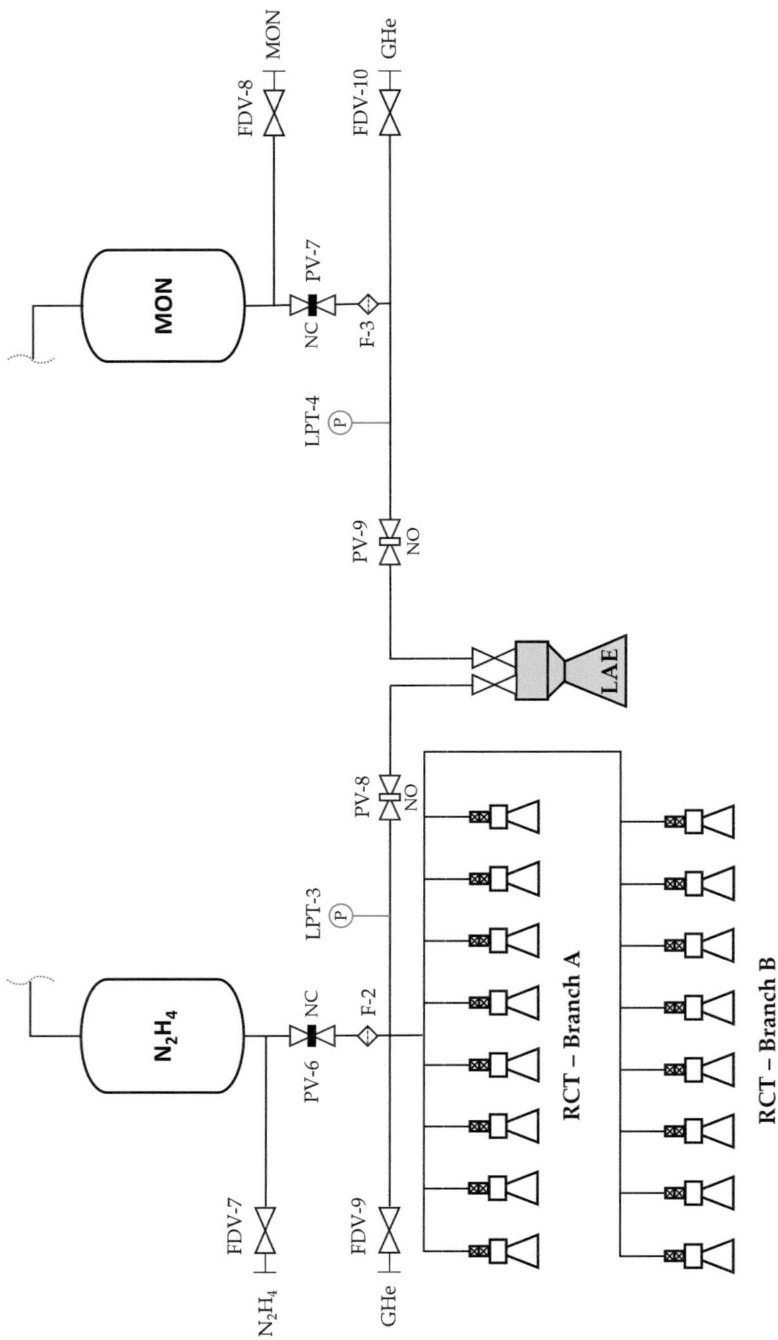


**Fig. 9.11** Principle flow schematic of a Dual Mode PIA

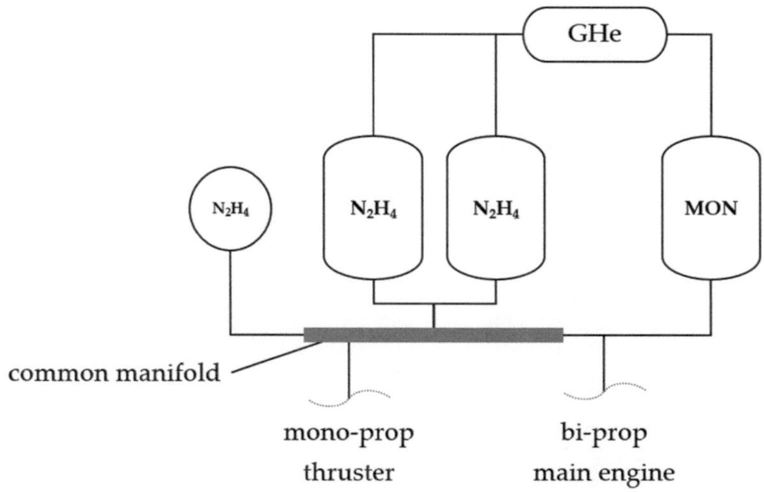


**Fig. 9.12** Flow schematic of the Messenger space probe utilising a Dual Mode CPPS [11]

It is therefore used for attitude control, and angular momentum management tasks. But it is also used for an important manoeuvre, called *propellant settlement*, in which the 4.4 N thruster in direction of the LAE are fired for *settling burns*. This is required to settle the propellant in the main tanks at the outlet prior to main engine burns [60]. A similar approach was adopted for NASA's NEAR mission. This was necessary, due to the lack of a propellant management device in the main tanks, see Sect. 9.6.

### Ariane 5 EPS And ESM

An example for a large pressure regulated propulsion system is the upper stage of Ariane 5G, called EPS, which stands for Etage á Propergols Stockables (in engl. Storable Propellant Stage). Depicted in Fig. 9.13, the structural design features a circular plate in which four tanks are placed in a symmetrical configuration. Since the tanks are exposed to space they are wrapped in multi-layer insulation blankets (short MLI) with a golden hue. This is needed to keep the propellant warm and prevent it from freezing. However, due to the proximity to the hot rocket motor, high temperature MLI is needed to protect the propellant from the engine's radiated heat—visible as white triangular blankets. The upper stage is loaded with 3.2 t of MMH and 6.6 t of $N_2O_4$ and propelled by the Aestus engine with a thrust of 29.6 kN and specific impulse of 324 s. It has a dry mass of 1,275 kg which results in total mass of 11 t and a dry mass index, $\sigma$, of 13%. The engine alone weighs 115 kg. With a height of 3.35 m and a maximum diameter of 3.94 m, it is a rather compact upper stage.

A further example is the European Service Module, ESM. It is the service module of NASA's crewed Orion capsule, providing electrical power and thrust as well as

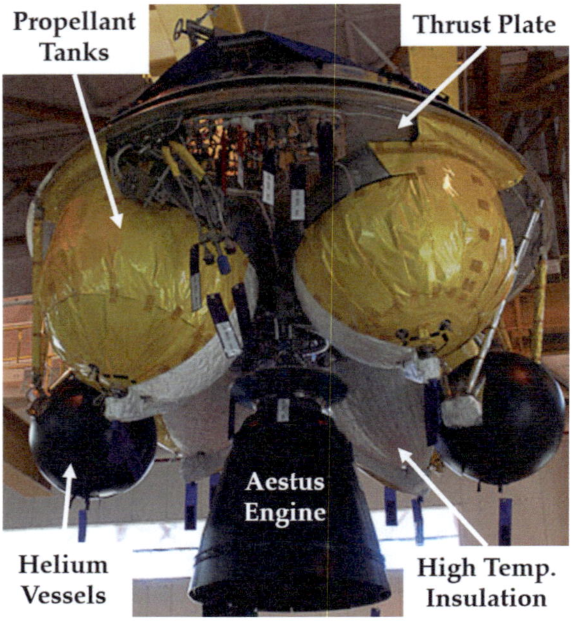


**Fig. 9.13** Upper stage of the retired launcher Ariane 5G, EPS (Etage á Propergols Stockables), accommodating 9.7 t of propellant, NTO-MMH. Credit: ESA/CNES/Arianespace/Photo Optique du CSG, S. Martin

consumables like air and water to the astronauts. Its propulsion system consists of a single orbital main engine (OMS) from AerojetRocketdyne (AJ10), eight auxiliary thrusters (R-4Ds) and 24 reaction control thrusters, in two redundant strings. Since Orion is crewed, ESM needs to be *human rated*.[17]

The auxiliary thrusters are required for two purposes. Firstly, to provide thrust vector control due to misalignment between the OME's thrust vector and the spaceship's centre of mass (CoM). Secondly, it serves as back-up in case of a main engine failure. The peculiarity of its tank design is depicted in Fig. 9.14. A single helium tank supplies two propellant tanks that are connected in series. The advantage of this system is that in total two sets of flow control equipments related to the propellant isolation can be saved, thus, minimising the number of failure cases.

### Small Cryogenic Propulsion Systems

Intuitive Machines is a private US company that plans to provide a commercial delivery service to the Moon. The lander to provide this service, dubbed 'NovaC',

[17] The term human rated generally means that the vehicle transports or houses humans and complies to stringent safety requirements, e.g. two-fault tolerance and higher safety margins.

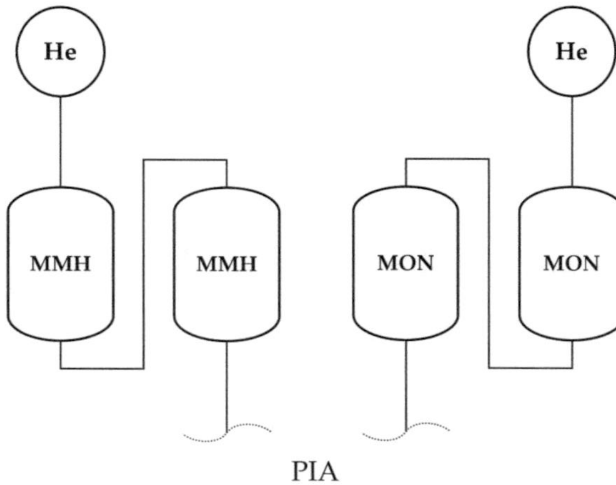


**Fig. 9.14** Schematic of ESM tank configuration

debuted in Feb. 2024. Although only partially successful[18] it is worthwhile to mention it since its propellant system uses the cryogenic propellant combination methalox. Table 9.2 gives the spacecraft configuration according to the companies FCC filing [61]. Although, not the first company to utilse this propellant combination, this honour belongs to the Chinese commercial space company LandSpace.[19] Intuitive Machines is the first to utilise it for a deep space mission with a mission duration of several days. As discussed in Sect. 8.2.2, duration matters when it comes to cryogenic propellant due to propellant boil-off. The main engine of NovaC, dubbed VR900, can achieve 3100 N. The specific impulse is not disclosed but given that the system is pressure-fed, it is to be assumed that the $I_{sp}$ will not exceed 315 s, but rather stay below this value.

The reaction control thrusters are not propelled by methalox. According to the filling these are cold gas thrusters with Helium as working medium. The reason behind using methalox instead of well-proven storable propellant combinations is apparently *strategic*. The company aims for more than the Moon and eyes for Mars and beyond very much like SpaceX. The intention of both companies is to establish

[18] The propulsion system worked nominally, the failure was with a sensor, called LIDAR (Light Detection And Ranging), that measures with help of a laser distance and relative velocity to the ground and the lander's altitude. The operation team was able to use a second LIDAR on-board, that was provided by NASA for experimental purpose only. The team managed to write the necessary software patches within 60 minutes that allowed the incorporation of the experimental LIDAR into the GNC (Guidance navigation and control) loop. This audacious act of engineering was unfortunately only partially successful. The lander had a too large transversal velocity that made it tip over to its side.

[19] It is noteworthy that Landspace's Zhuque-2 (Vermilion Bird—one of the four symbols of the Chinese constellations) designed and achieved an orbital launch of a methalox launcher prior to SpaceX and Blue Origin's BE4. The latter debuted as first stage engine of Vulcan Centaur in January 2024. In SpaceX's defence, it should be noted that the company has implemented methalox on a much larger scale, namely in the gigantic Starship, which brings its own challenges in development.

**Table 9.2** Publicly available mass constituents of NovaC [62]

| Property | Mass (kg) |
|---|---|
| Dry mass | 624 |
| $LCH_4$ | 845 |
| LOX | 422 |
| Helium | 17 |
| Payload | 100 |
| Wet-Mass | 2,008 |

the technology that can utilise Methane as propellant since this chemical element is believed to be wide spread in our solar system or can be readily extracted by ISRU, Sect. 4.5.

## 9.4 Pressure-Regulated PMS—Electric Propulsion

The mass flow rate of electric propulsion systems is measured in mg/s or sccm (standard cubic centimeter per minute), which is in stark contrast to chemical propulsion systems discussed in the previous section. Such low flow rates present different, but no less complex challenges.

The challenge of the electric propulsion feed system is the controlled reduction of the propellant pressure by almost four orders of magnitude in a highly accurate, stable and reliable manner to meet the thruster inlet requirements, which are primarily pressure, mass flow and temperature. This feat is only feasible by a sophisticated pressure and flow control assembly which is subject of this section. For illustrative purposes and to align with the nomenclature used in most publications and product data sheets, it is assumed that xenon is used as the blowing agent without loss of generality or implication of preference.[20] In case of another agent, the variable $X$ must be substituted accordingly.

Figure 9.15 shows a simplified block diagram of the xenon Feed System (XFS). The schematic commences with the xenon Tank Assembly (XTA), followed by the Propellant Supply Assembly (PSA), and then the Electric Thruster Assembly (ETA). The three blocks will be discussed in the following paragraphs. Two important points should be noted. Firstly, there are multiple ways to segment the feed system, and there are various names for the equipment. Some are named after the dominant flow control equipment, which leads to different designations for the same physical

[20] The reader will find it helpful to compare the information provided here with the product data sheets.

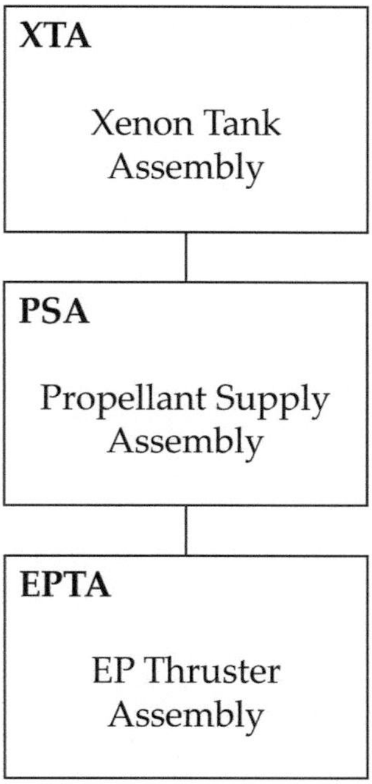


**Fig. 9.15** Block diagram of a xenon Feed System (XFS)

device. Secondly, the segment names suggest closed compartments or units, such as the FCU, but this is not always the case. Instead, it should be thought of as a functional segmentation.[21]

### 9.4.1 Xenon Tank Assembly—XTA

The XTA stores sufficient propellant to supply the electric propulsion system for all planned manoeuvres and needs to foresee extra propellant for contingency manoeuvre. Its schematic, in its simplest form, is depicted in Fig. 9.16. The XTA is designed around a central tank (or multiple tanks), which is monitored for pressure (P) and temperature (T) using a transducer. The only active control mechanism is the heater with which the temperature of the tank can be raised to prevent the high pressure gas from excessive cooling as this could cause a phase change, see Sect. 9.6.1. The tank is framed by two service valves of which service valve 1 (SV1) is a venting

[21] This is because there is no unified standard regarding this subdivision: XTA, PSA and ETA. While XTA is relatively clear, the boundary between PSA and ETA is not. It partially depends on the design (e.g. integrated) or very profane reasons like the workshare within an industrial consortium. The most important thing is that there is a clear interface agreement and that responsibilities can be clearly assigned.

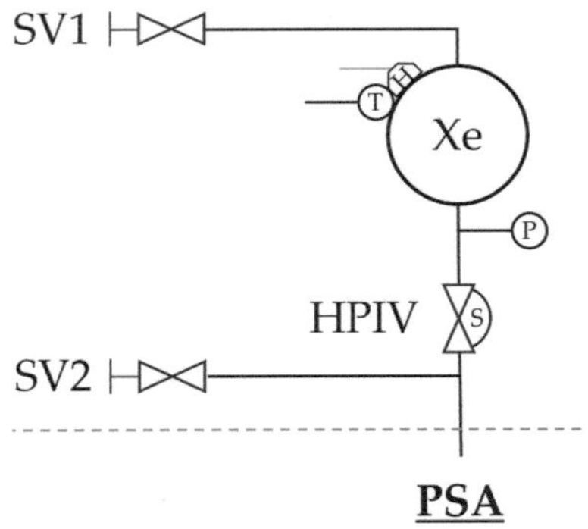


**Fig. 9.16** Flow schematic of an XTA design

valve and SV2 a fill and drain valve.[22] Some XTA designs foresee merely a single valve. The use of two valves eases the ground operational procedures for fuelling and venting during test campaigns and launch preparation. The tank is isolated from the PSA by a dedicated high pressure isolation valve (HPIV) that is nominally closed. The main function of this valve is to reduce leakage and pressure build-up in the tubing downstream towards the PSA during flight. A second reason is that this valve can be used for high pressure tests of the PSA without the need to load the tank. This is needed to perform integrity and performance checks of the assembled components following integration and welding. The proof pressure test, for instance, is an important mechanical design test.

Isolation valves belong typically to the solenoid valve family, which require a constant voltage to remain in the non-nominal position, here *open*.

However, several design requirements factor into the decision whether a solenoid or latching valve should be used and a case by case analysis is necessary. The argument for a solenoid as an isolation valve is that the valve has a bias towards its nominal position, closed, in the current example. Power, and thus active commanding is needed to leave this position and to open the valve and maintain it open. A latching valve on the other hand is *bi-stable* and requires merely a short power command to switch its state, from closed to open and back thanks to an integrated permanent magnet. Latching valves, therefore, do not have a nominal state. From operational point of view (a paranoid perspective by nature), these differences need to be considered to assess potential failure modes, e.g. 'fail-to-close'.

[22] Service valves (SV) and the aforementioned fill and drain valves (FDV) fulfil the same functions. The use of different terms for the same functionality is often the result of 'silo thinking' and working in the respective disciplines. This restricts the exchange of information and runs counter to a coherent use of technical language.

## 9.4.2 Propellant Supply Assembly—PSA

It is the PSA's task to reduce the propellant pressure from storage level down to values close to engine inlet pressure. Close is of course a relative term and an example shall shed some light on the basic principle. We have seen that xenon is stored at a pressure of 180 bar. In case of NASA's NEXT thruster, a GIT, this value needs to be reduced down to about 0.1 bar for the discharge chamber, 0.034 bar for the cathode and 0.08 bar for the neutralizer [13]. In the sequence of pressure reduction, this means, that the PSA has the task of doing the rough work: reducing the storage pressure down to a *set point* of about 2.5 bar (±10%). In the second step, the fine regulation will take place, which is discussed further below.

This 'rough' pressure down-regulation can be achieved in two ways, either mechanically or electrically. The mechanical pressure regulator (MPR) is a *passive* device that works according to its initial calibration, which cannot be changed in flight. The electrical pressure regulator (EPR) is an *active* regulator that enables a variable pressure set-point. There are currently two EPR variants in use: the bang-bang regulator and the proportional pressure regulator, Fig. 9.17. The following paragraphs will discuss the functionality, limitations and performances of the three pressure regulation devices.

### Mechanical Pressure Regulator—MPR

The MPR for the electrical propulsion system works like in the pressure-regulated chemical propulsion system. In both cases the gas needs to act against the force of a spring, $F_s$, and pressure conversion is achieved via a force balance. The resulting pressure ratio of outlet to inlet is inversely proportional to the area ratio:

$$\frac{p_{out}}{p_{in}} = \frac{A_{in}}{A_{out}}. \tag{9.5}$$

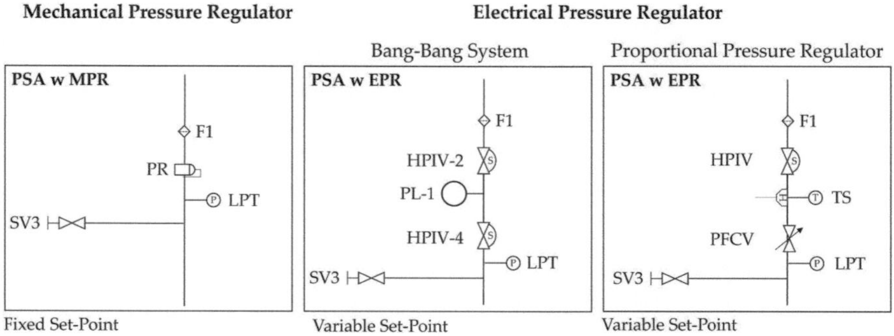


**Fig. 9.17** Variants of propellant supply assemblies [14]

**Table 9.3** Mechanical pressure regulators for xenon. *Source* MOOG Inc

| | | Single stage | Two stage |
|---|---|---|---|
| Inlet pressure, MEOP – Minimum | (bar) | 186 – 6 | 152 – 6.07 |
| Regulated outlet pressure | (bar) | 2.55 | 2.55 |
| Xenon Flow, min – max | (mg/s) | 4 – 60 | 4 – 33.2 |
| Cycles | (–) | 12,000 | 40,000 |
| Mass | (g) | 450 | 994 |

Because of this straightforward mechanism, that does not require an electrical interface, MPRs form a robust and reliable pressure regulation option. The specifications of a single and a two stage MPR are provided in Table 9.3. It should be noted that in both cases, there is a minimum threshold for the inlet pressure below which the MPR can no longer regulate in a predictable way, if at all. Engines are not operated below this value but the tank can still be drained for passivation purpose.[23]

### Bang-Bang Regulator—EPR

Historically, the Bang-Bang system was employed in electric propulsion systems prior to the MPR, as it was perceived to be a relatively low-risk approach and was capable of accommodating the extensive throttle ranges required for the first EP missions [15]. The Bang-Bang module consists of a high pressure isolation valve (HPIV) a small tank called *plenum*, a low pressure isolation valve (LPIV) and a low pressure transducer (LPT). The measured value of the LPT plays a pivotal role in the active control loop, serving as the control variable. The HPIV is the actuator in this control loop and it must be a solenoid to fulfil its task. The valve is actuated in pulse width modulation (PWM). For instance, an on-off duty cycle of 25% means that the valve is 25% open (on) and 75% closed (off) within a cycle period of $\Delta T$ in case of a nominally closed valve, which is the baseline. This ultimately changes the mass flow since only a fraction of the fluid can pass through the valve over a cycle period. Consequently, less fluid expands into the extra volume provided by the plenum, leading to the desired reduction in pressure. This principle has two side effects that need to be considered, pressure *oscillations* and *Joule-Thompson cooling*. Both will be explained in the following. The right schematic in Fig. 9.18 depicts the PWM and the sequential operation of both solenoid valves with respect to the applied voltage. The left depicts the corresponding pressure response. Note that the indicated times

[23] Recently introduced space debris mitigation standards mandate passivation for all spacecraft, including rocket stages. These rules encompass any component containing energy, thereby posing a risk of explosion and contributing to debris proliferation. In the context of propulsion systems, this necessitates the depletion of tanks to reduce the remaining pressure (i.e. energy density) and thereby reducing the risk of explosion in the event of debris impact.

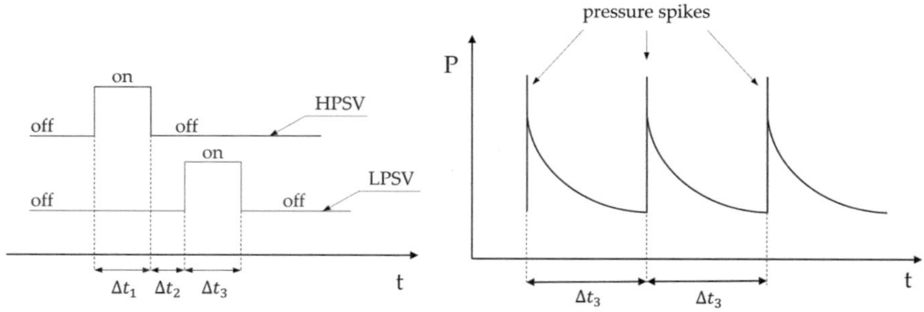


**Fig. 9.18** Single Bang-Bang Cycle, time span not to scale

are not to scale. In fact, $\Delta t_1$ and $\Delta t_2$ are much shorter than $\Delta t_3$. The PWM allows to adjust the mass flow and the pressure by modulating the frequency and duration of pulses.

It can be seen that the pressure follows a sawtooth profile, also referred to as ripples in the literature. The ripples start with a spike in pressure followed by an exponential decrease as the plenum gets depleted. These ripples would translate into undesired thrust oscillation or even cause unstable discharge behaviour and must be therefore damped. The Deep Space 1 mission,[24] for instance, had a thrust accuracy requirement of $\pm 3\%$ [16]. This damping is achieved by the plenum tanks which in case of Deep Space 1 had a volume of 3.7 liter [16]. The plenum acts as a low-pass filter, smoothing out high-frequency contributions and producing a sufficiently steady pressure profile for the ETA [17].

The second effect to be considered is the Joule-Thompson cooling effect due the rapid expansion of the fluid into the plenum. This cooling can readily lead to a saturated liquid/vapour mixture (i.e. a two-phase flow). Active temperature control is required to maintain xenon in gaseous form. This is achieved by a heater downstream of the plenum. An adjacent sensor carefully monitors the temperature which serves as control variable.

The opening times of both solenoids are determined by a number of parameters, including the xenon supply pressure, temperature, plenum pressure, and even solenoid temperature [16]. For each new mission, this increases the tuning effort of the cycle times, which is equivalent to more engineering work in design and verification phase. Such work includes control equipment characterisation for the new environmental conditions and operational requirements as well as calibration and software development.[25] The reward, however, is a flexible regulator whose pressure-set point can be adjusted via software in flight.

---

[24] Launched on October 4th 1998 by a Delta II launch vehicle, NASA's Deep Space One was the first spacecraft with a throttleable ion propulsion system used for primary propulsion. It was a technology demonstration mission that paved the way for EP in the US, while Hall Effect thruster were in use in the Soviet Union at least a decade ago.

[25] A high degree of standardisation is only possible for standard missions, as is the case with satellite constellations.

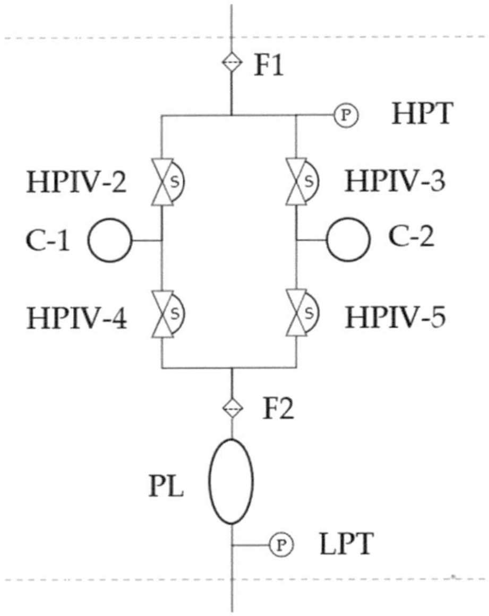


**Fig. 9.19** Redundant Bang-Bang pressure regulation of Smart-1 [18]

An elaborated variant of this principle was implemented in SMART-1, Fig. 9.19. The BPRU (Bang-Bang Pressure Regulation Unit) consists of two strings for redundancy. Each string consisted of the elements discussed above. However, instead of a single plenum, SMART-1 engineers introduced a cavity for each string and a plenum for both. This staggered tank design approach worked in a similar way as discussed above. The first solenoid valve (upstream) in the string lets only a slug of gas into the cavity before it closes. The second valve (downstream) does the same, letting only a small amount into the plenum. The gas expands each time into a larger volume, which reduces its pressure [18].

Irrespective of the detailed layout, there is the option of operating the bang-bang pressure regulator in an open or closed loop. In the first case, the measured pressure by the LPT is used solely for housekeeping telemetry purposes to characterise the health state of the system. In the latter case, the software development and operational effort become more complex.

The operation logic needs to be translated into a mathematical model (i.e. algorithm) and then into software. In a general desire to keep the onboard computer slim and standardised, execution of the SW is outsourced into a dedicated electronic unit.[26] As a consequence supplier of pressure regulation assemblies started to develop their own propulsion driving electronics (PDE) to offer a *turn-key* solution.

A general concern of this pressure regulation approach is that the huge number of oscillations can cause wear-out of the solenoid valve. In case of SMART-1,

[26] This means dedicated development and test effort for both the software and HW. System primes that do not manufacture EP systems but procure it are therefore reluctant in adopting this extra effort and to risk the project.

**Table 9.4** Performances of a PFCV for EPPS [20]

| Inlet pressure, MEOP – Minimum | (bar) | 186 – 2.8 |
|---|---|---|
| Regulated outlet pressure | (bar) | 2.28 – 2.55 |
| Xenon Flow, min – max | (mg/s) | 0 – 30 |
| Cycles | (–) | $10^6$ |
| Mass | (g) | 115 |

the solenoid performed about 700,000 cycles during its lifespan [18]. A significant amount of engineering design is required to implement the bang-bang system which is why there is a tendency towards other options [15].

Proportional Pressure Regulator—EPR

The second type of electrical pressure regulator is the proportional pressure regulator approach, which is named after its main flow control equipment, the proportional flow control valve (PFCV). Like the MPR, pressure is regulated via a force balance. But unlike the former one, PFCVs can vary the counter pressure, thus enabling active and continuous pressure control making it a viable alternative to mechanical regulators and the Bang-Bang method. There are two design types. The first uses a solenoid valve capable of varying the position of a flow restrictor, which consists of a spool assembly inside the valve.

The inlet pressure has to overcome the magnetic force induced by the coil and exerted by the spool. The PFCV can be thought of as the equivalent to an MPR but with a variable spring strength. Precisely speaking, it is a proportional *electro-mechanical* valve. This enables the active throttling of either the operating pressure or the propellant flow rate of the system, while also providing propellant isolation capability [21]. Performance figures are provided for an off-the-shelf product in Table 9.4. In the second approach, a piezo actuator is used to generate the required resistance force [19]. Due to their electric nature, both types can vary the pressure set-point in-flight.

## 9.4.3 Electric Thruster Assembly—ETA

The electric thruster assembly (ETA) is the last block in Fig. 9.15. It consists of a Flow Control Unit (FCU). In some definitions this block contains also the Power Processing Unit (PPU) and the thruster boom for thrust vector control. In this section we will focus on the FCU.

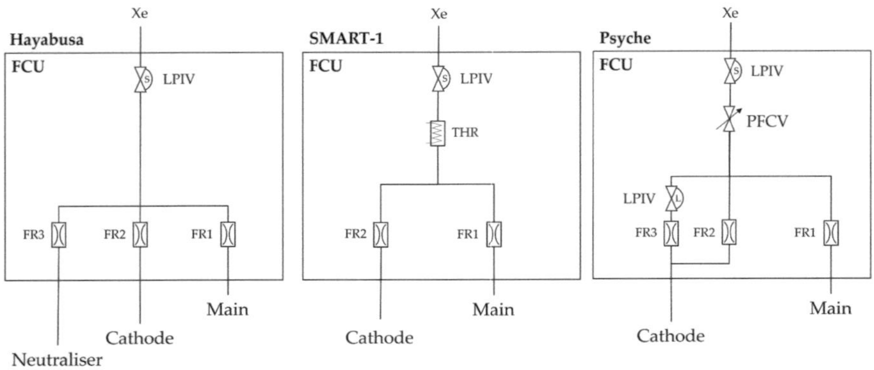


**Fig. 9.20** Three FCU types. Left: Hayabusa [22]. Middle: SMART-1 [26]. Right: Psyche [24]

## Flow Control Unit—FCU

The main task of the FCU is to split the low pressure flow into two in case of HET, or three in case of GIT, flow streams that meet the specific inlet requirements. Figure 9.20 shows three FCU examples for three missions: JAXA's Hayabusa, ESA's experimental satellite SMART-1 and NASA's deep space probe Psyche.

The FCU of Hayabusa is the least complex, it foresees three pre-calibrated flow restrictors to achieve flow splitting. The flow restrictor itself is a small device consisting of a ladder of capillary tubes that together exert a resistance to the flow. The flow split can be adjusted by adjusting the magnitude of the resistance of the three orifices in parallel, similar to an electric current. Despite its simplicity, it is necessary to characterise the device in terms of temperature and pressure. The result is a look-up table which the PPU uses to control the temperature of the restrictor. In case of Deep Space 1 the PPU also adjusted the plenum tank pressure [23].

The second FCU design belongs to ESA's SMART-1 mission which was propelled by a Hall Effect Thruster (PPS-1350). Besides the flow restrictor the FCU of SMART-1 features a *thermothrottle* upstream. This thermotrottle consists of a porous sintered metal plug that works according to the *thermo-viscous* principle: an increase in temperature leads to an increase in the viscosity of the liquid and consequently to a reduction in the mass flow rate[27] through the porous element [25].

The third depicted FCU design was implemented in NASA's Psyche mission, launched October 2023. It features a proportional flow control valve downstream the isolation valve, which was necessary due to the deep throttle requirements typical for interplanetary missions. This need was discussed in Sect. 7.2.6, where we said that the main way to throttle down an electric propulsion system is to reduce the discharge current, $I_d$. The depicted latch valve is closed later in flight to reduce the

[27] A detailed discussion of the fluid physics within a thermal throttle, including a method for predicting the performance of thermal throttled systems, is given in [27].

electron production rate, thus controlling the discharge current. By closing the latch valve, the relative ratio of the cathode flow to the main flow drops from 9% to 5% [24].

In conclusion it shall be highlighted that new pressure regulation concepts foresee an integrated design of the PSA and the FCU. This saves testing effort and reduces the susceptibility to errors. Such concepts are already in use for satellite constellations but it remains to see the applicability for deep space missions.

## 9.5 Pump-Fed PMS

The objective of the feed system of an electric propulsion system is to reduce the pressure in a controlled manner by two orders of magnitude: from 300 bar down to approximately 2–3 bar. In contrast, the objective of a PMS for a high thrust propulsion system is to increase the pressure in a controlled manner by also two orders of magnitude. However, the challenge of increasing the pressure is significantly more demanding than the reducing the pressure. This can be compared with the task of rolling a one-ton ball up a hill with the task of rolling it down.

The reason why engineers need to do this in the first place is because of the thrust equation derived in Sect. 7.1.2, which states that the thrust of a thermal engine is proportional to the chamber pressure and the nozzle throat area:

$$T \propto p_0 A_t. \tag{7.21}$$

Hence, to design compact high thrust engines like RD-181, RS25 or Raptor, it is imperative that the feed system is capable to supply propellant with very high pressure levels in excess of the required inlet pressure of 200–300 bar, Fig. 5.8. A passive feed system in which the propellant is stored at a high pressure and then drops in the feed lines as for a blow down system, Fig. 9.4, is not an option. It would lead to unrealistically heavy tanks, same for a pressure-fed design. Therefore, an *active* feed system is needed with *turbomachinery* equipment, most notably pumps, which is why this feed system class is also referred to as *pump-fed*. The power source of the pump is in general thermal but smaller engines may utilise electrical energy. Currently, only two engines have been implemented that use battery-powered pumps: Rocketlab's Rutherford engine with a thrust of $T_{sea} = 25$ kN and Astra's Delphi engine with a thrust of $T_{sea} = 29$ kN. As we will see below, the energy required to pump tons of propellant is considerable and the propulsion system mass fraction $\mu_{ps}$ of battery-powered pumps is poor with increasing propellant mass. Besides small launcher, electrically powered pumps are frequently discussed for planetary lander and large orbital systems with high thrust requirements. In particular, the relatively simple and smooth throttle capability renders them an attractive option for the landing of large masses on the Moon and dwarf planets. This is arguably a niche in which this technology will excel in future.

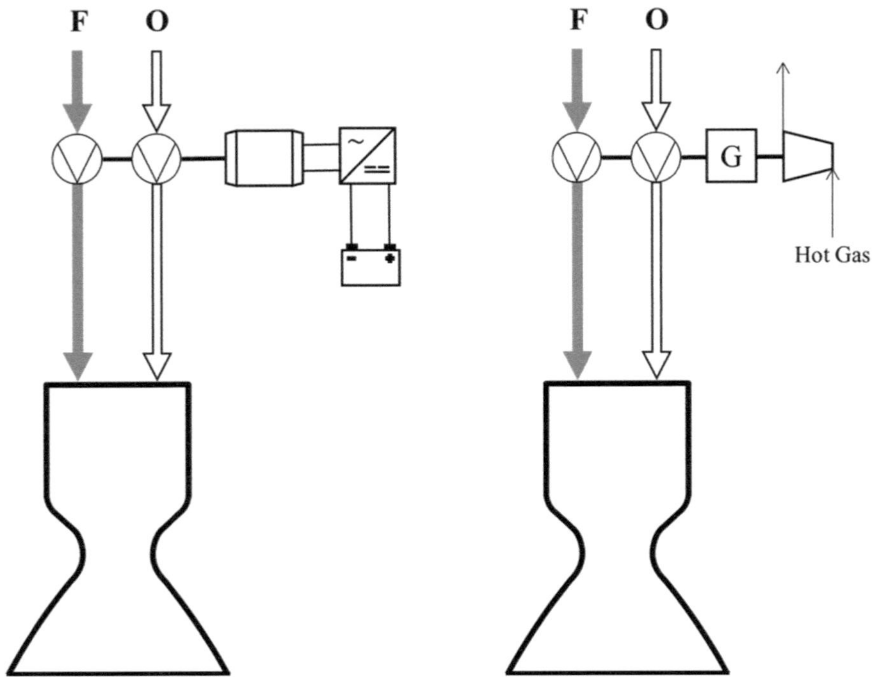


**Fig. 9.21** Schematics of electrically-powered and thermo-mechanically powered pumps

The vast majority of pump-fed feed systems, however, are driven by turbines and the assembly of pump and turbine is called *turbopump*. The schematic representation of the two options is presented in Fig. 9.21. Traditionally, turbomachines are part of heat engines whose operation and efficiency are best characterised by thermodynamic cycles. It is this historic analogy that led to the term *engine cycle* for pump-fed feed systems.[28] Rocket engineers distinguish between three distinct cycle architectures:

- Expander Cycle
- Gas Generator Cycle
- Staged Combustion Cycle

These three cycles are available in a number of sub-variants. To discuss them all would rightly require a textbook in itself, and is beyond the scope of our objectives. We, therefore, will emphasise on the working principle of these cycles and the prime technological differences on architecture level.

The first rocket motor to be designed and operated was, in fact, a gas generator, namely the engine of Aggregat 4. Despite this historical order, we commence our discussion with the expander cycle. Subsequently, we will examine turbopumps, which are the hidden champions of rocketry.

[28] The two terms, pump-fed feed system and engine cycle, are used synonymously in literature about high thrust engines.

### 9.5.1 Expander Cycle

The first expander cycle engine, debuted in 1962, was Rocketdyne's RL10A-1. It was furthermore the world's first hydrolox powered engine [30]. This remarkably successful engine laid the foundation for the RL10 engine family, which continues to be in service and is anticipated for use in several upcoming space programs. Since then several other expander cycle engines have been designed in various variations of which the three main types will be discussed in the following:

- Classical closed Expander Cycle
- Expander Bleed Cycle
- Expander Tap-Off Cycle

As we have seen in Sect. 8.2.1, large portions of the engine, in particular the combustion chamber, need to be cooled to prevent the walls from structural disintegration. As discussed, cooling is achieved by channelling off some portion of the propellant (in general the fuel) and divert it through cooling jackets and passages within the upper part of the engine, the thrust-chamber. The expander cycle makes use of this necessity by diverting the resulting hot gas into the turbine where it expands and drives the pumps via a shaft. This hot gas exiting the turbine can either be dumped off into the low (or zero pressure) pressure environment via a separate duct with a small nozzle or it could be redirected back into the combustion chamber. In the first case the cycle is called to be *open*, it literally bleeds propellant—hence the name *expander bleed cycle*. In the second case the cycle is *closed*. The Japanese LE-5 engine family is an example of a bleed cycle. The open cycle type has the advantage that the turbine power is larger than for the closed type due to the deeper pressure drop. Whereas for a closed cycle, the pressure of the propellant used to drive the turbine, still needs to be sufficiently high, so that this propellant can be injected into the combustion chamber. On the other hand, open cycle engines face the disadvantage that the hot propellant that is dumped overboard is missing in the combustion chamber which consequently reduces the reaction mass flow in the thrust term, $\dot{m}c_e$, and the achievable temperature in the combustion chamber.

The majority of the engines with an expander cycle type are of the closed type. Prominent examples are the US RL-10 family, the Franco-German Vinci, and the Chinese YF-75D. The generated turbine power for both types (open and closed) depends strongly on the enthalpy gain of the hot gas before entering the turbine and in consequence on the amount of energy that can be absorbed by the coolant, i.e. heat capacity. Cryogenic fuel is particularly well suited for this cycle type. To increase this amount, the combustion chamber of some expander cycle engines is more elongated than chemically required for the combustion process only. This can be clearly seen in the case of the Vinci thrust chamber, Fig. 9.22. The reason is obvious, the surface area for the heat transfer between chamber and coolant shall be increased. This working principle comes with a limitation that becomes clear when visualising the geometric relations: the available heat absorption capacity for the turbine medium scales with the surface, $r^2$, of the thrust chamber while the power the turbine has to supply

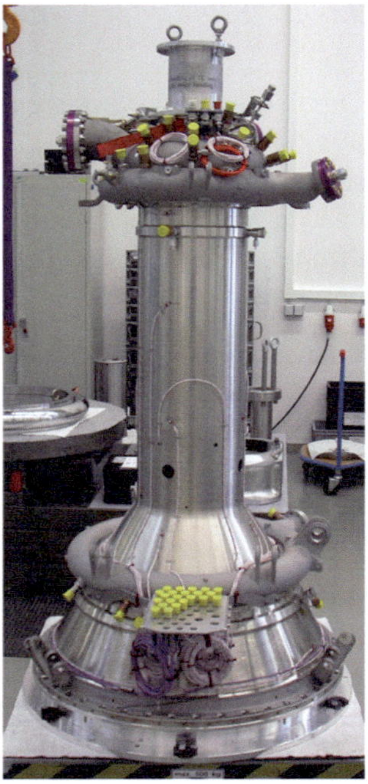

**Fig. 9.22** Elongated combustion chamber of Vinci. Credit: European Space Agency—ESA

scales with the volume, $r^3$, of the thrust chamber. The overall engine performance is, therefore, limited by the geometry of the engine. In other words, it is not possible to increase the supply pressure which is directly related to the turbine output power without increasing the engine geometry and ultimately its weight. Thus, engines based on the expander cycle are practically limited in size and thrust magnitude. A consequence is a relatively large thrust-to-weight ratio for higher thrust levels beyond $\approx 250\,\mathrm{kN}$ compared to other cycles [63]. The so far largest expander cycle engine was designed by the Japanese company Mitsubishi Heavy Industries, the LE-9 with a thrust level of 1471 kN and a specific impulse of 425 s at sea level[29] [31]. Figure 9.23 depicts the flow schematic of the LE-9.

Valves and other other flow control equipment are actuated pneumatically with helium. There is a trend, however, towards electric-driven valves [55] to save mass. Figure 9.24 (left) shows the simplified flow schematic without valves and other flow control equipment of the RL10-A series. Note that there is only one turbine for both pumps.

The turbine and fuel pump are directly connected by a drive shaft while the oxidizer pump is driven through a reduction gear due to the different rpm requirements. The

[29] The LE-9 engine powers Japan's new H3 rocket, successfully launched Feb. 2024. The intriguing element is that the launch system architects decided to use—for the first time—an expander cycle engine to power the first stage. Given the unflattering thrust-to-weight ratio of this engine type, makes the decision and the decision making-process even more intriguing.

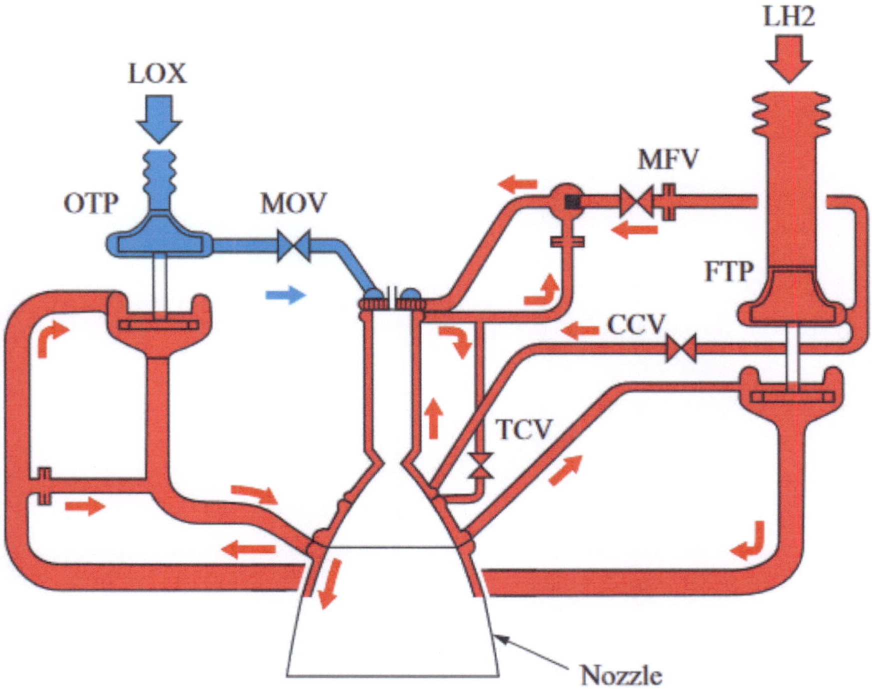


**Fig. 9.23** Flow schematic of the Japanese LE-9 [31], with Oxygen Turbopump (OTP), Main Oxygen Valve (MOV), Fuel Turbopump (FTP), Main Fuel Valve (MFV), Combustion Chamber Cooling Valve (CCV), Thrust Control Valve (TCV). Credit: JAXA/Mitsubishi Heavy Industries

figure shows also the corresponding ideal (i.e. isentropic) thermodynamic cycle as experienced by the fuel in an h-s diagram. The stations are as follows:

| | |
|---|---|
| *1*→*2*: | isentropic compression in the pump |
| *2*→*3*: | enthalpy gain through thrust chamber cooling |
| *3*→*4*: | isentropic expansion in the turbine, power output is larger for open (i.e. bleed) compared to closed cycles |
| *4*→*5*: | enthalpy increase through isobaric combustion |
| *5*→*6*: | isentropic expansion in the nozzle |

One method of enhancing the turbine power of this engine cycle type is to utilise a gas that has a higher enthalpy than the cooling-flow. This is what the second variant, the expander *tap-off* cycle does. Here, the hot gas to drive the turbine is directly taken from the combustion chamber. In consequence the thermo-mechanical load on the turbine is much more severe and material needs to be selected (and developed) that can sustain these high loads. This cycle type was initially tested in the United States between the years 1965 and 1969 (J2-S). It was not until Blue Origin's BE-3 that this type became flight proven in the companies suborbital and re-usable launch vehicle called New Shepard. The BE-3 has an estimated thrust level of 490 kN and

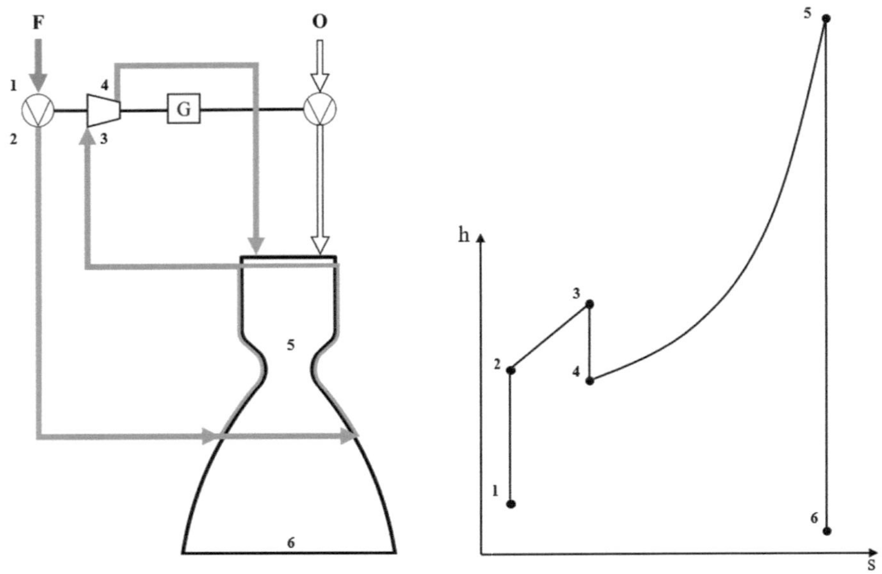


**Fig. 9.24** Left: simplified flow schematic of closed expander bleed cycle (e.g. RL-10A). Right: Loss-free h-s diagram of the fuel cycle

runs with hydrolox. Most expander cycle engines operate in a thrust range between 80 and 250 kN and are used as upper stage engines.

### 9.5.2 *Gas Generator Cycle*

The first modern rocket engine was designed and developed under the leadership of Dr Walter Thiel in Peenemünde (Germany) for the A4 programme in the late 30s and early 40s of the 19th century. The design team called it *der Raketenofen* the rocket oven, see footnote 8 on Sect. 8.2.1. The thrust requirement of 250 kN was achieved by a pump-fed feed system driven by hot gas that was generated in a dedicated combustion process by means of a so-called *pre-burner* [28]. The pre-burner is nowadays called *gas generator* and the complete engine cycle type *gas generator cycle*. Within the following 20 years, this principle was steadily further developed until it reached 1965 its peak thrust level of 6,770 kN with Rocketdyne's F-1, of which five propelled the mighty Saturn V.

Figure 9.25 shows a flow schematic of a generic gas generator cycle and the corresponding simplified h-s diagram of the fuel.

The gas generator cycle belongs to the group of *open* cycle combustion engines. A portion of both propellants is tapped off and diverted into the gas generator where it is burned to produce hot exhaust gas to drive the turbine. In order to limit the thermo-mechanical loads on the turbine blades, the combustion mixture ratio is fuel-rich.

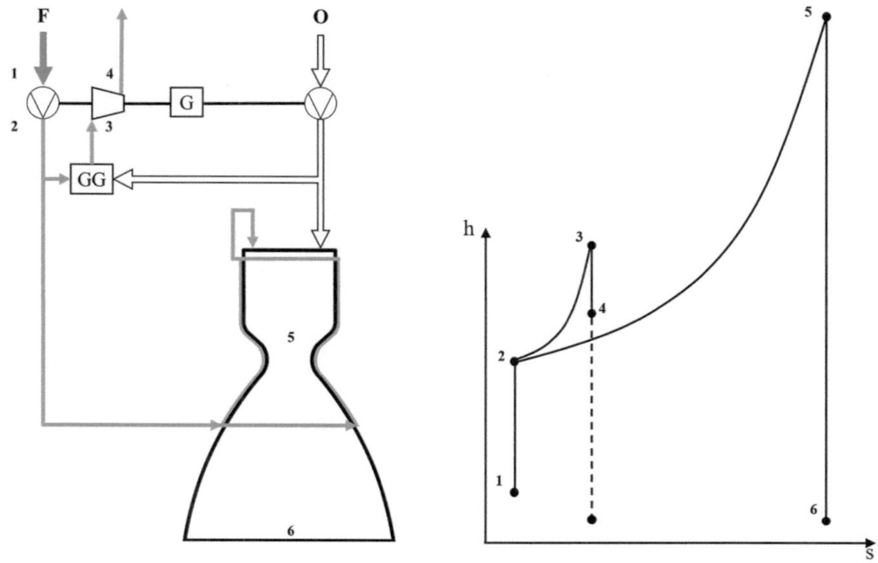


**Fig. 9.25** Generic gas generator cycle, flow schematic (left) and a simplified loss-free h-s diagram of the fuel (right)

Some engines like the F1 or the upper stage version of the Merlin engine, use the turbine exhaust gas to cool the lower portion of the engine nozzle beyond the throat by means of film cooling, see Sect. 8.2.1. This is clearly visible by dark streams made of dust particles, called soot, in the nozzle's exhaust gas caused by unburnt RP-1 of kerolox engines. The advantages of the gas generator cycle are its simplicity since it is an open cycle, its compactness, its good thrust-to-weight ratio, and foremost, its scalability. Engines with this cycle type span two orders of magnitude in thrust: from the Chinese YF-73 with 44 kN to the US F-1 with 6,770. On the other hand, the fact that the combustion in the gas generator is fuel-rich renders this cycle sub-optimal since chemical energy and reaction mass is 'wasted' when the hot not fully combusted gas is dumped out. This motivated engineers to design a closed cycle engine with even pre-burner and better performances, the staged combustion cycle.[30]

### 9.5.3 *Staged Combustion Cycle*

A closed cycle high pressure engine is offered by the *staged combustion cycle* (SCC). This cycle can be considered as the peak of rocket engineering as it offers excellent thrust-to-weight ratios and the highest efficiencies. Soviet engineering was partic-

[30] Nowadays, the environmental impact of fuel-rich gas generator cycles is an important consideration, especially if the fuel is a hydrocarbon like methane or kerosene, and particularly if the launcher is institutionally financed.

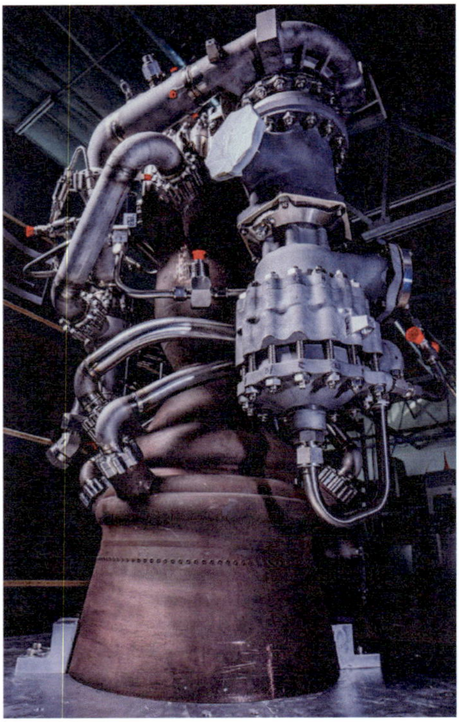

**Fig. 9.26** E-2 oxidiser rich kerolox SCC from Launcher. Courtesy John Kraus

ularly renowned in this field, with the NPO Energomash design bureau producing engines such as the RD-171M, the world's most powerful liquid rocket engine. These engines exhibited performance characteristics that US engineers did not believe possible until the fall of the Iron Curtain. The remarkable performances of the staged combustion cycle in general come with an architectural complexity. This engine type was first suggested by the Russian engineer and entrepreneur Aleksei M. Isayev.[31]

Figure 9.27 shows the flow schematic of an oxidiser rich staged combustion cycle. The latter misses the dashed line since all turbine exhaust gas is directed into the combustion chamber. Due to its high energy density, the propellant combination LOX-RP1 is a very good choice to propel the first stage of launchers, refer to discussion in Sect. 8.2.2. Figure 9.26 shows an oxidiser rich kerolox staged combustion cycle of the US company Launcher.

However, to develop a staged combustion engine for this propellant combination is not straightforward. As for the gas generator cycle, the pre-burner cannot be operated in stoichiometric conditions but needs to be either fuel or oxidizer rich. The latter case has been avoided since very hot oxygen gas of more than 1,300 K is extremely aggressive and ruthlessly erodes any metal on its way. A fuel rich pre-burner was

[31] Aleksei Mikhailovich Isayev (1971†) was together with Sergei Korolev (1966†) the founding father of Soviet rocketry [29].

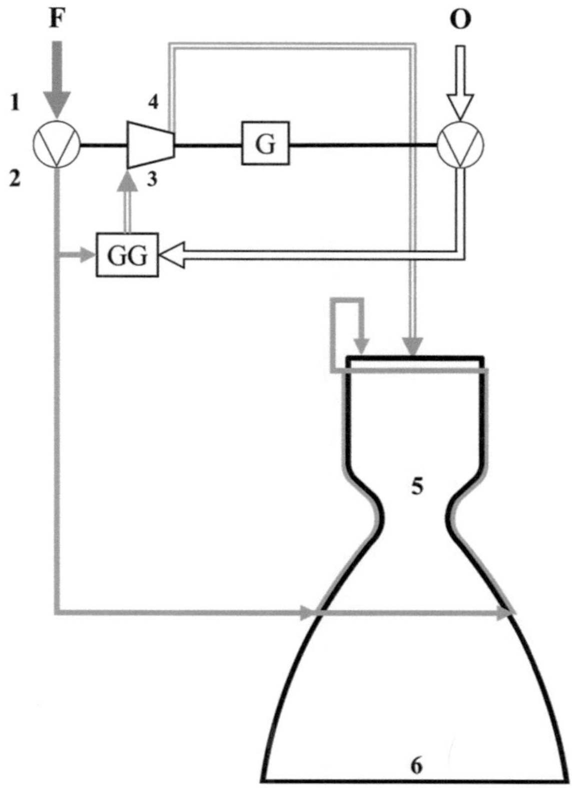


**Fig. 9.27** Generic flow schematic of oxidizer rich SCC with single shaft

not possible either, as incomplete combustion of heavy hydrocarbons in RP-1 lead to dust particles referred to as soot. Although soot is not aggressive in terms of erosion, it is extremely sticky and tends to clog, cooling channels, injectors and other narrow passages. Therefore, both paths led to a dead end and the reason why US engineers did focus on large gas generator cycles to power their launchers, like the F-1. Soviet engineers, however, were undeterred by the challenges posed by high-temperature oxygen and continued to pursue the potential of staged combustion technology in combination with LOX-RP1. They developed alloys that were able to withstand the aggressive oxidiser, but for almost 30 years their work was not recognised in the world, on the assumption that this cycle type was not feasible. When this ingenuity was finally revealed, it came indeed as a complete surprise. It led to powerful engines such as RD-171M, the world's most powerful liquid engine, and RD-180 which was sold for more than a decade to the US to power ULA's Atlas V launcher. Figure 9.27 shows an oxidiser rich staged combustion cycle in its most simple form.

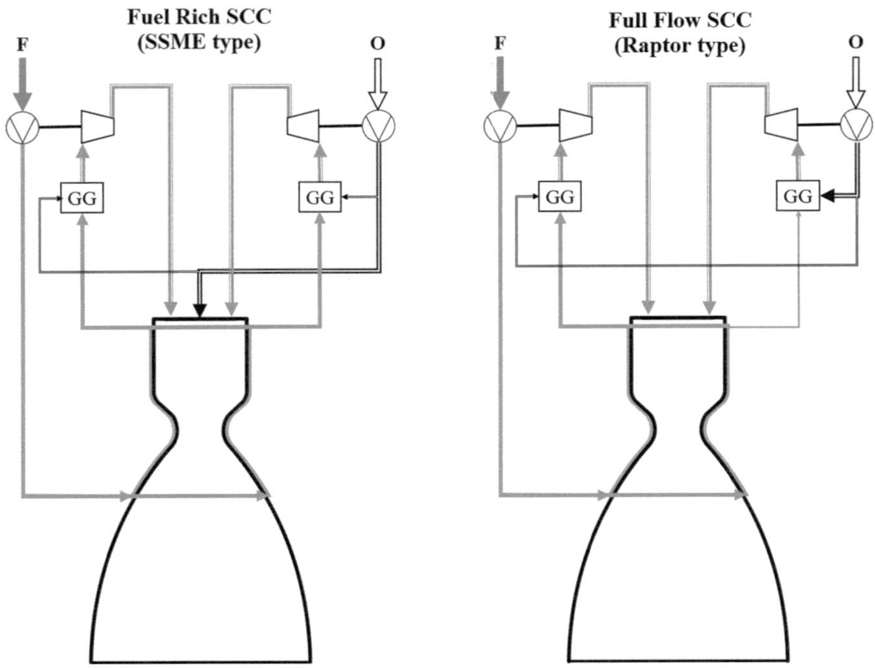


**Fig. 9.28** Generic flow schematics of staged combustion cycle (SCC) variants

Fuel Rich

The US was fully aware of the merits of staged combustion and did not give up on the technology itself. Instead, they shifted the focus to another promising propellant combination: hydrolox. Still convinced that alloys to withstand very hot oxygen do not exist, US engineers designed a fuel rich hydrogen staged combustion cycle. The result was the impressive RS-25, also known as the Space Shuttle Main Engine (SSME).[32]

Figure 9.28 (left) depicts the generic flow schematic of this cycle type. It is a dual shaft SCC, meaning that each pump is powered by its own turbine at its specific rpm without the need for a gear. Each pump has its own pre-burner. The use of hydrogen for a fuel-rich SCC brings significant challenges. We have seen in Sect. 8.2.2 that hydrogen is a highly volatile element and due to its tiny size capable of creeping through seals and joints. The danger is obvious, the escaped hydrogen could react with the surrounding air during the ascent and cause a catastrophic failure. US engineers coped with this challenge, but the RS-25 remained the only hydrogen-rich SCC engine to date.

[32] Following the retirement of the Space Shuttle, the RS-25 was revitalised and is used to power the core stage of NASA's Space Launch System (SLS).

Full Flow Staged Combustion Cycle

A third variant of the staged combustion cycle is the *full flow* SCC. The idea is to have two pre-burners, one running fuel-rich, Fig. 9.28 (left), and the other oxidizer-rich, Fig. 9.28 (right). Although, rigorously tested in the US and in the Soviet Union, this variant has not made it into operation until SpaceX has selected this type to propel its Starship transportation system. The result is the most efficient engine ever built, the Raptor engine. It uses the propellant combination methalox, which avoids the clogging issue due to the soot and the volatility issue that comes with hydrogen. The Raptor engine is the pinnacle of staged combustion technology, with a combustion pressure of 300 bar that outperforms the RD-181 by 40 bar.

The pressure upstream is in fact significantly higher. The turbopumps of the raptor engine (full-flow SCC), for instance, increase the oxidiser (LOX) pressure by 645 bar and the fuel (LCH4) pressure by 638 bar. The required power is 34 MW ($\eta_{\text{p,LOX}} = 0.8$) and 28 W ($\eta_{\text{p,LCH4}} = 0.8t$), respectively. The pressure losses caused by the injectors alone are 78.9 bar and 71.7 bar [32]. Due to the significance of turbopumps, we will delve a bit deeper into these compact power machines in the following section.

## *9.5.4 Turbopump*

The arguably most complex technical device of a rocket engine is the *turbopump*. This term is a compound of a turbine and a pump implying the fact that the pump is run by a turbine. This can be achieved either directly, then both run with the same angular velocity (rpm) on the same shaft or via a transmission gear. The term is specifically used to denote compact high-pressure turbomachines, as required in helicopters and rocket engines.

In essence, a pump is tasked with supplying a specified mass flow at a specified discharge pressure. For a fixed pump design, the two quantities are related to each other. We will discuss this relation after we describing each device.

The Physics of Turbopumps

From a fluid dynamics perspective, a pump exerts work on the fluid and thus increases the fluid's enthalpy, whereas a turbine receives work from the fluid and thus reduces the fluid's enthalpy. Both processes are so fast, that a fluid particle does not exchange noticeable heat with its environment, we can adopt the previously discussed assumed of an adiabatic process. The pump increases the fluid's enthalpy by forcing the fluid through a shrinking duct and thus volume. In Sect. 5.2 we said that the fluids enthalpy depends on both, the inner energy and its capability to change the volume of its boundaries. In other words to perform work against its surrounding, which is quantified as $pV$. By shoveling the fluid into a smaller volume (i.e. duct) work is performed on it, thereby increasing its $pV$ term.

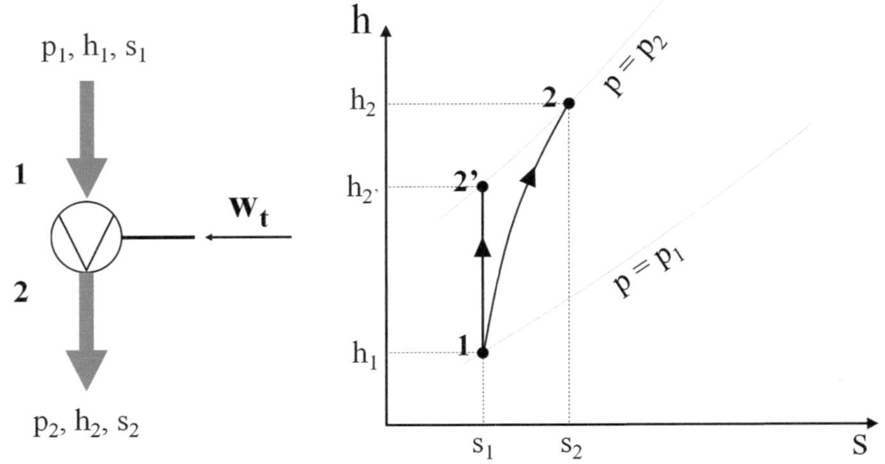


**Fig. 9.29** Flow schematic (left) and h-s diagram (right) of the compression process for an ideal and realistic case in a pump

The 'shoveling' is achieved by aerodynamically shaped rotating blades and can be quantified by an hydrodynamic force balance [33]. Figure 9.29 depicts the flow schematic of a generic pump and the related state change in the h-s diagram. In an ideal process with no entropy creation, the process would be a straight vertical line in the h-s diagram from state 1 to state 2'. Due to friction the process takes instead a 'detour' and goes from state 1 to state 2. This detour means in fact that more work needs to be performed by the pump than ideally required to reach the desired pressure $p_2$. An ideal processes in which no entropy is created is called *reversible* or *isentropic*. The reversible specific work exerted by the pump can be derived directly from Fig. 9.29. Ignoring the velocity change we receive:

$$w_{\mathrm{p,rev}} = h_{2'} - h_1 - \frac{c_{2'}^2 - c_1^2}{2} \approx h_{2'} - h_1 = \Delta h_{\mathrm{rev}}. \tag{9.6}$$

This value is lower than the actually performed specific work by the pump:

$$w_{\mathrm{p,real}} = h_2 - h_1 - \frac{c_2^2 - c_1^2}{2} \approx h_2 - h_1 = \Delta h_{\mathrm{real}}. \tag{9.7}$$

The ratio of both quantities gives the efficiency of the pump:

$$\eta_{\mathrm{p}} := \frac{\Delta h_{\mathrm{rev}}}{\Delta h_{\mathrm{real}}}. \tag{9.8}$$

The power delivered by the pump is called discharge power, $P_{\mathrm{dp}} = \dot{m}_{\mathrm{P}} \Delta h_{\mathrm{real}}$, the power needed to drive the pump is called shaft power, $P_{\mathrm{s}}$ and is supplied by the turbine.

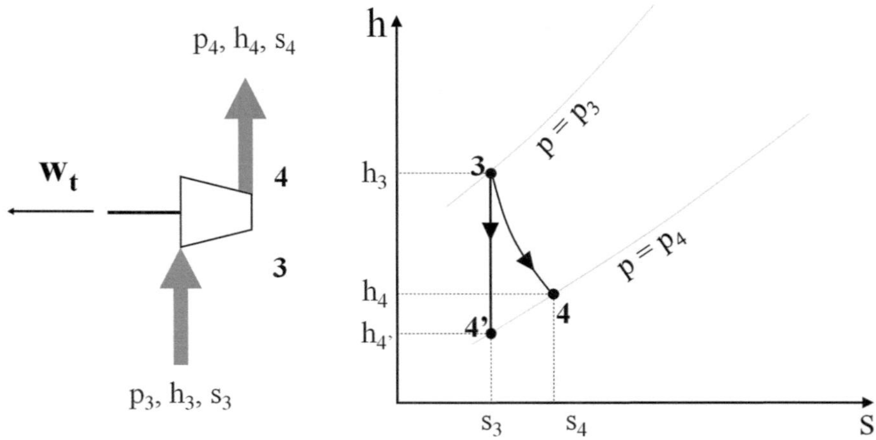


**Fig. 9.30** Flow schematic (left) and h-s diagram (right) of the expansion process for an ideal and realistic case in a turbine

We will determine this power following the same consideration as before. Figure 9.30 depicts the flow schematic of a generic turbine and the related h-s diagram.

The fluid transfers part of its enthalpy to the turbine via hydrodynamic forces [33]. These forces cause the turbine blades to rotate, and since the blades are mounted on a shaft, the shaft also begins to rotate. Like in case of the pump, the process is not ideal, $3 \rightarrow 4'$, but leads to an increase in entropy, $3 \rightarrow 4$. The specific work seen by the shaft, $\Delta h_{34'}$, is lower than what could have been delivered in an ideal process, namely $\Delta h_{34}$. Ignoring further transmission losses, the shaft power is then:

$$P_s = \dot{m}_T \Delta h_{34}. \tag{9.9}$$

Equation 9.9, though inconspicuous, is in fact very important for the turbine design and the efficiency of the engine cycle. The required shaft power could be achieved by either a larger enthalpy gain, $\Delta h_{34}$, or by a larger mass flow rate, $\dot{m}_T$. A larger enthalpy gain can be achieved by a combination of either hotter entry conditions of the pre-burner exhaust gas (i.e. large $h_3$) and/or a deep expansion (i.e. low $h_4$). Hotter entry gas could be in principle realised by a mixture ratio closer to stoichiometric conditions in the pre-burner or in case of an expander cycle by tapping off hot combustion gas, as realised in the BE-3 engine. Deep expansion can be realised, if the gas is dumped overboard, as done in all open cycles[33] To limit the thermo-mechanical loads on the turbine, temperatures beyond 1000 K must be avoided to avoid elaborated cooling mechanisms for the pre-burner and to avoid an overly complex design. This is the reason why the combustion in the pre-burner is either oxidizer rich or fuel rich but never close to stoichiometric. The only variable left to increase the discharge power

[33] This discussion highlights that its in fact the performance characteristics of the turbine that drives the rocket engine cycles.

of the turbine is the mass flow rate that is tapped from the main stream. This, however, brings the open cycle processes, gas generator & expander bleed cycle, in a dilemma. It means that more mass needs to be diverted from the main stream, which cannot be fully burnt and is expelled out of the engine while still containing valuable energy and mass. Both have the negative effect to reduce the thrust magnitude. Closed cycles do not suffer from this vicious circle.

### The Technology of Turbopumps in Rocket Engines

There is a manifold of turbomachinery designs which can be roughly classified in *axial* and *radial*. Axial devices are more effective but require also more space. They are used for instance in stationary gas turbines of power plants and to power jet engines. Radial devices have a higher power density and are always the first choice where volume and mass matter, e.g. helicopters and rocket engines. Figure 9.31 depicts the LOX turbopump (left) and the LH2 turbopump (right) of the Japanese LE-9 engine. Note the differences in relative size of pump and turbine for both turbopumps. Due to the lower density of LH2, the Impeller needs to be much larger relative to its turbine compared to the LOX case. The pump is of radial design whereas the turbine follows the axial design approach. Both turbopumps have a two stage turbine and a single stage pump with only one impeller. Further, both exhibit an *inducer* which is essentially for a low *head pump* meant to prevent *cavitation* at the impeller. Head is an important performance parameter used by the turbopump guild and is defined as the difference of the pump's discharge pressure and suction pressure normalised by the density:

$$H = \frac{p_d - p_s}{\rho} = \frac{\Delta p}{\rho}. \tag{9.10}$$

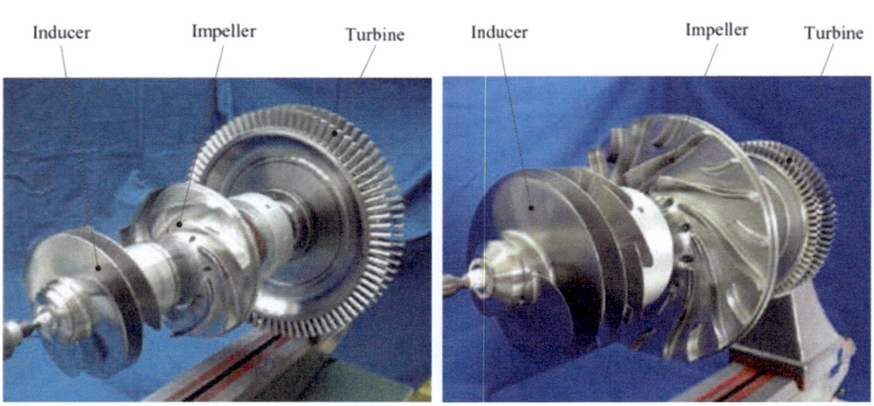


**Fig. 9.31** LE9 turbopumps, left for LOX and right for LH2 [31]. Credit: IHI

The Net Positive Suction Head (NPSH) is an important parameter of pumps and is closely related to cavitation. Simply put, if the inlet pressure drops below a specified threshold, the pump starts to suck more mass per second than supplied. This causes transient effects. As the fluid accelerates through the blades, the static pressure drops. This is a direct consequence of the Bernoulli equation, which relates the static pressure as measured by a co-moving observer with the fluid's kinetic energy density[34]:

$$p + \frac{1}{2}\rho v^2 = const. \tag{9.11}$$

This a normal effect and we have encountered it before, see Sect. 5.2. The pressure, however should not drop below the vapour pressure since this causes bubbles to form in the fluid—the fluid literally starts boiling. These bubbles have a lower pressure than the surrounding and when collapsing they form shock waves that can severely damage the pump's blades. This well-known phenomenon is called *cavitation* [33]. The lower the entry pressure, the less margin there is for the onset of cavitation—hence the importance of the pump inlet pressure. We will see below that tank pressurisation, too, plays a crucial role to prevent this failure case.

### Throttling

The capability to throttle a thermo-chemical engine is as relevant as for an electric thruster, Sect. 7.2.6. In case of a launcher the need is driven by the necessity to limit the vehicle's acceleration in the dense atmosphere when the mechanical load is largest, known as $Q_{max}$. The Space Transportation System (STS), for instance, throttled the engines down to 65% of the nominal thrust level. The other two cases, where deep throttle is necessary, are planetary landing and vertical landing of the reusable launcher stages. In the former case the engines must support in hovering above the ground until a hazardous-free landing spot is identified.

Since thrust is a function of the chamber pressure, Eq. 7.21, throttling is achieved by reducing this precise parameter. The chain of action is depicted in Fig. 9.32.

We have seen that the turbine's power output is a key variable within the chain. In an electro-mechanical design in which the turbine is driven by an electric motor, powering up and down is a relatively easy and smooth task. However, in a thermo-mechanical case its the exhaust gas, $m_{\mathrm{T}}$, which drives the turbine and which needs to be tuned. In all above cycles this is achieved by a single or several *thrust control valves* (TCVs). This valve controls the amount of propellant that is tapped-off the main stream, and less propellant is equivalent to a lower shaft power, Eq. 9.9. For LE-9, the TCV is depicted in Fig. 9.23. The effect is similar for a gas generator cycle. The valve reduces the amount of propellant flow $\dot{m}'_{\mathrm{T}}$ that is diverted to the pre-burner, thus controlling its energy output: $\dot{m}'_{\mathrm{T}} h_{34}$. The picture is different for a full flow

---

[34] The full equation is $p + \frac{1}{2}\rho v^2 + \rho g h$. We have omitted the last term as it is insignificant unless cases like hydroelectric power plants are subject of interest.

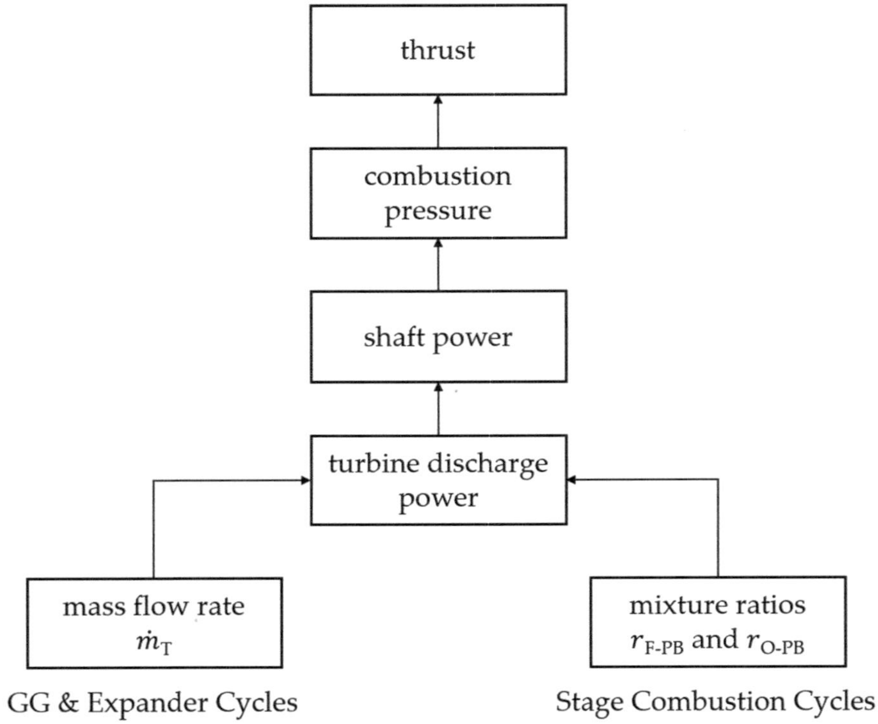


**Fig. 9.32** Principle chain of action to throttle a combustion rocket motor, with $r_{F,PB}$ $r_{O,PB}$ mixture ratios for the fuel rich and oxidiser rich pre-burner, respectively

staged combustion engine, since all propellant is routed through the two pre-burners. The two control parameters that are tuned are the mixture ratios of the respective pre-burners, $r_{F,PB}$ and/or $r_{O,PB}$. This is obviously considerably more complex.

In conclusion, we have discussed the static or steady-state picture of combustion engine cycles, with the exception of throttling. The transient cases of engine ramp-up and shut-down present the greatest challenge and require the greatest development effort. This is the point at which success or failure is determined.

## 9.6 Propellant Storage and Pressurisation Systems

A plethora of tanks exists with varying capacities from a few kilograms to hundreds of tons. These tanks must fulfil a multitude of system requirements, including axial, lateral and rotational acceleration. The latter is of particular importance in the case of spin-stabilised spacecraft. Material compatibility and long-duration storage can also become driving design requirements. For the sake of order and clarity, we will group these tanks into categories. The first category is that of storage pressure, which can be classified as high (180–300 bar), medium (20–30 bar) or low (3–5 bar). The general rule is: the higher the required inlet pressure the lower the propellant storage pressure.

#### Very High Storage Pressure

It is somewhat ironic that the propulsion system with the lowest thrust level stores its propellant at the highest pressure. This is because electric propulsion systems require the propellant in a *gaseous* state. Xenon for instance, has a very low density of 5.894 kg/m$^3$ under standard conditions, which is about 4.5 times denser than air. Fortunately, gas can be stored under high pressure with little expenses in terms of tank mass. We will discuss examples in Sect. 9.6.1. Pressurisation gas for pressure-regulated systems is also stored at extremely high pressure.

#### Medium Storage Pressure

Tanks in this pressure regime store *liquid* propellant as required for mono- and biprop systems. Satellite propellant tanks, for instance, operate in this regime but also ESA's contribution to NASA's lunar Orion spacecraft, the European Service Module, uses medium pressure storage tanks. The volume of these tanks varies between 10 and 2000 l. Most of them require a pressurant to maintain the tank pressure and ensure steady performance of the thruster—they are pressure-regulated. We will discuss the different types and designs in Sect. 9.6.2.

#### Low Storage Pressure

High thrust rocket motors require a huge amount of propellant at very high inlet pressure between 60 and 300 bar. Storing a liquid at this high pressure is prohibitively expensive in terms of mass. Instead, the propellant is stored at low pressure, 2–5 bar and the pressure rise is achieved via turbopumps, Sect. 9.5. Another feature of these tanks is that they become integral parts of the primary structure—hence the name *integral tanks*.[35] We will see in Sect. 9.6.4 that cryogenic propellant combinations offer a unique way to omit an additional pressurant thanks to auto-pressurisation.

All tank concepts are characterised by generic and specific performance parameters. The two most important generic parameters are :

- expulsion efficiency
- tank mass fraction

[35] Structural engineers distinguish between primary, secondary and tertiary structure. The primary structure carries the load. These are the main loads that the vehicle experiences and that endanger its overall structural integrity.

The *expulsion efficiency* quantifies the capability of the tank design to expel the loaded propellant. The factor refers to the issue that for several reasons not all loaded propellant can be expelled and a so called *static residual* remains within the tank.

$$\text{Expulsion Efficiency} = \eta_{\text{EE}} = \frac{\text{Tot. Expelled Propellant Volume}}{\text{Tot. Loaded Propellant Volume}}. \tag{9.12}$$

The tank mass fraction is the structural index of the tank, $\sigma_{\text{tank}}$, defined as the ratio of loaded propellant mass to tank dry mass, Eq. 9.13. It quantifies the mass efficiency of the tank design.

$$\text{Tank Mass Fraction} = \sigma_{\text{tank}} = \frac{\text{Tank Dry Mass}}{\text{Tot. Loaded Propellant Mass}}. \tag{9.13}$$

### *9.6.1 Gaseous Propellant—High Pressure Low Volume*

Electric propulsion systems are propelled with heavy inert gases such as xenon, krypton and argon. Fortunately, gas can be stored at very high pressure, thus saving volume and mass. xenon, for example, undergoes a phase change and becomes *supercritical*—a fluid state in which the distinction between gas and liquid is indistinguishable and the fluid exhibits both the properties of a gas and those of a liquid. For simplicity, we will refer in the following to the stored fluid simply as gas. Figure 9.33 shows the density profile of xenon as function of pressure for constant temperature.

The point beyond which a gas becomes supercritical is called the *critical point*. For xenon that corresponds to a temperature, $T_{\text{cr}}$ of 289.73 K (16.58 °C) and pressure, $P_{\text{cr}}$ of 58.42 bar.

Conversely, this means that the temperature of the xenon tank needs to stay above $T_{\text{cr}}$ to prevent the xenon from liquefying during the draining process thereby causing

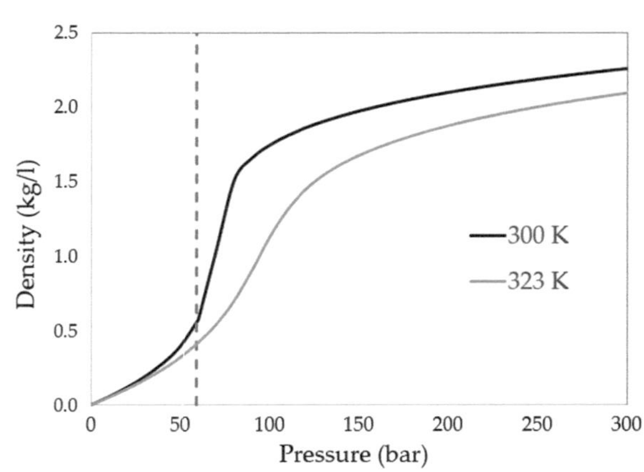


**Fig. 9.33** Xenon density as function of pressure for 300 and 323 K

a pressure drop. This is the task of the thermal control system. The other two commonly used noble gases, krypton ($T_{cr} = 209.46\,K$) and argon ($T_{cr} = 150.86\,K$) are less demanding, which reduces design and operation effort.

The high pressure gas is generally stored in *composite overwrapped pressure vessel* (COPVs) due to the mass advantage compared to all-metal tanks. They are often referred to as *vessel* to distinguish them from tanks that store a liquid under much lower pressure. They are composed of two elements: a metallic *liner* and a fibre composite *outer shell*. The main function of the liner is to form a tight barrier between the stored gas and the outer shell in order to prevent leakage. Different metals qualify as liner. Titanium alloys are commonly used as liner material due to their mass efficiency, that is 50% better compared to aluminium alloys and almost 100% to stainless steel [34]. The liner is then overwrapped with either a glass fibre or carbon fibre composite. The first is often referred to as type 2 and the second as type 3 COPV. From a mechanical perspective, the liner is considered non-structural, while the outer shell is considered structural. This means that the liner does not bear loads, while the outer shell must support the internal pressure load and externally induced loads, such as dynamic loads, bending, and torsion. To further reduce mass, advanced COPVs have a polymer-based liner, known as type 4. The next logical step is to design an extremely tight fibre composite for the tank and to fully omit the liner, this is the type 5 COPV.

This what the aforementioned private exploration company, Intuitive Machines, has utilised for their methalox propelled lunar lander, Nova-C. This makes the company the first to employ this technology in space[36]—however not as a pressure vessel but as a propellant tank.

Table 9.5 provides a list of tank specifications. A margin of 1.5 needs to be guaranteed between MEOP and design burst pressure. Figure 9.34 shows that in terms of storage efficiency xenon outperforms krypton and argon. The other two have significantly lower densities and thus require larger or more tanks with even higher pressure. However, price becomes an important selection criteria in large production volumes. This is the case for SpaceX's mega broadband constellations, Starlink. The designer omitted xenon first for the cheaper krypton and than moved on to the even cheaper argon [35]. Figure 9.35 shows two designs of xenon tanks. The left tank belongs to NASA's retired deep space probe, DAWN, to VESTA and Ceres. Note that the tank is almost completely covered with heaters. The right tank is a computer-aided design (CAD) model of the largest xenon tank currently in development with a storage capacity of 900 l of xenon. It is called L-XTA and is funded by the European Space Agency (ESA).

[36] https://x.com/Int_Machines/status/1688563888512528384.

**Table 9.5** Tank specifications for electric propulsion systems

| Designation | Volume (l) | $m_T$ (kg) | MEOP (bar) | $P_{bs}$ (bar) | $m_{prop}$ (kg) | $\sigma_{tank}$ (%) | Origin |
|---|---|---|---|---|---|---|---|
| 80386-101 | 32 | 6 | 172 | 259 | 66 | 9.7 | NG |
| 80412-1 | 50 | 7 | 150 | 225 | 102 | 6.8 | NG |
| 80458-201 | 54 | 12 | 198 | 285 | 111 | 11.0 | NG |
| S-XTA-60 | 60 | 12 | 187 | 281 | 123 | 9.5 | MTA |
| 80458-101 | 120 | 19 | 198 | 285 | 245 | 7.8 | NG |
| S-XTA-120 | 120 | 15 | 187 | 281 | 246 | 6.0 | MTA |
| 80458-1 | 133 | 20 | 198 | 285 | 272 | 7.5 | NG |
| LXTA-600 | 600 | 68 | 187 | 281 | 1230 | 5.5 | MTA |
| LXTA-900 | 900 | 85 | 187 | 281 | 1845 | 4.6 | MTA |

*MEOP* Maximum Expected Operating Pressure, $m_T$: tank mass
$\sigma_{tank}$ tank mass fraction, $P_{bs}$: burst pressure = 1.5 × MEOP
$m_{prop}$ max. stored propellant, NG: Northrop Grumman, MTA: MT Aerospace

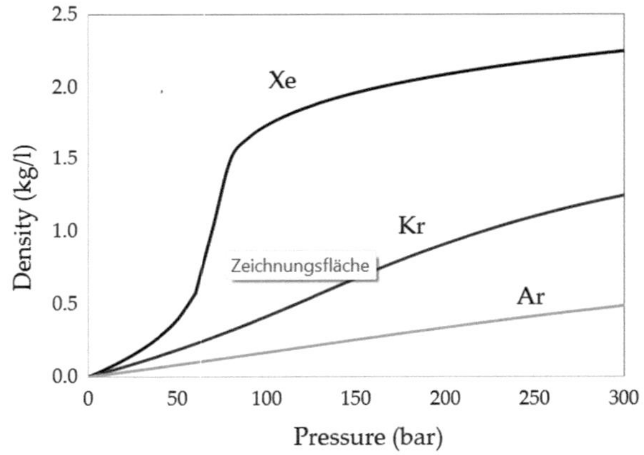


**Fig. 9.34** Xenon density as function of pressure for 300 and 323 K

## 9.6.2 *Liquid Propellant—Medium Pressure Medium Volume*

Medium pressure tanks in the range of 20–30 bar are used for low thrust propulsion systems as typically employed in satellites and space probes. There are two types of tanks in-use: the *positive expulsion tanks* (PET) and *surface tension tanks* (STT). The first are in general the standard solution for mono-propellant systems whereas the latter are used for bi-propellant systems. A pressurisation system is optional in the first case but the rule in the latter,[37] Fig. 9.36.

[37] The Globalstar tank (OST 31/0) discussed in Sect. 9.2 is such an exception.

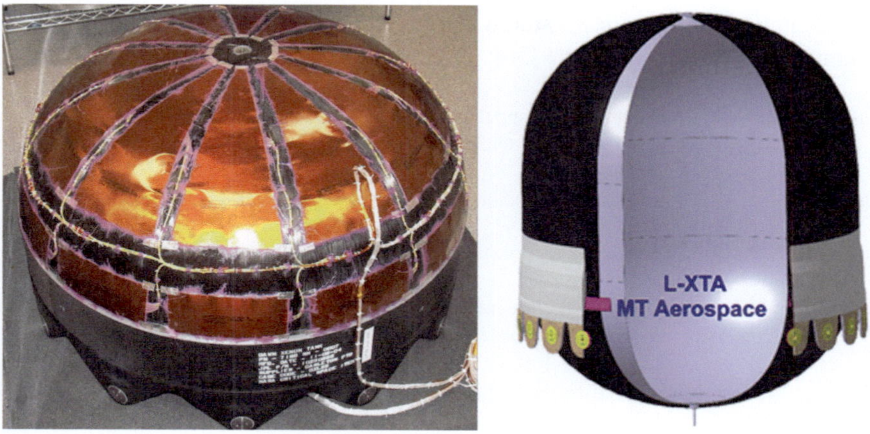


**Fig. 9.35** Left: Dawn's xenon tank covered with heater patches. Credit: NASA/JPL. Right: Largest xenon tank (900 l) in development. Credit: ESA/MT Aerospace

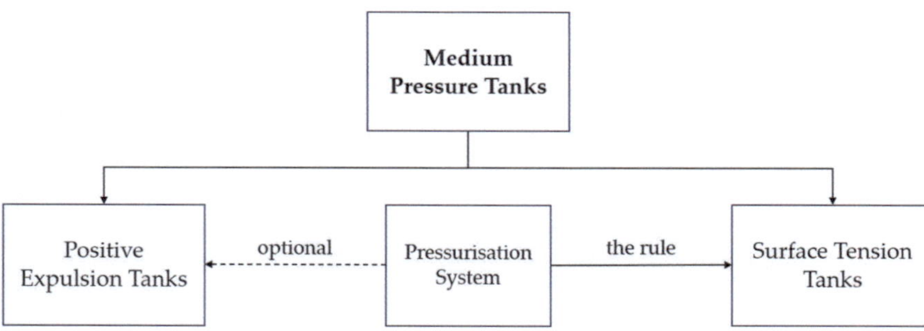


**Fig. 9.36** Relation pressurisation system and PET, STT

#### 9.6.2.1 Positive Expulsion Tanks—PET

*Positive expulsion tanks* (PET) have the striking advantage of carrying their own pressurant. They operate in blowdown for which no additional pressurant gas and hardware is required. PET make use of the principle that liquids are incompressible[38] and gases are not. Compressed gas is stored together with the liquid propellant in the *same* tank but in different compartments separated by a physical barrier. The space

[38] In fact everything is compressible, even a solid is compressible and it forms strange states like degenerated matter under extreme gravitational forces like in white dwarfs and neutron stars. Therefore, the term incompressible must be read in technical context. The density change of a liquid at 30 bar compared to 1 bar is negligible, therefore incompressible.

in which the pressurant gas is stored is referred to as the *ullage* volume. The main requirements for the barrier are [36]:

| | |
|---|---|
| *impermeability*: | to both, the gas and the propellant to minimise the formation of gas bubbles in the propellant tank |
| *flexibility*: | to allow the propellant to be expelled |
| *inert*: | chemically non-reactive to the propellant so as not to dissolve |
| *durability*: | capable to perform many expulsion cycles, if re-filled, relevant for re-usable vehicle as was the case for the space shuttle |

To achieve this physical barrier, there are two approaches: either by a *bladder* or by a *diaphragm*. Both expulsion device concepts are well proven and highly reliable. We will present the two separation concepts briefly in the following.

## Bladder Tank

Bladder tanks have a *balloon* made of an elastomeric membrane within the metallic tank shell, which in principle can either be filled by either the propellant or the pressurant [36]. In modern tanks, the propellant is stored in the balloon and surrounded by the pressurant as depicted in Fig. 9.37. The propellant leaves the elastomeric balloon via the perforated stand pipe in the centre. The design of the balloon must prevent *creasing* during the expulsion process as this would trap propellant in the manifold and reduce the amount of usable propellant. The engineering challenge of a tank

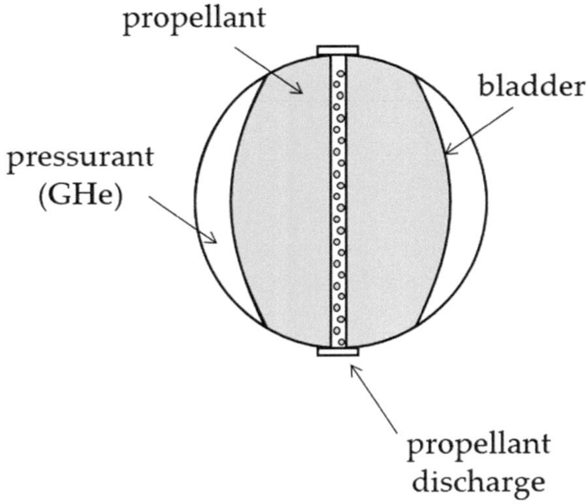


**Fig. 9.37** Schematic of a bladder tank with inward expulsion

designer is to develop a balloon that offers maximum propellant expulsion capacity without the danger of creasing and propellant trapping.

### Diaphragm Tank

As the name suggests, the physical barrier between propellant and gas is achieved via an *elastomeric diaphragm*. The diaphragm is made of a flexible *rubber* material and has undergone an evolutionary process in three steps leading to a thruster friendly *silica free* material called *SIFA*. It was developed under ESA guidance and became industry standard [37]. Figure 9.38 shows the schematic of pressurant loading and propellant depletion. The initial state (left image)is called the reversed position and the diaphragm is not allowed to touch the tank dome. Furthermore, excessive stretching of the diaphragm must be prevented to avoided damage of the material [37]. The loading procedure must take measures into account to prevent the excessive stretching and damage of the material [37].

Probably the most famous membrane tank was used for the space shuttle's RCS. A special feature of this tank was its durability. It was completely emptied and refilled after every flight and the material was therefore designed to withstand many operating cycles [37].

Both tank concepts, bladder and diaphragm, are predominantly used for monopropellant propulsion systems operated in blowdown mode, Sect. 9.2. Positive expulsion tanks have three striking advantages that make them the preferred solution for in-space propulsion systems:

1. PETs have limited *sloshing* characteristics. Sloshing refers to the uncontrolled movement of propellant within the tank which poses a challenge for spacecraft (SC) attitude control. In microgravity, the propellant floats in a rigid tank and is subject to inertia, adhesion forces and surface tension. When the SC changes its orientation, the propellant behaves like a pendulum that swings from side

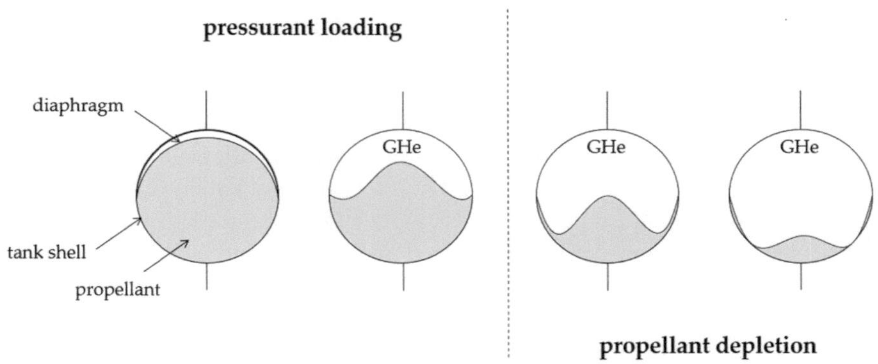


**Fig. 9.38** Pressurant loading and expulsion process of a diaphragm tank [38]

to side, making it difficult to control the attitude of the SC. This behaviour is largely mitigated in a PET for high frequency sloshing, due to frictional effects in the expulsion device. However, the behaviour varies during tank depletion, and characterisation is necessary to prevent *coupling* with the attitude control frequency.
2. Propellant expulsion capability is not dependent on the SC *orientation* in a PET. The propellant is always pressed against the outlet so that it can be drained any time without the need of a settlement manoeuvre.
3. The separation of the propellant and pressurant by a membrane ensures that thrusters are always supplied with gas-free propellant.

We will see in the next section that these three advantages are in fact not a given. A disadvantage of positive expulsion tanks is that they lack scalability. The tank mass fraction, $\sigma_{\text{tank}}$, becomes prohibitively large with larger propellant mass and other solutions must be found, as will be discussed in the next section.

#### 9.6.2.2 Surface Tension Tanks—STT

The surface tension tank (STT) does not suffer from the scalability issue of PETs. STTs are in general pressurised and thus require additional hardware, a pressurisation system as discussed in Sect. 9.3. The STT lacks a separation device (a membrane), which means that propellant and pressurant are not hindered from mixing in *micro-gravity* environment, $10^{-2}$ g – $10^{-4}$ g. The challenge for the tank designer is to achieve a gas-free propellant supply despite this mixing. The solution engineers came up with takes advantage of the dominant forces acting in this environment, *adhesion* forces and *surface tension*. Specific geometric devices have been invented to move the liquid propellant from the center or aft of the tank to the outlet [39]. These devices are known as *Liquid Acquisition Device* (LAD) or *Propellant Management Devices* (PMD).

Figure 9.39 shows the spatial location of the liquid propellant and gas within the tank with and without PMD. The left picture shows a stable stratification as it forms under noticeable gravity or acceleration: the propellant settles in the bottom of the tank due to its higher density compared to the gas (principle of Archimedes). This picture remains basically the same also during launch but changes completely under microgravity when no PMD is used. Due to surface tension and adhesion forces the propellant will be ‘attached’ to the walls pushing the gas into the centre. This is a non-critical state. However, the spacecraft’s attitude change will disturb the balance, and floating gas bubble could eventually migrate into the propellant outlet. If the thrusters are ignited gas enters then the feed lines and pushed into the combustion chamber causing combustion instabilities and in the worst case engine failure. The picture changes drastically with the use of an PMD (on the far right). The propellant wets the walls and the PMD and the gas is confined to the centre of the tank.

Several PMDs have been developed and successfully applied in space. They can be grouped in *communication* devices and *control* devices. Vanes, for instance, are

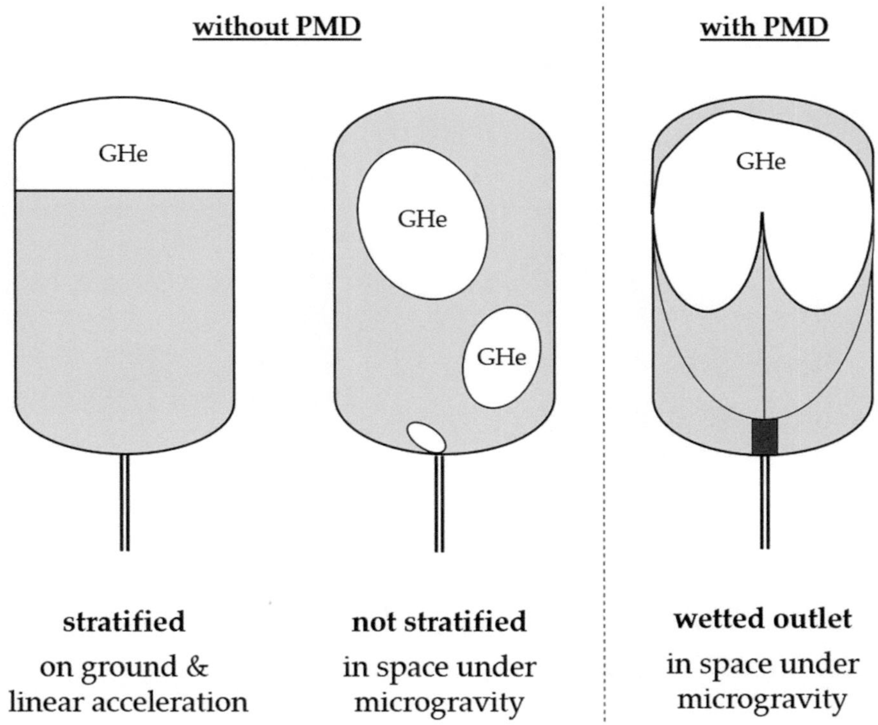


**Fig. 9.39** Spatial location of liquid propellant and gas within a tank without and with PMD

communication devices. Their task is to deliver propellant from within the tank to the outlet. The idea of the control device is to trap propellant in a small compartment close to the outlet, so that the outlet is always covered with propellant and enough propellant is trapped to ensure safe engine ignition. Sponges are typical control devices. Due to the tapered design of the panels, narrow spaces are created to which the propellant adheres. This design pushes any gas bubbles to the outer edge and guarantees that the propellant surface moves inward during depletion [41].

Figure 9.40 depicts the different PMD solutions. Most applications use a combination of communication and control devices to assure reliable operation under all circumstances. The PMDs mentioned so far have been used almost exclusively for storable propellants and not yet for *cryogenic propellants*. That is mainly due to the low surface tension of cryogenic propellants. Only *screen-channel PMD* have been successfully applied to the Centaur upper stages (LOX-LH2) [39]. A further complication when applying PMDs to cryogenic propellant is related to the very low boiling point, Table 8.8, which means that the vapour pressure is low. This could lead to an increased vapour generation in the vicinity of the outlet due to parasitic heat flow.

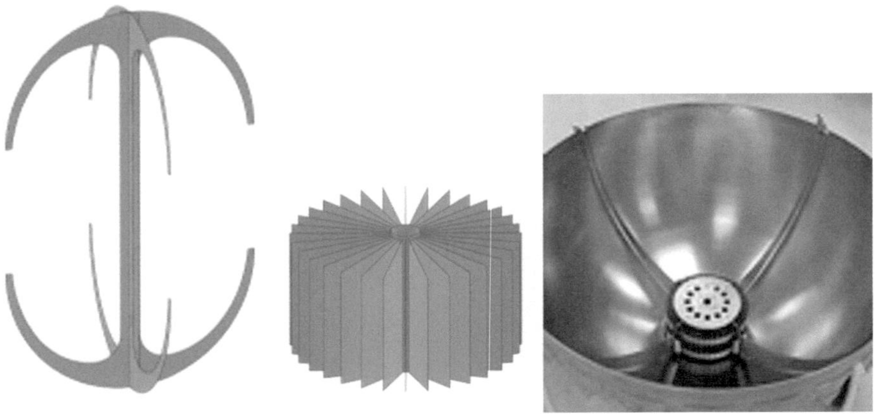

**Fig. 9.40** PMD solutions, vanes (left), sponge (middle), combination (right). Courtesy: PMD Technology

**Table 9.6** Specifications of surface tension tanks according to manufacturer's specifications

| Designation | Volume (l) | $m_T$ (kg) | MEOP (bar) | Size (mm) | | Form factor | Origin |
|---|---|---|---|---|---|---|---|
| 80353-1 | 59 | 3.9 | 27.6 | 483 ⌀ | | Sphere | NG |
| 80599 | 95 | 6.6 | 22.0 | 562 ⌀ | | Sphere | NG |
| 80420-1 | 151 | 8.1 | 21.9 | 562 ⌀ | 813 h | Cylinder | NG |
| OST 21-0 | 235 | 16.0 | 22.0 | 756 ⌀ | | Sphere | AG |
| OST 25-0 | 282 | 21.0 | 22.0 | 753 ⌀ | 643 h | Cylinder | AG |
| OST 25-3 | 331 | 22.7 | 19.5 | 753 ⌀ | 953 h | Cylinder | AG |
| 80340-1 | 504 | 17.1 | 19.0 | 990 ⌀ | | Sphere | NG |
| OST 23-0 | 769 | 31.7 | 17.5 | 1153 ⌀ | | Sphere | AG |
| OST 22-X | 1,108 | 49.0 | 19.5 | 1,146 ⌀ | 1,087 h | Cylinder | AG |
| OST 24-0 | 1,207 | 52.5 | 19.5 | 1,146 ⌀ | 1,456 h | Cylinder | AG |
| OST 26-X | 1,450 | 61.0 | 19.5 | 1,141 ⌀ | 1,683 h | Cylinder | AG |
| 80554-1 | 1,754 | 61.2 | 17.2 | 1,251 ⌀ | 1,927 h | Cylinder | NG |
| ESM | 2,100 | 110.0 | 25.0 | 1,145 ⌀ | 2,651 h | Cylinder | AG |
| 80576-2 | 2,310 | 63.9 | 17.9 | 1,158 ⌀ | 2,515 h | Cylinder | NG |

*NG* Northrop Grumman, *AG* ArianeGroup, *h* height

Table 9.6 shows a list of surface tension tanks and their system relevant specifications. These tanks are commonly made of titanium alloys (Ti6AlV st) which offers high specific tensile strength leading to a low tank mass fraction of about 5%. Anther important tank parameter is the MEOP (maximum expected operating pressure). This quantity is highly relevant for the PMS as it defines the starting point of the pressure drop, see Fig. 9.4. The higher this figure the higher the margin for the subsequent pressure drop.

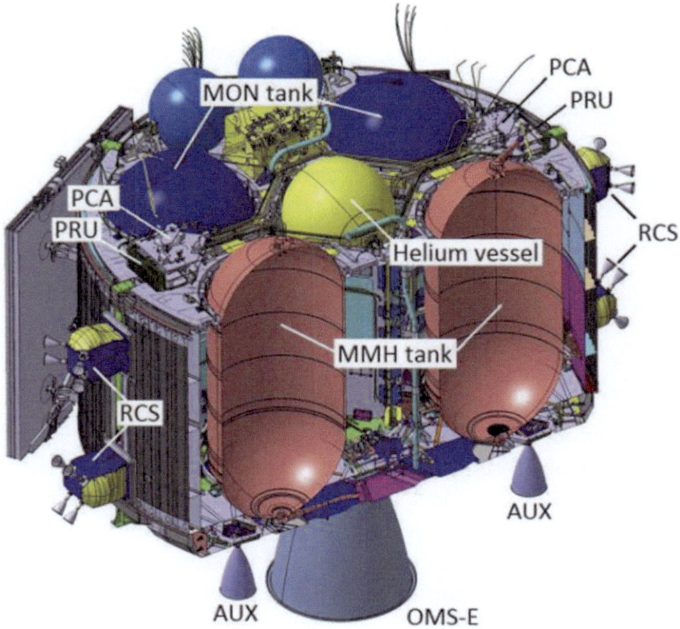


**Fig. 9.41** Cross section of ESA's European Service Module (ESM) for NASA's Orion Spacecraft [42]. Credit: European Space Agency—ESA/NASA/Airbus Defence & Space

Figure 9.41 shows the four ESM tanks mounted within the support structure of the European Service Module. The pressurisation tank is visible in the middle of the propellant tank arrangement. In the next section, we will see that large tanks no longer require a supporting structure but become part of the load-bearing structure themselves.

The PRU (pressure regulation unit) is an electronic interface that regulates the pressure while the PDE (Propulsion Drive Electronic) actuates all flow control equipment as well as the thruster and is connected with the Orion Data Network (ODN) to the Vehicle Management Computer (VMC) of the Crew Module [40].

### 9.6.3 Tank Pressurisation

In contrast to blowdown systems, pressure-fed systems can maintain their operational pressure throughout the entire operation, thus enabling constant specific impulse and thrust. This simplifies mission planning considerably. In order to maintain the operational pressure in the propellant tank, it is necessary to pressurise the tank with external gas. This gas is usually an inert gas like helium or nitrogen to prevent chemical reactions with the propellant. For illustrative reason, we adopt helium as pressurant gas throughout the book.[39] The helium gas is stored at high pressure, typically up to 310 bar with current developments aiming for 400 bar.

In the following, we will look at the typical task of determining the required amount of pressurising gas. The following variables are known:

$P_V$ pressure inside helium vessel,
$P_T$ operational pressure inside propellant tank,
$V_T$ volume of propellant tank.

We are looking for the following unknown quantities for a given storage pressure:

$m_{He,V}$ initially stored helium mass inside the vessel,
$V_V$ volume of helium vessel.

A heuristic approach will be employed to provide a preliminary estimate, which then needs to be refined through the use of a detailed hydrodynamic simulation and taking the compressibility of the selected pressurant gas into account. We start by calculating the required vessel volume. To do so, we take advantage of the fact that the process in which helium is released from its high pressure vessel into the propellant tank can be considered as *isothermal*. It means that the process of helium expansion is sufficiently slow that the accompanied temperature drop is compensated by heat entering both tanks—this is the opposite of adiabatic. The thermodynamic equation that describes an isothermal process with our denotation is:

$$\bar{P}_V V_V = P_T V_T. \tag{9.14}$$

It is important to note that the pressure inside the helium vessel changes during the expansion process. Starting with $P_{V,0} = 180$ bar, it drops down to about $P_{V,f} = 50$ bar at end of life. The mean vessel pressure is then $\bar{P}_V = 115$ bar. Rearranging Eq. 9.14 yields

$$V_V = \frac{P_T V_T}{\bar{P}_V}. \tag{9.15}$$

With the vessel volume, $V_V$, it is possible to calculate the *residual* helium mass left in the helium vessel:

$$m_{He,V} = \rho_{P_{V,f}} V_V. \tag{9.16}$$

The helium that has entered the comprised tanks to displace the propellant can be calculated in the same way using the helium density at tank operating pressure:

$$m_{He,T} = \rho_{P_T} V_T, \tag{9.17}$$

[39] Note that nitrogen cannot be used in combination with hydrogen as the two would react to form ammonia. Furthermore, a nitrogen based pressurisation systems is considerably heavier than one with helium.

$\rho_{P_{V,f}}$ helium density in storage vessel at the end of life,
$\rho_{P_T}$ helium density in propellant tank at end of life.

Finally it shall be considered that this is not the only helium mass to be considered in the budget. After fueling the propellant tank on ground, helium is added to pressurise the propellant tank so that it reaches launch pressure level, which is typically 12 bar. This helium mass is hidden in $m_{\text{He,T}}$ and did not come from the helium vessel. We call it initial helium ullage mass:

$$m_{\text{He,u}} = \rho_{\text{He},P_L} V_u, \tag{9.18}$$

$m_{\text{He,u}}$ helium for initial pressurisation in ullage volume,
$\rho_{\text{He},P_L}$ helium density at launch pressure,
$V_u$ ullage volume, typically 5–10% of tank volume.

### 9.6.4 Liquid Propellant—Low Pressure High Volume

A general rule in the structural design of rockets and large spacecraft is that if the most dominant mass contributor is the propellant, it is better to have no tanks. The rationale behind this approach is that storing the propellant in separate tanks, which are subsequently attached to the *load-bearing* structure, is not a mass-efficient solution.[40] Instead, an *integral tank* design is preferable. This means that the cylindrical part of the tank becomes the load-bearing structure of the rocket. It is, therefore, inevitable to discuss structural design aspects of large propellant dominated stages as is the case for launcher with a focus on upper stages.

Figure 9.42 shows the partially exploded view of two Ariane 6 configurations including payload. The payload has usually a smaller diameter and is safely placed beneath the fairing. Stages are connected through cylindrical shells called interstages. If stacked stages have different diameter, these interstages have the form of truncated cones. The payload adapter between the upper stage and the payload is in the majority of cases a truncated cone as can be seen on the left side.[41] The most important load to consider for a first mechanical design concept is the *load path*. The structure that carries the load path is referred to as *primary structure*—for a launcher this is the outer *shell*. In other words, it is the skin that bears the load. Two distinct shell structure designs have evolved in aerospace engineering, *stiffened and unstiffened*. Both will be discussed in the following. In order to stay within the scope of this book, we will

[40] The first stage of Saturn 1B was the sole modern stage to employ this structural approach. It clustered tanks within the stage shell structure.

[41] Like its predecessor Ariane 5, Ariane 6 has the capability to launch two payloads in one go separated by a tall launch adapter, called 'sylda'. The payload that is deployed first is placed on top of the sylda, the second payload is placed within the sylda.

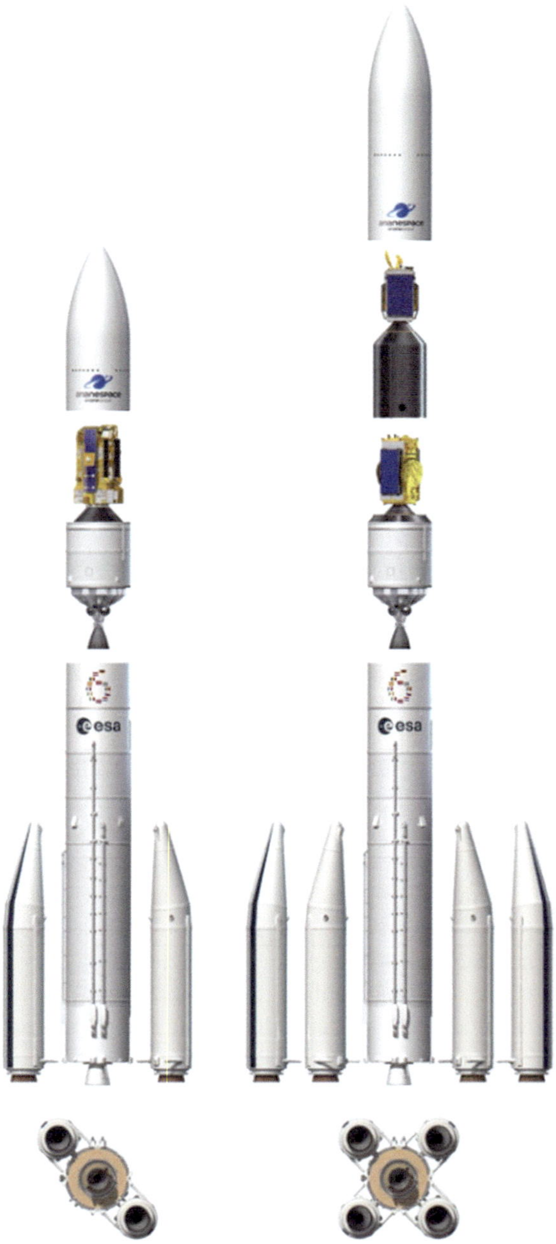

**Fig. 9.42** Expanded view of Ariane 6 launcher. Credit: European Space Agency—ESA/D. Ducros

adopt a phenomenological[42] approach rather than a physical-mathematical one. This means we present the design solutions first before discussing the main load cases during launch.

### 9.6.4.1 Stiffened and Unstiffened Metal Shells

Aerospace vehicles, like airplanes and launchers, are in principle pressurised vessels—hence the name space vessel. The former contains passengers and/or freight, the latter large amounts of propellant. A truss structure, though highly mass efficient and successfully applied in many engineering disciplines, is not suitable for aerospace vehicles. The volume should be maximized and remain undisturbed Shell designs have evolved that are capable to carry load *and* to offer undisturbed volumes. In addition, they are characterised by a very high load-bearing capacity and a comparatively low structural mass. We can distinguish among two types:

1. unstiffened isotropic shells,
   - steel balloons,
   - welded barrels, i.e. cylinder segments,
   - monocoque, either as composite or 3D printed,
2. stiffened shells, by,
   - mechanically fastened stringer, called skin stiffener,
   - integrally machined stringers of iso- and orthogrid shapes, i.e. stiffened skin.

Figure 9.43 depicts the above listed design options. Sandwich structure, though important for payloads, will be omitted in the context of this discussion.

#### Steel Balloons

The concept of steel balloons[43] implies a very *thin-walled* cylindrical design that takes advantage of internal pressure to stabilse the structure. Without this pressure, the construction is structurally not stable and would collapse. The thickness of the shell varies between an incredible 0.36 and 0.94 mm [44]—hence the name balloon. This concept was applied to the first US ICBM (intercontinental ballistic missile), the SM-65, of which a human rated derivative, the Atlas LV-3B served as launch vehicle in the Mercury programme. The balloon is built from corrosion-resistant stainless steel in extra full-hard (EFH) grade. To improve formability, shells that require non-cylindrical shapes, such as truncated cones for transitions or domes, were made from

---

[42] The phenomenological method of investigation is the most important method for gaining knowledge in many scientific disciplines, such as zoology and anthropology. But also engineers, who get their hands on technology of a competitor, for instance, have to apply this method.

[43] This concept was first developed by Karel Bossart (1975†), a Belgian-born engineer who emigrated to the United States [43].

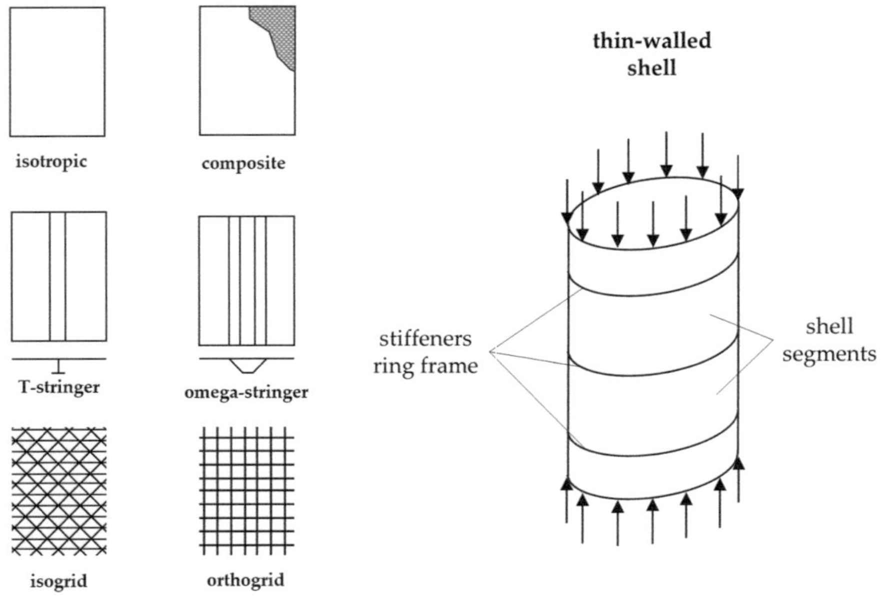


**Fig. 9.43** Design options to for thin-walled shell structures

less hard grades (e.g. 1/2–3/4). While shells are effective at handling typical launch loads, they are incapable to cope with local loads caused by external hardware such as strap-on boosters, electrical cables, or piping. Therefore, machined reinforcing rings must be placed at these locations to provide additional support [43]. The balloon concept for entire rockets was also used for the commercial Atlas family of launch vehicles and was retained until the Atlas III. This design approach is still in use today, notably on the Centaur upper stage, which has been in service for 60 years. The latest version flies on ULA's Vulcan rocket. Figure 9.44 shows an early version of the Centaur from 1962 during assembly. The tank is later insulated with a spray foam to reduce boil-off losses during ground operation and flight.

### Carbon Fibre Structure

Carbon fibre reinforced plastic (CFRP) is being discussed in the launcher industry as a replacement of steel or aluminum based alloys due to their many times higher tensile strength-to-weight ratio. The European Space Agency, ESA, initiated the ICARUS project (Innovative Carbon Ariane Upper Stage) to demonstrate that carbon fibre can handle the cryogenic temperatures of the propellant. However, since SpaceX's maxim of subordinating design decisions to cost savings rather than performance improvements has become widely recognised, other space agencies and aerospace companies have also turned to traditional and cost-effective steel for launch vehi-

**Fig. 9.44** Centaur upper stage with steel balloon shell in assembly hall. Credit: NASA

cles.[44] The aforementioned lunar lander mission, NovaC, of Intuitive Machines being a refreshing exception. It remains to see, where CFRP elements justify the related cost and add value to the product or enable missions otherwise not feasible.

### Stiffened Shells

Stiffened shells may be considered as the opposite extreme to balloon tanks. They are stable under their own weight, which facilitates ground handling enormously, especially for large structures, like a first stages. Integrally machined stringers of iso- and orthogrid shapes are the most common stiffening methods, Fig. 9.43.

Milled from an aluminium block consisting of a high-quality aluminium-lithium alloy, 95% of the material is lost, which is one of several reasons why this form of manufacturing is very costly. Isogrid shapes are more popular among structural engineers than orthogrid shapes, primarily due to their suitability for numerical modelling. This is because they can be modelled as isotropic shells. However, both find application in aerospace structures. Atlas V and Delta IV core stages were stiffened by isogrids while Space Shuttle's external tank and Vulcan's booster have orthogrid shapes. Figure 9.45 depicts the inner part of the LH2 tank of the Space Transportation System. The eight segment barrel design separated by ring frame stiffeners is clearly visible. The grid free space between the barrels is required to weld the seg-

[44] In fact, prior to the so called NewSpace era, performance gain at any cost was top priority for all stakeholder. This was very much in line with the business model of a state-funded ecosystem in which the survival of the company depended on government contracts.

**Fig. 9.45** Inner view STS main tank [45]. Credit: NASA

ments together and is sometimes referred to as *weld land*. Besides launcher, also capsules like Dragon, Orion and Starliner utilise gridded structures. An exception is SpaceX's Falcon 9 rocket that utilises stringers and frames to stiffen the structure. This design choice is fully in-line with the company's philosophy, since externally attached stringers are cheaper than milled iso- and orthogrids. The Delta Cryogenic Second Stage (DCSS) is an example for an upper stage with isogrids for both cylindrical tank segment, LH2 and LOX tank. Sub-variants of the DCSS power NASA's lunar exploration rocket Space Launch System (SLS), namely the Interim Cryogenic Propulsion Stage (ICPS) on Block 1 and the Exploration upper Stage (EUS) on Block 1b, Fig. 9.46.

ICPS is very similar to its parent, DCSS, both have a diameter of 5.1 m and are propelled by a single RL10-C2 engine with an extendable nozzle. This is required to increase the nozzles expansion ratio $\epsilon$, see Sect. 7.1.1. However, the EUS is larger with a diameter of 8.4 m and is powered by four RL10-C3 engines with fixed nozzles, whereby only three would be needed to perform the trans-lunar injection manoeuvre the additional engine is operated in hot redundancy. A peculiarity of the DCSS upper stage family is its suspended oxidiser tank attached by a composite inter-tank truss assembly to the LH2 tank cylinder. It is noteworthy that the main load path flows through the LH2 cylindrical tank section, the aft adapter and then into the inter-stage (not shown here), thus bypassing the truss assembly. Therefore, as long as the upper stage is attached to the main stage, neither the truss structure nor the cylindrical LOX tank segment experience *axial compression*. Instead, they are stretched as they are pulled in flight direction. This changes only when the upper stage engine(s) ignite and begin pushing and thus compressing all elements above.

Another example is the upper stage ESC-A (Etage Supérieur Cryotechnique de type A), of the now retired Ariane 5 ECA. Due to height limitations of the integration hall, the engineers had to adapt. This resulted in a compact design in which the LOX tank intrudes the LH2 tank. The lower tank dome is known as *reveres bulkhead* (Fig. 9.47).

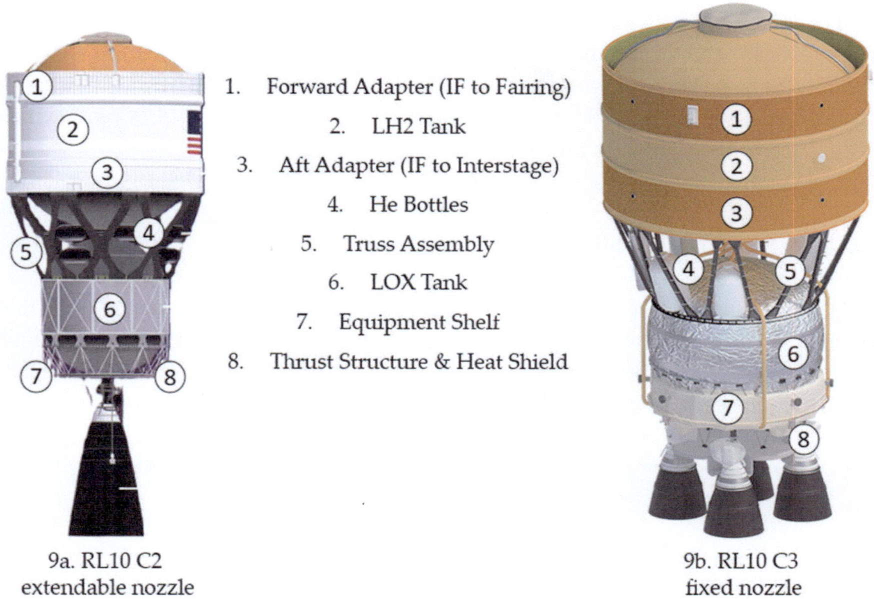


**Fig. 9.46** Left: Interim Cryogenic Propulsion Stage (ICPS). Right: Exploration Upper Stage (EUS). Credit: NASA

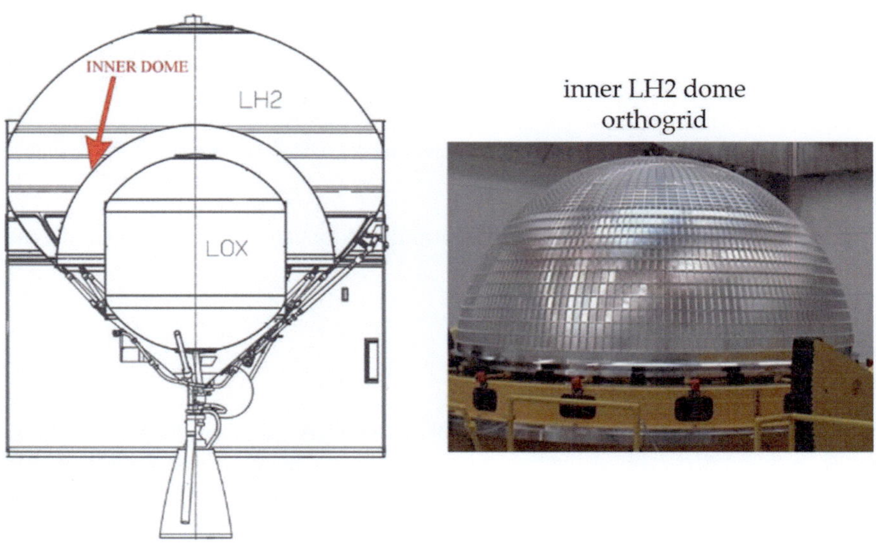


**Fig. 9.47** Reversed LH2 inner dome of the upper stage ESC-A of Ariane 5 ECA [46]. *Credit* European Space Agency—ESA/DLR

The load caused by the LH2 inertia during launch makes this design particularly vulnerable to structural failure, which is the reason why it was decided to stiffen it by an orthogrid. This compromise resulted in an unprecedented unfavourable upper stage dry mass index of 30%. Another notable design feature is the LOX tank, which is attached to the LH2 tank by a truss structure, akin to DCSS.

#### 9.6.4.2 Structural Loads

A spacecraft experiences a range of different types of *thermo-mechanical loads* that need to be accounted for. It starts with ground-handling and transportation and ends with in-flight separation. The most severe mechanical loads occur during launch[45] but also in-space loads need to be considered. Prominent loads cases are:

*static & quasi-static loads*: caused by axial compression due to axial acceleration/internal pressure/ aerodynamic pressure

*thermal loads*: caused by aerodynamic heating and cryogenic propellant temperature,

*dynamic loads*: caused by bending and torsion due to lateral acceleration/propellant sloshing/engine ignition and shutdown/interaction with propellant feed system causing the notorious pogo oscillations[46] / gust loads due to wind shear,

*vibration loads*: a result of engines that either transmit high frequency loads through the structure up to the payload or caused by acoustic loads that reflect back from the ground during lift-off/aerodynamic noise due to launcher interaction with the atmospheric.

The corresponding requirements are formulated for calculable material and structural properties:

- stiffness,
- strength,
- fracture and fatigue.

When it comes to the structural integrity of thin-walled shell designs, it is the failure mode *buckling* that gives structural engineers a headache. Buckling is an instability mode that originates from axial loading or external pressure. It is primarily a *shape failure* and not material failure. Meaning that it can occur even when stresses in the structure are below the failure threshold of the material. The phenomenon of buckling is complex and its occurrence does not necessarily lead to catastrophic failure. The structure rather deforms and takes on a new deformed-state, which could still be

[45] In-space manufacturing and assembly is a hot topic research field.

able to carry the load and continue the mission. However, further loading leads to unpredictable deformations, causing a complete loss of the carrying capability of the primary load path.

The theoretical limit for the *critical buckling load* in case of an ideal and perfectly isotropic cylinder was derived a century ago by the pioneering work of Timoshenko in 1910 [48], Lorenz in 1911 [47] and Southwell in 1913 [49]:

$$F_{\mathrm{id}} = \frac{2\pi E t^2}{\sqrt{3(1-\nu^2)}}, \tag{9.19}$$

$F_{\mathrm{id}}$ ideal buckling limit,
$E$ modulus of elasticity,
$t$ wall thickness,
$\nu$ Poisson's Ratio, transverse contraction strain to longitudinal extension strain.

Note that this equation is independent from the cylinder's radius $R$ and length $L$, which is counter intuitive. In particular, the slenderness of the cylinder, defined as the ratio of radius to thickness, $\lambda = R/t$, is expected to play a decisive role for the critical limit. Shortly after, it was discovered that the ideal limit is far from reality. The drop in allowable load is called *knockdown* and the related factor *knockdown factor* (KDF):

$$\rho = \frac{F_{\mathrm{id}}}{F_{\mathrm{exp}}}. \tag{9.20}$$

The Dutch mechanical engineer Warner Tjardus Koiter proved in his PhD thesis in 1945 that the theoretical explanation for this discrepancy is due to unavoidable *geometric imperfections* [51]. However, due to the complexity of the impacting factors, the only feasible way to deduce a quantitative approach with sufficient margin is through a large number of experiments. These costly campaigns started in the 30s and intensified in the 50s and 60s yielding several empirical formulae for the knockdown factor:

$$\rho = 1 - 0.902 \cdot \left(1 - e^{\frac{1}{16}\sqrt{\frac{R}{t}}}\right) \qquad \text{NASA SP-8007 [59]},$$

$$\rho = 3.87 \cdot \left(\frac{R}{t}^{-\frac{1}{2}}\right) \qquad \text{Russian guideline [60]}.$$

Similar experiments have been established for composite cylinders but in contrast to isotropic shells no clear dependency on the body's slenderness could be observed [50].

Finally, it shall be noted that the synopsis provided here is far from complete, but merely dips into this intriguing matter.

### 9.6.4.3 Pressurization of Large Systems

Pressurisation and control of ullage pressure is arguably a more challenging task for large systems than for small ones. The main reason is the pace of tank depletion during engine operation. While small engines burn propellant in the order of hundred grams per second larger engines require dozens to hundreds of kg per second[47] The pressure drop within the propellant tank is therefore much faster and the pressurisation system has to keep up with the pressure loss. This is particularly critical in the phase shortly after ignition. The reason is the very low ullage volume of 5%. A small ullage volume means that the proportional changes are large and the pressurisation system needs to keep up with the ullage increase. A discrepancy between the gas supply by the pressurisation system and the propellant surge by the turbopump could lead to underpressurisation thereby placing the turbopump at risk, if the Net Positive Suction Head (NPSH) is not maintained due to cavitation.

Another critical phase is re-ignition after a long coasting phase. The propellant and the ullage gas—consisting of propellant vapour and pressurant—mix and condensation of the vapour could occur and lead to a pressure drop. The case is even more severe for autogenous pressurisation systems. These events need to be prevented, especially before engine ignition.

#### Autogenous Pressurisation

Cryogenic propellants have an inherent advantage over storable propellants in that they allow for autogenous pressurisation. Due to the very low evaporation temperature, propellant vapour can be tapped-off from the feed system after it passes the pump and engine cooling. The tapped vapour is then fed back into the tank via a pressure regulator as pressurant. The advantages are obvious, this 'hack' allows to save mass and cost of a separate tank pressurisation system with He and associated work in terms of integration and testing. The principle was already used for the Saturn rocket, the external tank of STS and the upper stage Centaur [53]. Almost all full cryogenic launch systems use autogenous pressurisation, with the ICPS and EUS on SLS being an exception, Fig. 9.46. A popular example is SpaceX's Starship. It uses autogenous pressurisation of both the LOX and the LCH4 tank.

[47] The 400 N MON-MMH apogee engines of ArianeGroup requires 135 g/s while the upper stage engine, Vinci, of the same company consumes a total of 39.6 kg per second to achieve 180 kN (34 kg/s LOX and 5.6 kg/s LH2).

### 9.6.5 *Propellant Mass Estimation and Management*

Precise propellant budgeting becomes an important task for satellites and space probes when mission duration is in the order of years. Spacecraft operators, however, need to accurately know how much propellant is left and how much $\Delta v$ capacity is available. This is particularly important when it comes to mission extension and has even legal consequences, if national space debris mitigation rules are not met, e.g. de-orbiting. NASA's Polar mission is such a negative example [3]. The knowledge on how much propellant is consumed during a manoeuvre is strongly related to the question of how much $\Delta v$ has been performed and can be still achieved. Hence, the question how much propellant is left, is of high relevance for manoeuvre planning. Proper budgeting of in-orbit propellant mass is therefore of vital interest to the operator. Three methods have proven themselves:

- propellant bookkeeping[48]
- pressure-volume-temperature (PVT) analysis,
- thermal gauging.

These methods do not exclude each other and are often used in combination. We will present their working principle, their strengths and their weaknesses briefly in the following.

#### Propellant Bookkeeping

The task may sound trivial but the technical problem is not. The acceleration level can be measured by the inertial measurement unit but computing the thrust magnitude requires knowledge about the mass, which varies along the manoeuvre. Thrust and specific impulse can be used to determine the mass flow rate. Knowledge about the $I_{sp}$ can be gained through a predetermined functional relation with the engine inlet pressure by the thruster manufacturer. The inlet pressure can be measured on-orbit, albeit with uncertainty as all in-flight measurements. All these parameters are fed into an on-board model to compute the consumed propellant and to subtract it from the previously computed value. Looking closer at the matter reveals the complexity. The specific impulse is an illustrative example. The thruster's $I_{sp}$ is a function of the mixture ratio, which is expressed in the mass flow ratio of fuel and oxidiser. The mass flow, however, is a function of tank pressure. A concern in that respect is that for a bi-prop system one of the two tanks could heat up more than the other, either because of environmental reasons or internal dissipation of equipment or payload.[49] This

[48] In analogy to the famous navigation method that relies solely on a map and compass without the help of the sun and stars, the bookkeeping method is also known as dead reckoning.

[49] Thermal control engineers need to be aware of this thermal imbalance and verify the design in that respect. To ensure that the respective analyses are not overlooked, it has become good practice in systems engineering to establish standard requirement specifications, which must be actively reviewed to judge their applicability.

causes both its temperature and pressure to rise. If the heated tank contains MON, the pressure increase is exacerbated by the fact that MON, due to its lower vapour pressure, is more volatile than MMH. The consequence will be an off-nominal mass flow of the heated tank (higher than nominal) leading to an oxidiser rich mixture ratio (higher than nominal),[50] that alters the propellant estimate and forecast. This is one example of the level of detail the model has to have for precise propellant bookkeeping.

In addition, there are several uncertainties to consider but the most dangerous is the *systematic error*. This can be either rooted in the underlying model in terms of a bias or due to in-flight calibration, which in turn may be affected by accumulated errors [3]. In general, the absolute error in the bookkeeping method accumulates. It is small at the outset of the mission but increases towards the end, thereby leading to an underestimation of the residual propellant. It is therefore advisable to combine this method with the PVT analysis.

#### Pressure-Volume-Temperature Analysis—PVT

The second method to estimate the residual propellant is the *Pressure-Volume-Temperature (PVT) Analysis*. This method goes back to the source of propellant, the tank. Since tank volume and original filling ratio are known, residing propellant mass can be calculated with the knowledge of tank pressure and temperature. The method takes an indirect approach and estimates the ullage volume occupied by the pressurant from the ideal gas state equation:

$$pV = nRT. \tag{9.21}$$

With $n$ is the number of moles, a fluid specific value, and $R$ is the universal gas constant 8314 J/K/mol. For this, it requires the temperature $T$ and partial pressure of the pressurant $P$. The method takes advantage of the simple fact that the drained propellant volume $V$ is replaced by the pressurant. The assumption of thermal equilibrium is of significant importance to the algorithm and is well-founded, provided that sufficient time elapses between the burns.

The PVT method is not limited to liquid propellant. It has been successfully applied to the xenon tank of the SMART-1 probe [54].

#### Thermal Gauging

The third method uses the *thermal inertia* of the propellant inside the tank. For this, the tank is regularly heated to prevent it from freezing, due to inevitable heat losses to the cold space. The process is cyclic; if the measured temperature falls

[50] If the deviation is too large a dangerous situation for the thruster could occur. The inlet conditions could be outside the allowed thruster operation box and it is said that the thruster runs hot.

below a predefined lower threshold, the attached heaters are commanded ‘on’ by the power subsystem, Fig. 9.35 left. The heaters are deactivated, when the temperature exceeds an upper threshold. The thermal response after each heating cycle, in terms of absolute heat-up and cool-down time, is a key indicator of the remaining thermal capacity of the propellant. A smaller amount of propellant results in a lower thermal inertia, which in turn leads to a greater temperature sensitivity for the same heater power. This is the reason why this method delivers higher accuracy towards end of life (EOL) than begin of life (BOL)—contrary to the previously mentioned methods. The method leads to a thermal model with four parameters, starting from an energy balance equation [3]:

$$m_f c_v \Delta T = Q_{\text{heater}} + Q_{\text{Sun}} + Q_{\text{Earth}} - Q_{\text{loss}}, \tag{9.22}$$

$m_f$ mass of the fluid,
$c_v$ heat capacity at constant volume.

It is solved via a Monte-Carlo method, making the thermal gauging method the most demanding in terms of on-board computational effort. The in-flight thermal response experiments are conducted with dedicated heaters and not with the standard TCS heaters. These dedicated heaters are capable to boost the temperature temporarily by 10 °C to cause a stronger thermal response thereby improving the signal quality. Furthermore, the tanks need to be thermally well insulated to prevent disturbances to the signal profile caused by heat leakage or heat gain from external sources.

In conclusion, it shall be highlighted that signal processing of temperature and pressure sensors by the data handling system plays a decisive role in propellant mass estimation. Analogue signals need to be digitized which causes conversion losses. This is also the case for flow control equipments. Hence, accurate on-ground characterisation and in-flight of sensors and equipments is essential to ensure proper modelling and nominal operation.

## References

1. Stuhlberger, J. (n.d.). In-space propulsion. In *AIAA International Air and Space Symposium and Exposition: The Next 100 Years*. https://doi.org/10.2514/6.2003-2588
2. United Launch Alliance. (2010). Atlas V Launch Services User’s Guide.
3. Owens, B., Cosgrove, D., Sholl, M., & Bester, M. (2010). On-orbit propellant estimation, management, and conditioning for the THEMIS spacecraft constellation. In *SpaceOps 2010 Conference*. https://doi.org/10.2514/6.2010-2329
4. Sholl, M., Leeds, M., & Holbrook, J. (2007). THEMIS reaction control system: From I and T through early mission operations. https://doi.org/10.2514/6.2007-5579
5. Cattani, B. M., Soares, T., Morales Serrano, S., Briot, D., Val Serra, S., & Thiry, N. (2021) The impact of space debris mitigation requirements on mission design choices: An overview from ESA Clean Space. https://conference.sdo.esoc.esa.int/proceedings/sdc8/paper/131

6. Bombardieri, C., Traudt, T., & Manfletti, C. (2015). Experimental study of water hammer pressure surge in spacecraft feedlines.
7. Anflo, K. P. (2013). Hot-firing tests using a low temperature derivative of LMP-103S. In *5TH European Conference for Aeronautics and Space Sciences (EUCASS)*
8. Sawai, S., Sato, E., Uesugi, K. T., Furukawa, K., Mishima, H., Morishima, K., Nonaka, Y., & Kondo, M. (n.d.). Development of ceramic based 500 N class bipropell... In *56th International Astronautical Congress of the International Astronautical Federation, the International Academy of Astronautics, and the International Institute of Space Law*. https://doi.org/10.2514/6.IAC-05-C4.3.01
9. Fujii, G., Kawatsu, K., Daimon, Y., Michigami, K., Kushiki, K., Sawai, S., & Sakai, S. (2021). Verification of the transient behavior of the SLIM propulsion system (Smart lander for investigating moon). https://doi.org/10.2514/6.2021-3582
10. Devey, R., Gaydon, S., Statham, G., & Rose, L. D. (n.d.). Development and pre-flight testing of the Olympus propulsion subsystem. In *25th Joint Propulsion Conference*. https://doi.org/10.2514/6.1989-2511
11. Tam, W., Dommer, K., Wiley, S., Mosher, L., & Persons, D. (2002). Design and manufacture of the messenger propellant tank assembly. https://doi.org/10.2514/6.2002-4139
12. Mayer, A., Wieling, W., Watts, A., Poucet, M., Waugh, I., Macfarlane, J., & Valencia, F. (2018). European Fuel Blend Development for In-Space Propulsion.
13. Soulas, G., Domonkos, M., & Patterson, M. (2003). Performance evaluation of the NEXT ion engine. https://doi.org/10.2514/6.2003-5278
14. De Wijper. (2017). Development of the next-generation fluid management technologies for LEO, GEO & Navigation Satellites. In *EPIC Workshop 25*th, Bradford
15. Snyder, J., Randolph, T., Hofer, R., & Goebel, D. (n.d.). Simplified ion thruster Xenon feed system For NASA science missions.
16. Ganapathi, G., & Engelbrecht, C. (2002). Performance of the Xenon feed system on deep space one. *Journal of Spacecraft and Rockets, 37*. https://doi.org/10.2514/2.3573
17. Harmann, H. P., & Rothaus, S. (2013). $\mu$FCU - A miniaturized flow control unit for Xenon. https://api.semanticscholar.org/CorpusID:114400827
18. Koppel, C., Marchandise, F., & Estublier, D. (2004). Robust pressure regulation system for the SMART-1 electric propulsion sub-system. https://doi.org/10.2514/6.2004-3977
19. Bushway, E., & Perini, R. (2000). Proportional Flow Control Valve (PFCV) for electric propulsion systems. https://doi.org/10.2514/6.2000-3745
20. Data Sheet of Moog, Inc. (2020). Specification of Electric Propusion Proportional Flow Control Valve (PFCV)
21. Pehrson, D. M. (2007). Continuing development of the Proportional Control Valve (PFCV) for electric propulsion systems. IN *IEPC 2007–346, 30th International Electric Propulsion Conference*, Florence, Italy, Sept. 17–20.
22. Kuninaka, H., Nishiyama, K., Funaki, I., Yamada, T., Shimizu, Y., & Kawaguchi, J. (2007). Powered flight of electron cyclotron resonance ion engines on Hayabusa explorer. *Journal of Propulsion and Power - J PROPUL POWER, 23*, 544–551. https://doi.org/10.2514/1.25434
23. Ganapathi, G., & Engelbrecht, C. (2002). Performance of the Xenon feed system on deep space one. *Journal of Spacecraft and Rockets, 37*. https://doi.org/10.2514/2.3573
24. Lenguito, G., Neff, K., Barbarits, J., Snyder, J. S., & Chaplin, V. (2019) Versatile Xenon flow controller for extended hall-effect thruster power range. In *IEPC-2019-303. 36th International Electric Propulsion Conference University of Vienna • Vienna, Austria*
25. Edwards, C., & Mundy, D. (2000). Development and testing of the GOCE proportional Xenon feed system. In R. A. Harris (Ed.), Spacecraft Propulsion (Vol. 465, p. 571).
26. Koppel, C., Marchandise, F., Estublier, D., & Jolivet, L. (2004). The smart-1 electric propulsion subsystem in flight experience. https://doi.org/10.2514/6.2004-3435
27. Starling, D. A. (1996). *Propellant feed control for ion engine*. Monterey, California: Naval Postgraduate.
28. Dornberger, W. (1995). Peenemünde: die Geschichte der V-Waffen. Ullstein.

29. New Mexico Museum of Space History. (2024). https://www.nmspacemuseum.org/inductee/alexei-m-isayev/
30. Ha, D., Roh, T. S., Huh, H., & Lee, H. J. (2022). Development trend of liquid hydrogen-fueled rocket engines (Part 1: Performance and operation). *International Journal of Aeronautical and Space Sciences*
31. Azuma, N., Kojima, M., Kobayashi, T., Okita, K., Motomura, T., & Niiyama, K. (2019). The development status of LE-9 engine turbopumps for H3 launch vehicle. https://doi.org/10.2514/6.2019-4430
32. Piunti, M., Shytani, A., Persico, F., & Pasetti, S. (2022). Preliminary design of a Raptor-like engine [Phdthesis]. https://doi.org/10.13140/RG.2.2.18529.17760
33. Schobeiri., M. T. (2012). *Turbomachinery flow physics and dynamic performance*, 2nd edn. Springer
34. Le Floc'h, C., Aufaure, J., Salome, R., & Tschofen, J. (2000). New high pressure tank for Xenon storage. In R. A. Harris (Ed.), Spacecraft Propulsion (Vol. 465, p. 543).
35. Aria Alamalhodaei, A. (2023). SpaceX's acquisition of Swarm is paying off with new Starlink thrusters. TechCrunch
36. Porter, R. N., H., & Stanford. (2005). No . 32-899 propellant expulsion in unmanned spacecraft. https://api.semanticscholar.org/CorpusID:202548129
37. Tam, W., Kawahara, G., Wlodarczyk, K., Gutierrez, H., & Kirk, D. (2018). Review of ATK diaphragm tanks-an update.
38. Tao, R., Yang, Q.-S., Liu, X., He, X.-Q., & Liew, K.-M. (2017). Investigation of intelligent reversible diaphragm using shape memory polymers. *Journal of Intelligent Material Systems and Structures, 29*, 1045389X1774097. https://doi.org/10.1177/1045389X17740977
39. Hartwig, J. (2016). A detailed historical review of propellant management devices for low gravity propellant acquisition. https://api.semanticscholar.org/CorpusID:138574725
40. Berthe, P., Over, A., Picardo, M., & Byers, A. (2017). Orion European Service Module (ESM) development, integration and qualification status. https://doi.org/10.2514/6.2017-5144
41. Jaekle, D. E., Jr. (1991). Propellant management device conceptual design and analysis: Vanes. AIAA-91-2172.
42. Cellarier, D. (2020). Modelling avionics interfaces and generating ICDs for the propulsion subsystem of the MPCV-ESM. ESA Presentation
43. Dawson, V.P., & Bowles, M. D. (2004). Taming liquid hydrogen: The centaur upper stage rocket 1958–2002. The NASA History Series. NASA SP-2004-4230
44. Atlas-Centaur AC-12 flight performance for Surveyor 3. (1968). NASA Technical Reports Server (NTRS). https://ntrs.nasa.gov/citations/19690000964.
45. Wu, K., Cerro, J., & Wallace, M. (2015). Hardware-based non-optimum factors for launch vehicle structural design. *SAWE International Journal of Weight Engineering, 74*, 22.
46. ECSS Buckling Handbook (2010). ECSS-E-HB-32-24A, http://www.ecss.nl/wp-content/uploads/standards/ecss-h/ECSS-HB-32-24A24March2010.pdf
47. Lorenz, R. (1911). Die nicht-achsensymmetrische Knickung dünnwandiger Hohlzylinder. *Physikalische Zeitschrift, 12*(7), 241.
48. Timoshenko, S. (1910). Einige Stabilitätsprobleme der Elastizitätstheorie. *Zeitschrift für Mathematik und Physik, 58*(4), 337.
49. Southwell, R. V. (1913). On the general theory of elastic stability. *Philosophical Transactions of the Royal Society, 213*(A 501), 187.
50. Wagner, R. (2019). Robust Design of Buckling Critical Thin-Walled Shell Structures [Phdthesis]. https://doi.org/10.13140/RG.2.2.15095.16801
51. Koiter, W. T. (1945). Over de stabiliteit van het elastisch evenwicht. Amsterdam: Thesis Delft.
52. Almroth, B. O., Burns, A. B., & Pittner, E. V. (1970). Design criteria for axially loaded cylindrical shells. *Journal of Spacecraft and Rockets, 7*(6), 714–720.
53. Pressurization of cryogens - A review of current technology and its applicability to low-gravity conditions. (1992). In *Joint Propulsion Conference and Exhibit.*

54. Koppel, C. R., Rathsman, P., Borrajo-Pelaez, R., Demairé, A., & Estublier, D. (2017). Enhancement of the PVT method for xenon gauging of electric propulsion by the use of an EMA method. In *IEPC 2007-346, 35th International Electric Propulsion Conference, Atlanta, Georgia*, Oct. 8–12, 2017
55. Fuhrmann, T., Mewes, B., Kroupa, G., Lindblad, K., Dorsa, A., Matthijssen, R., & Underhill, K. (2019). FLPP ETID: TRL6 reached for enabling technologies for future European upper stage engines. https://doi.org/10.13009/EUCASS2019-208
56. Pavon, S., Tregubow, V., Peukert, M., & Lescouzéres, R. (2012). Engineering validation model for the Exomars bipropellant propulsion subsystem. https://api.semanticscholar.org/CorpusID:198186179
57. Henry, C. (2002). An introduction to the design of the cassini spacecraft. *Space Science Reviews, 104*, 129–153. https://doi.org/10.1023/A:1023696808894
58. Gamgami, F., Rohrbeck, M., Wollenhaupt, B., & Andersson, B. (2016). A dual-mode propulsion system with arcjets as an alternative propulsion system for the SGEO platform, 3125075, Space Propulsion Conference 2016. Rom, Italien.
59. Wiley, S., Dommer, K., & Mosher, L. E. (2003). Design and development of the MESSENGER propulsion system. https://api.semanticscholar.org/CorpusID:113052626
60. Tam, W., Dommer, K., Wiley, S., Mosher, L., & Persons, D. (2002). Design and manufacture of the messenger propellant tank assembly. https://doi.org/10.2514/6.2002-4139
61. Intuitive Machines ODAR - Version 1.0, Attachment D, Intuitive Machines-1 Orbital Debris Assessment Report (ODAR) IM-1-ODAR-1.0 https://apps.fcc.gov/els/GetAtt.html?id=265886&x=
62. NSSDCA ID: IM-1-NOVA https://nssdc.gsfc.nasa.gov/nmc/spacecraft/display.action?id=IM-1-NOVA
63. Gamgami, F., Sippel, M., & Dumont, E. (2010). Statistical analysis and classification of rocket motor efficiency, thrust to mass ratio and structural index [Techreport]. DLR. https://elib.dlr.de/75246/

# Part IV
# Reference Mission to Ceres

# Chapter 10
# Preliminary Mission and System Design for C-One

**Abstract** This chapter brings the basic technology discussed previously to life by applying it to an unmanned mission to Ceres, *C-One*. The mission aims to transport a 30 t payload from low Earth orbit to the surface of Ceres and land it smoothly, with a mission duration of no more than five years, preferably three. This ambitious mission requires the use of almost all the technologies discussed earlier, serving as an instructive case study of how these technologies integrate to design a comprehensive system.

## 10.1 The Art of Feasibility Studies

The architecture of a challenging mission, like our mission to Ceres, depends on numerous interrelated parameters. The art of *systems engineering* is to breakdown the complexity by establishing clear mission phases and system boundaries in order to define clear subsystems with well defined interfaces. The result is an evolving *baseline* and the starting point is referred to as *Design Departure Point* (DDP).[1] The DDP focuses on the main design driver with the objective to establish a preliminary mission and design baseline. Preliminary, for two reasons: (a) the set of requirements is incomplete and immature (b) the analysis effort is moderate. The term *baseline* refers to a unique and agreed configuration.

The main objective of early design phases is to assess mission feasibility with the aim to consolidate the requirements. Mission requirements are frequently revised until a feasible mission is achieved, or the evolution of the requirement reveals that the mission is no longer aligned with the intended objectives and constraints, e.g. too low payload mass, too long transfer time, too risky new technology developments, too small margins or simply too costly—hence the name *feasibility study*. This phase

[1] The DDP is a useful concept motivated by the mathematical discipline numerical analysis that solves complex problems in an iterative manner. It usually starts with a simplification of the original mathematical or physical problem with the aim of obtaining an easily solvable problem—at the best with an analytical solution. This solution is then fed into a sophisticated solver which maps the full problem for improvement.



F. Gamgami, *Space Propulsion and Spaceship Design*, Springer Praxis Books,
https://doi.org/10.1007/978-3-031-71336-1_10

is characterise by trade-offs on architecture, system and subsystem level. It is not uncommon that the study leads to more than a single concept with different emphasis: (a) high performance high risk and (b) low performance and low risk.

Depending on the complexity of the mission—and invested effort—a first design iteration lasts between some months to a year. This phase is characterised by a high amount of architectural trade-offs and moderate analyses effort. With the consolidation of the requirements base and a mature baseline architecture, the level of detail increases and is referred to as the detailed design phase. This phase is characterised by larger and more complex models.

Another aspect of establishing a preliminary baseline design is identifying exemplary equipment. The goal is not to make a final decision but to demonstrate that suitable equipment exists and to use it for sizing purposes. Therefore, we will look for existing equipment (so-called off-the-shelf) and adopt it accordingly.[2] A quote of an experienced systems engineer shall be repeated at this point, who asked in a heated discussion:

'Do you want a feasible or a consistent design?'

While it may seem contradictory, this statement underscores a common challenge encountered by projects in their early phases: it is not advisable to seek a feasible *and* consistent design during the conception of a mission and system. The effort would be firstly prohibitively high and secondly not target-oriented, which in early design phases is *feasibility*. Furthermore, the effort would be largely in vain since requirements are not consolidated yet and many lower level requirements not even formulated. The focus, therefore, should always be on technical and programmatic feasibility instead of full consistency. Experience has proven that inconsistencies resolve or become even obsolete in the course of the program due to changes in the requirement base and alternative design choices and are not hidden show-stoppers.[3]

The previous chapters have provided us with the knowledge of the physics and technology of space propulsion for the construction of a DDP for C-One.

Preliminary Baseline Mission

A transfer from LEO directly to Ceres with only electrical or only chemical propulsion is not advisable. The main reason is that we start the journey deep in Earth's gravity well. One of the defining characteristics of electric propulsion, however, is its

[2] Space projects frequently face the question of whether to: (a) use existing equipment, known as off-the-shelf; (b) modify existing equipment to meet new requirements, known as delta-development; or (c) develop completely new hardware and software. The latter has the advantage of yielding a fully compliant component or subsystem but bears the considerable risk of cost overruns and schedule delays. On the other hand, using off-the-shelf equipment is limited and leads to sub-optimal performance. The dawn of additive manufacturing promises to alleviate this dilemma in hardware manufacturing. Several commercial space companies, such as Relativity Space, are currently working on this breakthrough, making an on-demand rocket motor a real possibility.

[3] It goes without saying that major inconsistencies are not acceptable, like incompatible propellant combinations or insufficient tank size. They can, however, be clearly identified as they directly challenge feasibility.

low thrust. While the escape from Earth's gravity field is not impossible as demonstrated by Deep Space 1 and Smart-1, it is a challenging endeavour for a solar electric propulsion system, particularly if a large payload mass is transported [1].

To understand the matter better and to highlight the solution, it is helpful to draw an analogy to launchers. The mission profile of a launcher can be divided into two phases: firstly, to gain height quickly.[4] and secondly, to gain orbital speed. In simple terms, the first task is performed by the booster stage and the second phase by the upper stage. Booster stages are characterised by high thrust and lower specific impulse. Upper stages by lower thrust and higher specific impulse. Architects of deep space missions drew inspiration from this task-sharing approach, as both face a similar issue: the need to overcome the deep gravity well before cruising to the target orbit. Therefore, a combined architecture in which the escape burn is performed chemically, i.e. impulsive acceleration, and the remaining cruise phase with a high efficient solar electric propulsion system, i.e. continuous acceleration, is the smarter approach. It has been adopted by several missions Hayabusa, Dawn, Psyche and is suggested in several studies.

Based on the qualitative discussion above, it is reasonable to establish a mission architecture consisting of four phases, in line with systems engineering principles:

*Phase 0*: Launch and In-Orbit Assembly
*Phase 1*: Earth Escape
*Phase 2*: Powered Interplanetary Flight to Ceres
*Phase 3*: Smooth Landing

The design process must reflect the resulting requirements imposed by the phases on each other. This is similar to the boundary value problem in mathematics, where the solution sought must fulfil the conditions at the start and the end points. Applied to C-One, this means that launchers must be available with the capability to launch the respective spaceship element into the desired orbit. It is, however, not feasible to discuss a launch and deployment scenarios without knowledge of the system architecture and the design of the individual stages. This is true for each phase: it requires information about the succeeding phase. This dilemma is solved by an iterative approach starting backwards, in this case with phase 3. We will refer to this approach in the following as *Backward Engineering*.[5] Figure 10.1 depicts the recommended iterative design approach for C-One.

The design process starts therefore with phase 3, the final element of the C-One spaceship. This will lead to a payload mass requirement for phase 2, which in turn will lead to mass requirement for phase 1. The elements of each phase must be

[4] A launcher must satisfy two competing requirements. On the one hand, it must overcome gravity rapidly to minimise gravity losses: on the other hand, it must not be too fast to limit the aerothermodynamic loads.

[5] This method is rooted in numerical mathematics where it is called *Backward Integration*. It is a method in which differential equations are solved by working backward from a final known condition. In programming it is referred to as 'Backward Recursion', but it should not be confused with 'Reverse Engineering'. The latter involves analysing a finished product to understand its design specifications and manufacturing processes.

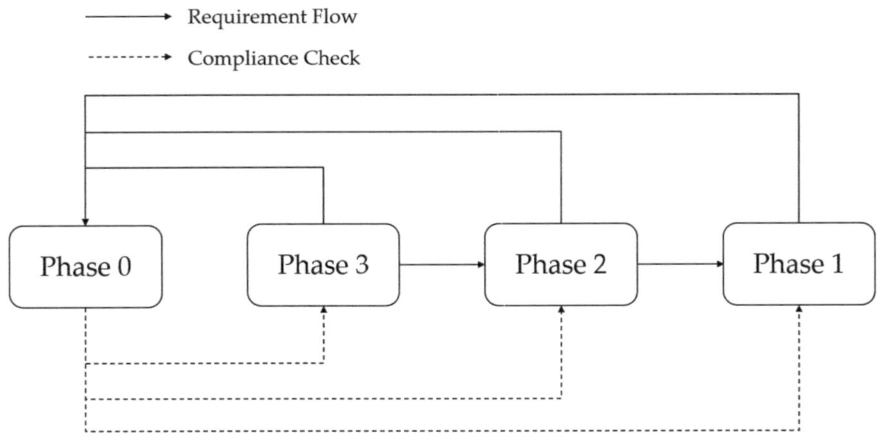


**Fig. 10.1** Design flow logic of C-One mission phases

transported from ground to space, where they are assembled and prepared for the journey—represented by phase 0. The dashed lines represent compliance checks that need to be performed in the course of the design process in order to guarantee launch feasibility. The two decisive parameters are mass and volume—as in any logistic business. The launch and in-space assembly strategy will be established at the end of the chapter.

To prepare for the launcher compliance check, this chapter begins with a non-exhaustive overview of suitable space transportation systems.

Launch Systems

Any space mission starts with the launch. It is therefore important to establish an overview of the capability and availability of current and envisioned space transportation systems.[6] The modules and stages of C-One will require a medium—supper heavy lift launch vehicle (MHLLV and SHLLV). While the mighty Saturn 5 rocket has been considered for decades to be a peculiarity in the evolution of launch vehicles, due to its breathtaking size and initial thrust,[7] we are about to enter an era in which super-heavy launch vehicles will become the new normal. SpaceX has shown that well-established market principles such as economy of scale apply to the space business as well, and that the space market can be multiplied once the launch price drops sufficiently. In the mid 30 s there will be Chinese and US super heavy lift launcher

[6] The term 'Space Transportation Systems' (STS), is the umbrella term for all systems capable to deliver payload into space. It includes classical launchers, space planes, horizontal take off systems and even exotic systems like kinetic catapult systems. which is development by a commercial US company 'SpinLaunch'. We restrict ourselves to classical vertical take of systems, i.e. launcher. The fact that this term is also the official programme name of the NASA space shuttle system is sometimes a source of confusion.

[7] 34,500 kN sea level thrust, 110 m height and 10 m diameter.

**Table 10.1** Payload capacity of the operational and soon-to-be operational launcher

| Launcher | LEO | Remark | Type |
|---|---|---|---|
| Vulcan (VC6)[a] | 27 t | In operation | Medium |
| Long March 5B[a] | 25 t | In operation | Medium |
| Angara 5[b] | 25 t | In operation | Medium |
| New Glenn[c] | 45 t | In development | Heavy |
| Falcon 9 Heavy[a] | 63 t | In operation | Heavy |
| SLS Block 1[d] | 95 t | In operation | Super heavy |
| Starship[e] | 100 t | 2024 operational | Super heavy |
| Long March 10[a] | 70 t | 2026 operational | Super heavy |
| Long March 9[a] | 150 t | 2033 operational | Super heavy |

[a]ISS like orbit 400 km height 51.6°/41.47°, [b]200 km, 63.1°
[c]200 km 51.6°, [d]Orbit not specified, [e]Minimum performance

vehicles in duty, which predestines these countries to collaborate for space mega projects like C-One. Table 10.1 lists the payload capability of candidate launchers.

Compatibility with a launcher refers to several domains, of which the mechanical compatibility is the most limiting one. First, the launcher must be capable of lifting the payload into the desired orbit—this is expressed by a wet or launch mass requirement. If this requirement is met, the payload must also fit into the launcher's fairing—this is the volume requirement. Furthermore, the payload needs to comply with the launcher's static and dynamic load environment as well as limitations in centre of mass.[8] The loads and the volume constraints by the launcher impact the mechanical layout and the structural design of the payload and consequently the mass budget. In order to limit the scope of the current feasibility study to the present chapter, we will focus on mass and volume checks for the single spaceship elements. We will see that mass compliance of the payload does not necessarily mean volume compliance, and that it is often the latter that drives the selection process.

Figure 10.2 depicts the *usable payload volume* beneath the launcher's fairing for a selection of launchers of each launcher category; medium (Vulcan), heavy (New Glenn) and super-heavy (Starship). The figure shows that despite Starship's superior launch capability, its payload bay is 3 m lower than that of Vulcan Centaur. The higher launch capacity is reflected in the larger usable diameter of 8 m, surpassing Vulcan Centaur by 3.2 m and New Glenn by 1.65 m.

A last remark shall be made on the *payload launch adapter* (PLA). The PLA provides a load path between the payload and eventually the launcher's primary structure, which is the outer shell. Launch service providers offer a number of off-the-shelf (OTS) products but customised designs can be made on request to mission unique needs. Each service provider offers a wide range of PLAs. They have in general a conical form thus transitioning from the wide launcher diameter to the narrow payload diameter. It shall be noted that some PLAs might protrude into the

[8] In very rare cases, the launcher is adapted to the payload but this is the absolute exception and not the rule.

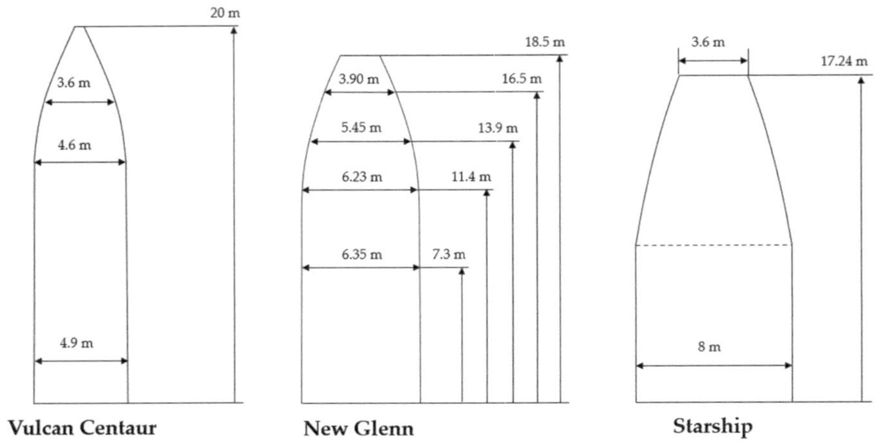


**Fig. 10.2** Static fairing envelope of Vulcan [2], NewGlenn [3] and Starship [4]

usable payload volume, indicated in Fig. 10.2 and could thus reduce this volume. Some launchers offer a PLA that coincides in diameter with the launcher itself. In this case, the payload's outer shell might not be protected by the fairing but instead exposed to the atmosphere like the SpaceX's dragon capsule on top of Falcon 9.

The *deployment and assembly strategy* will be defined once the physical dimensions of the spaceship elements are known. We, therefore, revisit Phase 0 at the end of this chapter when establishing the synopsis of the baseline mission architecture.

## 10.2 Phase 3 Landing on Ceres

Upon arrival at Ceres, the task will be to land the payload smoothly on the surface at a predefined location. This task will be accomplished by the *Ceres Landing Propulsion Module* (CLPM). Figure 10.3 depicts a schematic of the landing stack, consisting of the payload in form of a cylinder and the CLPM separated by an interstage. Together they form the Ceres Landing Module (CLM). The introduction of an interstage facilitates the design and assembly process of the two modules, allowing them to be designed and tested first separately in two different locations before being assembled into a single stack.

In order to establish a preliminary design of the CLPM's propulsion system, it is necessary to specify first the tasks it has to fulfill and derive the related technical requirements from then.[9]

Since the Moon and Ceres are both planetary objects without an atmosphere, *landing strategies* proven on the Moon could be adopted for Ceres. A proven strategy is to subdivide the landing sequence into three phases each with specific require-

[9] In a formal requirements engineering these tasks are functional requirements.

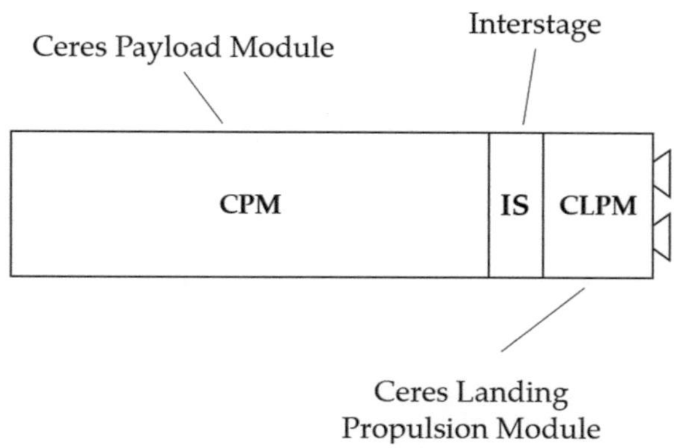


**Fig. 10.3** Schematic of Ceres Landing Module (CLM)

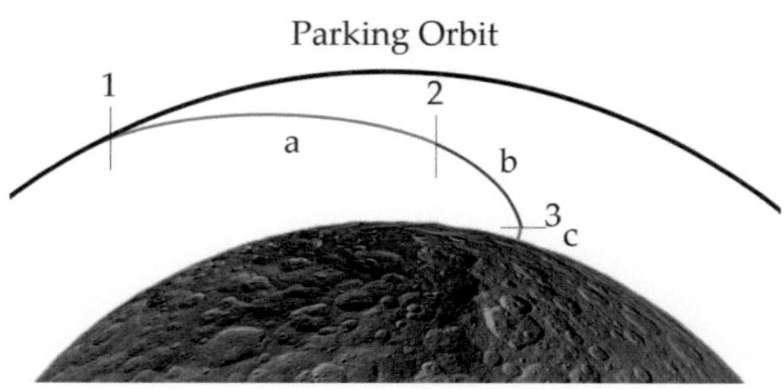


**Fig. 10.4** Descent sequence of phase 3

ments upon the propulsion system. The phases are equivalent to *three trajectory arcs*, depicted in Fig. 10.4. Starting from a safe parking orbit a Hohamnn transfer is initiated by a single retrograde burn (1) to bring the stack into a ballistic *intermediate orbit* (a). This orbit has a low pericentre but is still above the surface to avoid a collision course. At a certain height the engines are activated again (2) and the second phase, the *powered descent phase* commences (b), during which the flight path angle is continuously increased and the horizontal velocity reduced close to zero. The vertical velocity of the stack is then slowly reduced until the stack hovers over the surface (3) identifying the final landing spot. The last phase consists of a *controlled free fall* characterised by a vertical descent (c). The remaining impact energy is absorbed by mechanisms within the landing legs, for instance via a crushable structure.

Velocity Demand

The escape velocity of Ceres is 516 m/s and it can be regarded as an upper limit for the required $\Delta v$ to land on Ceres. To derive a more accurate value, a more detailed look at the landing sequence is needed. This will also help us to gain insight into the technology required. Drawing an analogy to successful lunar landing strategies is helpful.

For the Moon, the aforementioned parking orbit is usually at an altitude of 100 km $H_{po}$ or slightly above and the resulting pericentre height $H_{pc}$ of the intermediate orbit is between 30 km as for the Indian lander Chandrayaan-3 and 15 km as for the crewed Apollo Module [5]. The spacecraft Dawn orbited Vesta safely at an altitude of

470 km [6]. Vesta, also a dwarf planet, has an irregular shape causing a gravitational field with stronger higher harmonics[10] that perturb the orbit, Sect. 2.2, Fig. 2.5. Similar considerations led to the 100 km orbit for the Moon. Although more spherical than Vesta, the Moon has an inhomogeneous internal structure, which also causes strong higher harmonics in the gravitational field, with the result that orbits below 100 km are unstable and last only a few days without correction manoeuvres [7]. Since we know neither the topology of Ceres' gravitational field nor its internal structure, we assume that what is good enough for Vesta should also be good enough for Ceres. Adopting a circular parking orbit of 500 km altitude above the surface, CLM's orbital velocity will be then 255 m/s. During the landing process, this velocity needs to be dissipated completely. The velocity demand is actually higher than this value due to hovering over the ground in search of a safe spot to land—known as hazard detection and avoidance.[11] In case of Apollo, the final velocity demand was 23% higher than the initial circular velocity of the parking orbit [5]. Based on this similarity we should assume a velocity demand of 300 m/s for the complete descent phase. It is good engineering practice to consider margins. Given the qualitative approach and the uncertainties associated with Ceres' gravitational field 25% are necessary, which leads to 375 m/s.

The required propellant mass and the resulting total mass of the CLPM can be calculated from the set of equations derived in Sect. 6.1.2:

$$
\begin{aligned}
m_{\mathrm{PL}} &= m_{\mathrm{CLM}} \cdot \left[1 - (1+\sigma)\left(1 - e^{-\frac{\Delta v}{c_e}}\right)\right], \\
m_{\mathrm{CLM}} &= m_{\mathrm{PL}} + m_{\mathrm{CLPM}}, \\
m_{\mathrm{CLPM}} &= m_{\mathrm{p}} \cdot (1+\sigma).
\end{aligned}
\tag{10.1}
$$

To solve this set of equations, we need to know the dry mass index, $\sigma$, of the CLPM and the specific impulse, $I_{\mathrm{sp}}$. The performance ranges of the specific impulse depending on the selected propellant and pressurisation type are well-known. The dry mass index, however, is in principle unknown until a design has been established. This requires knowledge that is not available at this stage and a detailed quantitative structural analysis is out of scope for early feasibility studies—as discussed above. In Sect. 6.1.1, we discussed that the dry mass index, is a performance figure of space transportation systems—whether from ground to space or in-space. Therefore, we need to take an assumption, a so called 'best engineering guess' (BEG), which ideally is backed by a statistical databank or a strong technical analogy (like for the velocity demand for landing), or sometimes expert's guts feeling! The assumption needs to be verified a posteriori by detailed analyses.[12] It is further important to be aware of what

[10] Higher harmonics are a measure of the deviation from a spherical field.

[11] This is a very critical phase, as these hazards may include rocks on the surface within shadowed areas or areas with slopes exceeding design limits. All of this could cause the mission to fail at the last moment, rendering the entire program in vain.

[12] It must be possible to trace-back all assumptions and their reasoning. This is subject of *configuration management*.

$\sigma$ should comprise and what not. This question is less about right or wrong but rather about proper bookkeeping in order to verify properly the assumptions made at a later stage. In the current case, we shall assume that the propellant management system, all structural mass, the avionics required to control the landing and the interface to the payload are included.[13] Not all listed elements above scale with the propellant mass and the last two certainly do not. This is an inconsistency that we need to accept at this early design stage and re-visit it in the next project phase.

The specific impulse depends on the propulsion type and the chemistry of the propellant, Chap. 9 and Sect. 8.2. An electric propulsion system can be ruled out immediately since it cannot meet the thrust requirements for soft landing. A solid fuelled motor is also not suitable since throttleability beyond 'pre-programmed' thrust profiles is not possible. The remaining options are:

1. cryogenic propellants
   a. methalox
   b. hydrolox
2. storable propellants
   a. mono-propellant
   b. bi-propellant

The US company Intuitive Machines has demonstrated that a planetary lander employing methalox as propellant is indeed feasible, Sect. 9.3.3. Small scale pressure-fed systems with hydrolox have not flown yet, which, however, is merely a question of time. Both cryogenic propellant combinations suffer from boil-off losses and stringent thermal storage requirements, Sect. 8.2.2. The question for both is to what extent the architectural and operational disadvantages are justified by the performance gains. A higher dry mass index is to be expected due to lower propellant density and the required thermal insulation and constructive clearance.[14] Although a mission to the cold outer region of the solar system should favour this propellant combination. We will discard hydrolox from the trade-off due to its low average density and above discussed design implications (volume and mass) as well as the risk of strong propellant boil-off. Especially the latter poses a high risk in case of parasitic heat loads that have been overseen in the design or miscalculated. For the sake of an instructive comparison, we will keep the methalox combination and assume a benign specific impulse of 355 s [8].

Hypergolic propellant combinations have been in use since the dawn of space travel but due to their toxicity they will be replaced in future, by so-called green propellant combinations of similar performance. Both types have in common that they belong to the group of storable propellants. Table 10.2 shows the mass of the

---

[13] Landing legs, communication and power supply for the entire stack shall be provided by the payload module.

[14] The term clearance refers to a safety distance or a spatial provision that allows access, e.g. for maintenance work.

**Table 10.2** CLPM trade-off domain with respect to the dry mass index and specific impulse

| | Mono-prop | | **Bi-prop** | | Methalox | |
|---|---|---|---|---|---|---|
| $I_{sp}$ | 220 s | | **315 s** | | 355 s | |
| $\sigma$ | 0.4 | 0.2 | **0.4** | 0.2 | 0.4 | 0.2 |
| $m_p$ | 6.2 t | 5.9 t | **4.1 t** | 4.0 t | 3.6 t | 3.9 t |
| $m_{dry}$ | 2.5 t | 1.2 t | **1.6 t** | 0.8 t | 1.4 t | 0.8 t |
| $m_{CLPM}$ | 8.6 t | 7.1 t | **5.7 t** | 4.8 t | 5.0 t | 4.7 t |

Ceres Landing Propulsion Module (CLPM) fully loaded for two cases of dry mass index and three specific impulses reflecting three different propulsion systems.

The results show a variation of the CLPM wet mass from 4.7 t in the best to 8.6 t in the worst case. For comparison, Fregat, Soyuz' autonomous upper stage, has a propellant mass loading of 5.4 t (N2O4/UDMH) and a dry mass index of 0.2. The lower dry mass index is therefore not unreasonable for a storable bi-propellant propulsion system. The tasks of the CLPM is, however, different from that of an upper stage and we are advised to follow a more *conservative* approach.[15] A mono-propellant propulsion system is too massive and the benefit of a cryogenic propulsion systems does not justify the added design and mission complexity. The baseline configuration should be rather a classical storable bi-propellant system with a specific impulse of at least 315 s and a structural mass allocation of 1.6 t ($\sigma = 0.4$).

Finally we will assess the impact of the 25% margin on the velocity demand defined above. Without the margin, the required $\Delta v$ is 300 m/s. This corresponds to: $m_p = 3190$ kg and $m_{dry} = 1276$ kg, for the baseline configuration. Hence, a $\Delta v$ margin of 25% corresponds to a propellant margin of 22%. Additional margin on propellant would be double margin as both values are physically related.

Thrust Requirements

The main requirement for the propulsion system of a planetary landing system is the capability to *throttle* its thrust level. The reason lies in *continuously powered descent* strategy for landers.

This is equivalent to a gravity turn, a simple guidance law to reach orbit. However, the descent is more challenging than the ascent precisely because of the demand to throttle the engines from propulsion perspective.[16] This demand arose first during the lunar exploration era in the 60 s of the 20th century. The Apollo program led to the descent propulsion system (DPS) for NASA's lunar lander module. This system

[15] A conservative approach in engineering terminology means a cautious approach, which could mean to adopt the solution that offers higher technical margin or that has a higher technical maturity, e.g. flight heritage, tested etc.

[16] Furthermore, there is a need to continuously measure the relative distance and speed to the ground to feed the guidance algorithm, which in turn commands engine throttling. This has to happen autonomously, in the lander is uncrewed. Autonomous planetary landing missions are therefore, the most challenging missions in space engineering with moderate success rates compared to other missions.

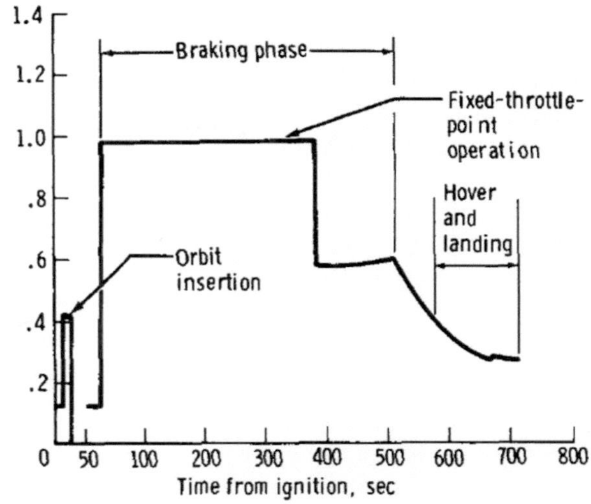


**Fig. 10.5** Normalised thrust profile of the Lunar 'Descent Propulsion System' (DPS) [9]. Credit: NASA

featured a maximum thrust of 47 kN with a throttle capability in the range of 60 to 10% of maximum thrust. In case of the lunar landing module, the thrust required to hover over the ground was about 11 kN. The minimum thrust requirement was 4.7 kN, which is slightly less than half the hovering requirement. Figure 10.5 shows the thrust profile as specified for Apollo's DPS normalised to the maximum thrust. To adopt this strategy, it is necessary to adequately benchmark the thrust. The benchmark figure should be related to the relevant physical property of the celestial body and the mass of the system. The hovering thrust is a suitable parameter as it fulfils both aspects.

Since hovering is the last stage in the complex landing sequence, we shall consider the very last instance when almost all propellant is consumed. This way we avoid an initial propellant loading bias in our similarity assessment.

We calculate first the ratio of the weight forces of both lander, the Ceres Landing Module (CLM) and that of Apollo shortly before touchdown assuming all propellant is consumed:

$$\frac{G_{\mathrm{C}}}{G_{\mathrm{L}}} = \frac{g_{\mathrm{C}} \cdot (m_{\mathrm{CLM,0}} - m_{\mathrm{CLPM,p}})}{g_M \cdot (m_{A,0} - m_{A,p})} \tag{10.2}$$

$$= \frac{0.284 \cdot (35.6\,t - 4.0\,t)}{1.626 \cdot (15.2\,t - 8.2\,t)} \tag{10.3}$$

$$= \frac{8.99\,\mathrm{kN}}{11.3\,\mathrm{kN}} = 0.79 \tag{10.4}$$

This yields us the desired benchmark figure, 0.79, which we can use to scale Apollo's nominal thrust, 44.7 kN, and to obtain CLM's nominal thrust, 35.5 kN. This value needs to be understood as a ballpark figure pending on confirmation by GNC analyses at a later stage.

Planetary landing in general puts upper and lower limits on the required deceleration profile. A high thrust engine results in a high vehicle deceleration which hampers the descent. Figuratively speaking, the CLM would hoover at high altitude not able to approach the surface, if not deep throttled or pulsed to reduce the thrust. Both engine operation options are driven by the engine's nominal thrust magnitude. Conversely, a low thrust engine generates a too low deceleration leading to a fast spiralling descent trajectory. In that sense a high thrust engine has some degrees of freedom to comply with the average deceleration requirement, which a low thrust engine does not possess.

When considering means for thrust magnitude modulation, throttling or pulsing, it is important to choose a general formulation of the propulsion requirement. This allows for deliberate openness in future detailed engine trade-offs.

Thruster Selection and Engine Configuration

Before establishing a baseline engine configuration for CLPM, two key questions need to be answered a) single or multiple engines and b) with redundancy or without? Especially the latter is a hot debated topic which is not without a psychological aspect, meaning fear of mission loss.[17] In view of the fact that C-One will be a multinational mission with a long development time and high cost, a *single point failure free* design will be certainly requested, which is equivalent with full redundancy down to engine level.[18] We, therefore, select a dual main engine configuration. In case of a failure, the remaining engine shall have sufficient thrust to complete the mission. The impact on system level of this decision that needs to be considered is a potential thrust misalignment and thus parasitic torques by the single engine. A remedy is to tilt the engines, but the CoM will move and in a fixed configuration thrust induced parasitic torques can only be zeroed for a specific moment. An alternative is to control the thrust vector by gimbaling the entire engine. This is achieved with electro-mechanical actuators (EMA), which is standard in the launcher industry [11].

---

[17] A multi-billion dollar project like C-One, which takes a decade to design and build, should not be in vain because of an engine or even valve failure! On the other hand, the crewed Apollo lunar descent module had only a single engine. The argument that Apollo was a child of the space race and the political will to win the race to the Moon supports actually the psychological dimension to the matter. Since Apollo, the zeitgeist has changed dramatically. Agencies have become more cautious and risk adverse.

[18] Note that redundancy is not always the answer to reliability or safety. A multi-engine configuration for instance faces the paradox situation that it is capable to absorb an engine-out failure but on the other hand it is more prone to more failure modes. Partially due to the introduced complexity within the propellant management system, partially due to the sheer existence of an additional propulsion string that contains elements that could fail for a particular reason. Simply put, the odds of failure increase with the number of involved units. On the other hand, it occurs that single point failures (SPFs) cannot be avoided in design. A prominent example is NASA's James Webb Space Telescope (JWST), launched 2021 by Ariane 5), which had in total 344 SPFs [10].

The long-term reliability of these mechanisms[19] under deep space conditions has not been proven yet and requires more scrutiny. A third method to achieve thrust vector control (TVC) is to utilise vernier thruster, a common method for orbital systems and we shall adopt it as baseline for TVC for CLPM.

We need, however, a thrust requirement these engines. Since the prime task of the thrusters is to compensate the torque induced by the main engine, knowledge about the CoM position relative to the thruster location is required. This information is currently not available. Based on other designs with the functional workshare shows that vernier thruster should have 10% the thrust magnitude of the main engine, in our case 3.5 kN. However, it should be noted, that in order to enhance the lever arm for torque control, it might be required to incline the thrusters, thereby reducing the in-axis thrust contribution—pending on the parasitic torque force.

Following the above mentioned off-the-shelf approach, we will look for suitable main engines and vernier thrusters that comply with the above stated thrust requirements. It shall be stressed that these are merely exemplary components and by no means a final selection.

The German rocket motor Aestus[20] is a pressure-fed engine that runs with storable propellant, NTO-MMH. It delivers 29.4 kN of thrust and boasts a very good specific impulse of 324 s. It has proven its merits as upper stage engine of Ariane 5's second stage (ES). The US AJ10-190 developed by Aerojet, is an equally powerful engine of similar performance, 26.7 kN but only 316 s specific impulse.

Different vernier thruster arrangements are conceivable as depicted in Fig. 10.6.

Off-the shelf engines with thrust in the order of the required magnitude, range between 400 N to 1.1 kN, as listed in Table 10.3. None of these engines supports a two engine configuration. The smallest of the three, S-400-15, requires an eight engine configuration to fulfil the minimum thrust requirement, the second, R-42, a four engine configuration. A configuration of three engines is sufficient with the 1.1 kN thruster of Nammo, which we will adopt as baseline for compactness reasons.

An example of an *inconsistency* can be discussed here. First, the required propellant oxidiser, though similar, is not exactly the same for all engines. Aestus is qualified for $N_2O_4$ as oxidiser while LEROS 4 requires MON-3. The other two engines have been qualified for both oxidiser types. This difference should not be a showstopper due to similar chemistry. In fact the data sheets of several engines mention that the engine is either directly qualified for both or it was qualified for one combination and performance was demonstrated for the other.[21] Further inconsistencies are related to

---

[19] Space engineers are generally very conservative when it comes to mechanisms. The foldable antenna of the Jupiter space probe Galileo, which failed to unfold, serves as a cautionary example. On the other hand, the James Webb Telescope had the most complex unfolding mechanism ever deployed in space, and it worked flawlessly. Overall, mechanisms are avoided unless proven essential.

[20] Developed in Ottobrunn, Germany, by Astrium AG, now Airbus Defence and Space.

[21] For instance Arianegroup's 200 N thruster, used on ATV and Orion ESM, has been qualified for MON-3/MMH and adequate performance was demonstrated for UDMH/$N_2O_4$. Aerojet has qualified immediately a large range of their thruster and engines for both oxidisers MON-3 and $N_2O_4$ in combination with MMH to maximise its applications and increase the accessible market.

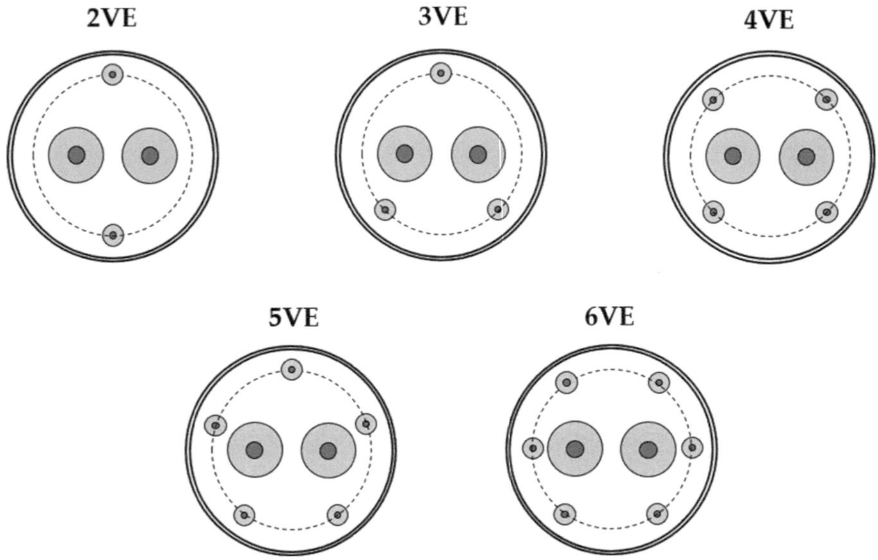


**Fig. 10.6** Options for CLPM vernier thruster configuration

**Table 10.3** Specification of potential main engines and vernier thruster, according to data sheet

| Property | Main engine | | Vernier thruster | | |
|---|---|---|---|---|---|
| | Aestus | **AJ10-190** | **LEROS 4** | R-42 | S-400 |
| Max Thrust | 29.4 kN | 26.7 kN | 1.1 kN | 0.89 kN | 0.44 kN |
| Specific Impulse | 324 s | 316 s | 321 s | 305 s | 321 s |
| Total Impulse | 32.5 MNs | 1.44 MNs | 13,6 MNs | 24,3 MNs | 13.46 MNs |
| Inlet Pressure (nom) | 17.7 | 16.6 | 15.4 bar | 16 bar | 15.5 bar |
| Oxidiser | $N_2O_4$ | $N_2O_4$/MON | MON | $N_2O_4$/MON | $N_2O_4$/MON |
| Fuel | MMH | MMH | MMH | MMH | MMH |
| Mixture Ratio (nom) | 1.9 | 1.65 | 1.65 | 1.65 | 1.65 |
| Engine Mass | 111 kg | 118 kg | 8.41 kg | 4.53 kg | 4.30 kg |
| Length | 2.2 m | 1.95 m | 1 m | 0.79 m | 0.669 |
| Diameter | 1.31 m | 1.17 m | 0.5 m | 0.39 m | 0.32 m |
| Expansion Ratio | 84:1 | 55:1 | 293:1 | 160:1 | 160:1 |

the inlet pressure and mixture ratio. Both can be handled by the PMS, Chap. 9 but as for the deviation in oxidiser type, this too will increase the technical design effort and most notably the verification effort by test. The decisive argument from an architectural point of view, however, is the mixture ratio. The Aestus engine falls off in that respect as it requires an O/F ratio of 1.95 compared to 1.65 of other engines. This difference complicates the tank selection and the PMS. We are therefore advised to insist on consistency in that respect, which excludes this engine.

In conclusion, the combination of two AJ10-190 engines and three LEROS 4 s appears to be a viable and prudent starting point for the preliminary design of the Ceres Lander Module.

Finally the CLPM needs to control also the attitude of the stack after separation from the solar electric cruise module. The set of thrusters is referred to as RCTs (reaction control thrusters). Similar to the vernier thruster, sizing of this system requires knowledge about the centre of mass (CoM) and the moment of inertia (MoI) of the stack and attitude requirements, like pointing accuracy and number of attitude changes. This is beyond the scope of our objective and the final design does not determine the overall architecture but rather adapts to it. We therefore, omit detailing this subsystem. Instead, we allocate 5% of the above computed propellant as a provision, which is rather conservative in absolute terms.

Tank Selection and Mechanical Architecture
We have seen that both, tanks and mechanical architecture, are closely related to each other: the latter needs to carry the first and the first defines the layout of the latter. We first start by defining the volume requirement for the tanks, summarised in Table 10.4

Note that the resulting storage volume of oxidiser and fuel are almost identical, as they differ merely by 1%. This is because the mass based mixture ratio (1.65) is almost equal to the ratio of propellant densities (1.67). This is not a coincidence but a deliberate design feature with a benign system impact: fuel and oxidiser tanks are interchangeable and most importantly do have the same filling ratio. Imbalance in the centre of mass (CoM) towards the launch axis due to propellant loading in a symmetric tank configuration cannot occur. Also tank depletion will be homologous causing no off-axial movement of CoM, if the tanks are not accommodated asymmetrically in the first place. ßß The here computed propellant volume is slightly larger than the largest satellites tanks currently available, such as the OST 26/X, which has a maximum volume of 1450 l. The options are to opt for many smaller tanks or to look for a larger tank that enables a dual tank configuration. The first is feasible and

**Table 10.4** Propellant properties and storage volume demand for the CLPM

| | |
|---|---|
| Propellant Mass for landing | 4083 kg |
| RCT allocation (5 %) | 204 kg |
| Total Propellant Mass | 4287 kg |
| Mixture ratio (O/F) | 1.65:1 |
| Oxidiser density (NTO) | 2669 kg/m$^3$ |
| Fuel density (MMH) | 1618 kg/m$^3$ |
| Oxidiser mass (NTO) | 1450 kg |
| Fuel mass (MMH) | 866 kg |
| Oxidiser volume (NTO) | 1841 l |
| Fuel volume (MMH) | 1868 l |

has been successfully applied before. Unless there are accommodation limitations, it is always mass efficient to select a dual tank configuration to reduce the tank mass fraction, $\sigma_{tank}$. A suitable propellant tank, that enables a dual tank configuration, is the MPCV tank used for Orion's ESM (European Service Module) with a capacity of 2100 l leading to a *filling ratio* of 89%. This is compliant with the maximum acceptable filling ratio during launch, which is in general 95% for surface tension tanks.[22] The general fear in this context is that parasitic heat could cause the tank to heat up, vaporising the propellant and increasing the pressure in the ullage beyond the burst pressure. The situation is even worse when the tank is pressurised before take-off. This is sometimes necessary for immediate manoeuvres, e.g. to avoid sun illumination of an optical instrument or for collision avoidance. The ESM tank provides a moderate *growth margin* of 6% in case later design stages reveal the need of more propellant—something a prudent systems engineer needs to expect[23] A further advantage of this tank compared to typical satellite tanks is its high MEOP of 25 bar. This is a critical quantity as it defines the starting point of the pressure drop due to tubing and fittings, Sect. 9.6.2.2. The pressure margin to the engine inlet pressure is a comfortable 8.4 bar, which is what the AJ10-190 main engine requires.

We now need to determine the required amount of helium for pressurisation, for which we can apply the set of equations derived in Sect. 9.6.3. The results are provided in Table 10.5. Like for the propellant mass, we need to distribute the required helium volume into off-the-shelf pressure vessels. There exist a range of suitable pressure vessels and the final selection depends on the mechanical architecture of the CLPM and dynamic loads during launch. A concept with only few (e.g. two) vessels will lead to higher local loads than a distributed approach with more but smaller vessels. Table 10.6 lists the potential options and their main characteristics. All pressure vessels consist of a titanium liner and are over-wrapped with an epoxy-based CFRP to withstand the high pressure. To accommodate the required 553 l of helium with the smallest vessel a total number of 14 would be required, and only two with the largest one.

Pending on the mechanical design, it is reasonable to opt for the configuration with the smallest number of tanks (ESM) as that would reduce the complexity of the propellant management system in terms of required tubing and flow control equipment. However, the mass penalty is about 40 kg and it needs to be traded against the PMS related savings (e.g. tubing and flow control equipment) and other savings in secondary and tertiary mounting structure—or simply accepted for the sake of a simpler feed system. Hence, this trade-off cannot be closed on this level but requires a more detailed analysis.

[22] https://www.space-propulsion.com/brochures/propellant-tanks/2100lt-mon-mmh-tank-mpcv-esm.pdf.

[23] Filling ratios of up to 97% are in principle acceptable, if the initial pressurisation level is sufficiently low—pending approval by the launch service provider.

**Table 10.5** Propellant tank pressurisation system for CLPM

| | | |
|---|---|---|
| Tank operational pressure | $P_T$ | 25 bar |
| Total propellant tank volume | $V_T$ | 4200 l |
| Initial vessel pressure | $P_V$ | 310 bar |
| Final vessel pressure | $P_f$ | 70 bar |
| Total required vessel volume | $V_V$ | 553 l |
| Residual helium in pressure vessel | $m_{He,V}$ | 5.82 kg |
| Helium inside propellant tank | $m_{He,T}$ | 16.1 kg |
| Helium in ullage volume | $m_{He,u}$ | 1.61 kg |

**Table 10.6** Options of helium pressure vessels and main specifications

| | PVG 40 L | PVG 75 L | PVG 120 L | **ESM** |
|---|---|---|---|---|
| Volume | 40 l | 75 l | 120 l | **300 l** |
| MEOP | 310 bar | 310 bar | 310 bar | **400 bar** |
| Single vessel mass | 8.5 kg | 14.4 kg | 23.5 kg | **81 kg** |
| Number of vessels | 14 | 8 | 5 | **2** |
| Total vessel mass | 119 kg | 115 kg | 118 kg | **162 kg** |

### *10.2.1 Preliminary Baseline Design CLPM*

The function of the Ceres Landing Propulsion Module (CLPM) follows two main requirements. They can be formulated in a partially-formal way as:

*CLPM-Func-REQ-1*: The CLPM shall land the payload on the surface of Ceres from an orbit of 500 km altitude

*CLPM-Func-REQ-2*: The CLPM shall provide attitude control for the Ceres Landing Module

Figure 10.7 summarises the above performed architecture relevant trade-offs related to the propulsion system of the CLPM and the path the discussion took. Table 10.7 lists the top level mission and system requirement for the Ceres Landing Propulsion Module (CLPM). Note that for any mission, the number of mission and system requirements is in fact small, but their analysis leads to many more requirements, called derived requirements. The process is called *requirement breakdown* and linking the requirements to each other is mandatory and ensures traceability. When formulating requirements, it is of utmost importance to refer to the source of the requirement and to justify it. The source is usually a higher level requirement or an interface requirement imposed by another subsystem or discipline. In all cases, justification is required. This is in general in form of a numerical analysis but could

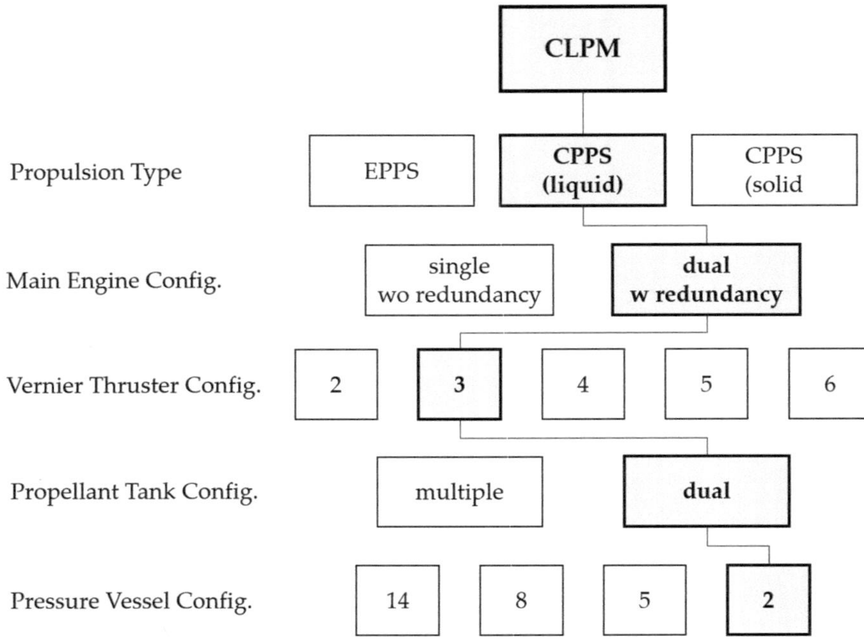


**Fig. 10.7** CLPM propulsion system relevant architecture trade-off space

**Table 10.7** Mission and system requirements for CLPM

| | |
|---|---|
| Payload mass | 30 t |
| Velocity demand | 375 m/s |
| Nominal thrust | 35 kN |
| Throttle range | 60–10% |

be any other form of reasoning.[24] It was also necessary to take design decisions that need to be verified (i.e. dry mass index) or challenged (i.e. propellant combination) in later design phases. Table 10.8 summarises both, the design choices and the derived requirements for the CLPM.

The discussion above led finally to a preliminary CLPM architecture with baseline equipment for the main components of the propulsion system. Table 10.8 summarises the baseline design. Note that the thruster specifications for the reaction control system are which is not untypical. The objective of this approach is to underscore the significance of these components while acknowledging that they have not yet been addressed and are currently outstanding—hence TBD (to be determined) (Table 10.9).

[24] The discipline that deals with this matter is called *requirements engineering* and is part of configuration management.

**Table 10.8** Derived requirements, design choices and assumptions for CLPM

| *Overall architecture* | | |
|---|---|---|
| Structural index, $\sigma$ | 0.4 | Assumption (TBC) |
| Max. structure mass | 1,633 kg | Derived requirement |
| Min. propellant mass | 4,083 kg | Derived requirement |
| Max. total wet mass | 5,716 kg | Derived requirement |
| CLM total mass | 35,716 kg | Derived requirement |
| *Propellant* | | |
| Fuel | MMH | Design choice |
| Oxidiser | MON | Design choice |
| Min. fuel volume | 1868 l | Derived requirement |
| Min. oxidiser volume | 1841 l | Derived requirement |
| *Pressurant* | | |
| Storage pressure | 310 bar | Design choice |
| Min. pressurant volume | 553 l | Derived requirement |
| *Engine configuration* | | |
| Nominal thrust | 35 kN | Derived requirement |
| Min. thrust | 3.5 kN | Derived requirement |
| Min. $I_{sp}$ | 315 s | Design choice |
| No. Main engines | 2 | Design choice |
| Min. No. vernier thruster | 2 | Design choice |

### 10.2.2 Launcher Compatibility Check CLM

The first question to be answered is whether the CLM shall be launched as a stack or separately. In the latter case, docking would be performed in orbit. The general guideline is: everything that can be done on-ground should not be done in space. This guideline stems from the fact that autonomous operations in space bear always the risk of failure. Regarding launch mass compatibility, even the payload alone (30 t) would require a heavy lift launch vehicle. It is therefore recommended to launch the stacked configuration in a co-manifested launch.

A volume check is would be hypothetical since the payload dimensions are not specified yet.

## 10.3 Phase 2 Powered Interplanetary Flight

Having discussed the CLM in the previous section, we now need to address how to reach Ceres. Subject of this second mission phase is the powered interplanetary transfer to Ceres and shall be accomplished by a Solar Electric Transfer Vehicle

**Table 10.9** CLPM baseline architecture

| *Propellant tank* | |
|---|---|
| Baseline tank | ESM prop. tank (AG) |
| No. tanks | 2 |
| Tank volume | 2,100 l |
| Tank diameter | 1,145 mm |
| Tank height | 2,651 mm |
| Tank dry mass | 110 kg |
| *Pressurant tank* | |
| Baseline tank | ESM press. tank (AG) |
| Storage pressure | 310 bar |
| Min. pressurant volume | 553 l |
| No. tanks | 2 |
| Tank volume | 300 l |
| Tank dry mass | 81 kg |
| *Main engine* | |
| Baseline main engine | AJ10-190 (AerojetRocketdyne) |
| No. main engines | 2 |
| Thrust | 26.7 kN |
| $I_{sp}$ | 316 s |
| Mass | 118 kg |
| Nominal inlet pressure | 16.6 bar |
| Length | 1955 mm |
| Diameter | 1168 mm |
| *Vernier thruster* | |
| Vernier thruster | LEROS 4 (NAMMO UK) |
| No. vernier thruster | 3 |
| Vernier thruster thrust | 1,100 N |
| Vernier thruster $I_{sp}$ | 321 s |
| Mass | 8.41 kg |
| Nominal inlet pressure | 15.4 bar |
| *Reaction control thruster* | |
| No. Reaction Control Thruster (RCT) | TBD |
| RCT thrust | TBD |
| RCT $I_{sp}$ | TBD |

(SETV). The intention of this section is to establish a preliminary design of the SETV. We start by determining the required velocity demand, $\Delta v$ for the transfer. The discussion shows nicely how phase 2 and phase 1 are intertwined.

### *10.3.1 Powered Interplanetary Flight to Ceres*

At the beginning of this chapter, we explained that it makes sense to use electric propulsion (EP) for interplanetary travel. We will elaborate this idea in this section further and shed some light why EP has become the first choice of rendezvous missions to asteroids and dwarf planets.[25] Table 10.10 shows a list of space probes that used or plan to use a solar electric propulsion system.

It is noteworthy that all missions performed a gravity assist manoeuvre en route to their destination. Hayabusa II gained 1.6 km/s during its swing-by at Earth [12]. In addition, all missions have in common that they need help by a chemical stage to reach the required escape velocity as discussed in the introduction. This is in general achieved by the launcher's upper stage, and if it is not capable enough an additional kick stage is used as applied for the Dawn mission, Sect. 2.2.3. After this initial kick, all space probes use their solar electric propulsion system to increase the semi major axis (SMA) and to align their trajectory with the target orbit.

Mission analysis and design for SEP based spacecraft is considerable more complex than for impulsive transfer missions with chemical propulsion system. The added complexity comes from the tight dependence of the flight trajectory on the spacecraft's acceleration profile. This means that on one hand knowledge of the spacecraft and its propulsion system is required by the mission analyst, who is responsible of the flight trajectory, while on the other hand detailed information on the trajectory parameters are needed by the propulsion and power engineers to design their systems. This intertwined relation demands an *iterative design process*.

To avoid a detailed and mathematically overloaded mission analysis as well as an intricate iteration process, we shall rather adopt a pragmatic approach, in-line with the purpose of this chapter and early design phases. This means that we draw reasonable analogies from adequate SEP missions. Psyche is a good *reference mission*. The orbital parameters of the asteroid Psyche and the dwarf planet Ceres are strikingly similar, Table 10.11, and it is safe to assume that a future Ceres mission, which is also based on solar electric technology, will follow a similar mission profile.

A closer examination reveals that Psyche's orbital energy is slightly higher than Ceres', which means a mission to Ceres should require a somewhat smaller velocity demand. On the other hand, Ceres has a higher inclination. Changes in inclination are

---

[25] In contrast, large planets (Venus, Mars, Jupiter etc.) require a high thrust manoeuvre capability for orbit insertion to shorten the capture duration. An EP-based capture is much more complicated, if possible at all. This is because the hyperbolic excess velocity must be reduced within a single orbital approach, or in other words, while the spacecraft is within the sphere of influence of the planetary body. If successful, the remaining EP based capture manoeuvre takes considerable time and is operationally complicated. It is therefore in general discarded.

**Table 10.10** Interplanetary missions with SEP

| | Hayabusa I | Hayabusa II | Dawn | Psyche |
|---|---|---|---|---|
| Mission | Itokawa | Ryugu | Vesta | Psyche |
| Launch date | May 03 | Dec. 14 | Dec. 07 | Oct. 23 |
| Distance at arrival | 1.40 AU | 0.94 | 2.47 | 2.7 |
| Inclination | 1.6° | 5.88° | 7.1° | 3.1° |
| C3 | – | 21 $km^2/s^2$ | 11.4 $km^2/s^2$ | 34 $km^2/s^2$ |
| DLA | – | – | 28.5° | 28° |
| Time of flight | 2.32 yr | 3.57 yr | 3.21 yr | 5.92 yr |
| Gravity assist | Earth | Earth | Mars | Mars |
| Thruster technology | GIT | GIT | GIT | HET |

**Table 10.11** Comparison of orbital parameters of Psyche and Ceres

| Orbital parameter | Psyche | Ceres |
|---|---|---|
| Perihelion | 2.53 AU | 2.55 AU |
| Aphelion | 3.32 AU | 2.98 AU |
| SMA | 2.92 AU | 2.77 AU |
| Incliniation | 3.1° | 10.6° |
| Orbital period | 4.99 yr | 4.60 yr |

energy-intensive and the above-mentioned savings could be cancelled out.[26] Psyche required a $\Delta v$ capacity of 6.25 km $s^{-1}$. In the context of a feasibility study it is safe to assume that the above described counter acting effects cancel each other to a large extend[27] and we adopt the mission profile of Psyche as a reference for C-One.

### 10.3.2 EP-Model for C-One

In order to establish a preliminary design for the SETV, we will apply the EP model discussed in Sect. 6.2 to C-One. Based on the top-level system requirements, we will identify again a design departure point for a future more detailed analysis.

Table 10.12 provides a concise list of performance and environmental requirements. In particular, the fact that the solar flux drops by almost a factor of ten is relevant to the performance of a solar electric propulsion system and, as we will see, a determining factor in the size of the solar array.

---

[26] There are several ways to account for a different inclination: either during the escape manoeuvre by adjusting the launch azimuth and launch date or during the powered interplanetary cruise.

[27] This is a *working hypotheses*, needs to be checked a posteriori.

**Table 10.12** Mission and system requirements for SETV

| | |
|---|---|
| Payload Mass (CLM) | 36 t |
| Velocity demand | 6875 m/s |
| Flight time (target–max) | 3–5 yrs |
| Min. $F_\odot$ | 150 W/m$^2$ |

Starting point of the conceptional design is the rocket equation for a solar-electric spacecraft introduced in Sect. 6.2. The algebraic set of equations that needs to be solved is:

$$\begin{aligned} m_0 &= m_{\text{pl}} \cdot \left[1 - (1+\sigma)\left(1 - e^{-\frac{\Delta v}{ce}}\right)\right], \\ m_{\text{p}} &= m_0 \left[1 - e^{-\frac{\Delta v}{ce}}\right], \\ m_{\text{eps}} &= m_{\text{p}} \frac{\alpha}{2\tau\eta_{\text{int}}} c_{\text{e}}^2, \\ m_{\text{dry}} &= \sigma m_{\text{p}}, \\ P_{\text{d}} &= \alpha m_{\text{eps}}, \\ T &= P_{\text{d}} \frac{2\eta}{I_{\text{sp}} g_0}. \end{aligned}$$

Note that $m_{\text{pl}}$ is now the mass of the CLM and $m_0$ the combined mass of SETV and CLM. The decisive parameter of the EP model are:

$\Delta v$: velocity demand, 6875 m/s as per requirement,
$\alpha$: ratio of mass-to-power generator, 0.03 kg/W design requirement (TBC),
$\tau$: flight duration, 3–5 yrs (TBD),
$\eta$: engine efficiency (TBD),
$I_{\text{sp}}$: specific impulse (TBD).

A mission to Ceres propelled by an SEP will experience a significant decrease in solar power generation capability by about 85% compared to the initial value at 1 AU, but the decrease is not reflected in the model above. This lack does not render the model unsuitable but asks for the right interpretation of the results. A meaningful interpretation in the context of an SEP system is that the model predicts the *average* power and thrust demand to accomplish the required $\Delta v$, refer to the discussion in Sect. 6.3.

Figure 10.8 depicts the continuous decrease of a 100 kW (BOL) solar power generator (SPG) from Earth to Ceres aphelion, 2.98 AU. The mean power output, averaged over distance, is merely 39 kW (dashed line) due to the exponential decrease of the flux. The average thrust level is consequently also much smaller than at BOL. It is, therefore, important to note that the power and thrust requirements for a mission to Ceres, as predicted by the EP model, represent *mission average* values of approxi-

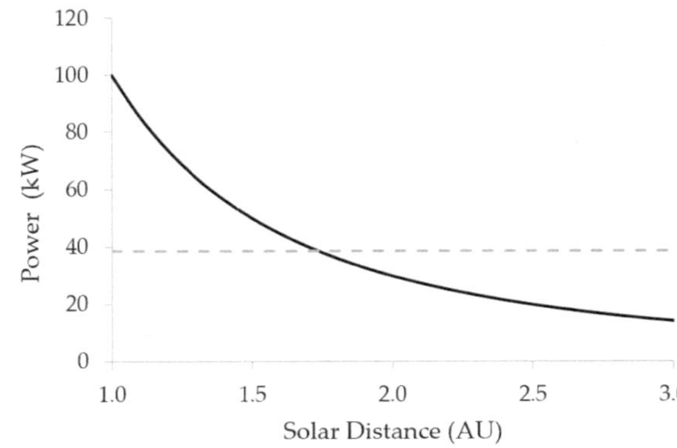


**Fig. 10.8** Maximum power point, $P_{mpp}$ variation from 1 AU to 2.98 AU

mately 40% of the initial capability at 1 AU! We will take this into account, when using this to define the DDP for the SETV.

In the following, we will perform a trade-off between HET and GIT technology to identify a propulsion baseline.

### 10.3.3 Thruster Trade-Off

A solar electric powered spaceship that transports a large payload into the asteroid belt needs high power thrusters. Restricting the trade-off to proven technology with a high technology readiness level (TRL) excludes Magentoplasmadynamic (MPD) thrusters and leads to the choice between Hall Effect Thruster (HET) and Gridded Ion Thruster (GIT). A detailed trade-off between the two types but also between the specific thruster candidates must include several factors, like the total propellant throughput and the maximum operating cycles. The latter case is important since outages are inevitable leading to a non-continuous thrust profile. For the sake of clarity we will focus on the main design and performance driving features to keep the discussion as comprehensible as possible.

Table 10.13 shows the effects of transfer time, engine efficiency and specific impulse on the discharge power. The first observation that can be made is that the more time the transfer is granted, the less power is required. The same applies to higher engine efficiencies. Conversely, the higher the specific impulse, the greater the power requirement. This is fully in-line with the findings made in Sect. 6.2.

In addition, we have seen in Sect. 7.2, that the two EP technologies tend to reign over different specific impulse domains: GIT tend to have larger $I_{sp}$ than HET in the range of 3000–4500 s, while the latter one ranges from 1800 to 3000 s. This will level the power demand since a HET based EP with an $I_{sp}$ of 2000 s will have a similar power demand than a GIT based EP with an $I_{sp}$ of 4000 s. The striking difference is in fact the achievable thrust magnitude, which is considerably larger for HET system expressed in the thrust-to-power ratio.

**Table 10.13** Required discharge power, $P_d$, as function of $I_{sp}$, mission duration, $\tau$ and thruster efficiency $\eta_T$ according to the EP model

| $I_{sp}$ | $\tau = 3$ yr | | $\tau = 5$ yr | |
|---|---|---|---|---|
| | $\eta_{T,HET} = 0.5$ | $\eta_{T,GIT} = 0.8$ | $\eta_{T,HET} = 0.5$ | $\eta_{T,GIT} = 0.8$ |
| 1600 s | 75 | 46 | 44 | 27 |
| 2000 s | 86 | 53 | 50 | 31 |
| 2400 s | 98 | 60 | 57 | 35 |
| 2800 s | 111 | 67 | 65 | 40 |
| 3200 s | 124 | 75 | 72 | 44 |
| 3600 s | 138 | 83 | 80 | 49 |
| 4000 s | 152 | 91 | 87 | 53 |
| 4400 s | 167 | 100 | 95 | 58 |

These competing effects require a detailed trade-off in the parameter space spanned by $I_{sp}$ & $\tau$, with real hardware performance figures. While this statement is unreservedly correct for smaller space probes like Dawn (GIT) and Psyche (HET), there is little debate for high payload mass missions like NASA's PPE (HET). This is because the pursuit of high-power GIT with a discharge power in excess of 20 kW is not currently being considered for a number of reasons, see Sect. 5.3.1. Instead HETs with discharge power levels even beyond 100 kW are subject of strong research interest. Given this technology trend, and the higher thrust required to propel the high payload mass within a short transfer, we baseline the Hall Effect Thruster technology for C-One.

Among the HET designs currently under consideration, the nested HET concept represents an intriguing proposition with distinctive benefits for deep space missions. The versatility of X3, introduced in Sect. 7.2.2, opens up new possibilities in engine operation. Unlike previous missions, where over 50% of the solar array was not utilized at the beginning of its life, Sect. 8.3.2, X3 can use almost all power generated and convert it into thrust, thus increasing the average acceleration level and reducing the time of flight. This is due to its quasi-continuous throttling capability, which renders it ideal for the specific needs of deep space missions. A single engine could in principle cover the complete required performance range. Table 10.14 shows an updated EP model with X3 thruster efficiency again for two flight time options.

As seen earlier, relaxing the flight time $\tau$ does not change the total mass $m_0$ significantly but it does have a considerable impact on the average discharge power and thrust.

DDP-3-3 matches the performance figures of X3 and will be adopted as baseline for a trnasfer time of 3 years. It requires an average discharge power of 55 kW which corresponds to 145 kW at BOL, according to the 40% rule as discussed above. While this number is three times higher than that of NASA's PPE it is only 50% higher than that of the Tiangong space station and therefore well within reach of near-term future SPG technology.

**Table 10.14** EP model as function of specific impulse, $\bar{\eta} = 0.66$

| | $\tau = 3$ yr | | | | | $\tau = 5$ yr | | | | |
|---|---|---|---|---|---|---|---|---|---|---|
| $\bar{I}_{sp}$ | DDP | $m_0$ | $m_p$ | $\bar{P}_d$ | $\bar{F}$ | DDP | $m_0$ | $m_p$ | $\bar{P}_d$ | $\bar{F}$ |
| s | | t | t | kW | N | | t | t | kW | N |
| 1800 | DDP-1-3 | 64 | 21 | 51 | 3.8 | DDP-13-5 | 63 | 20 | 30 | 2.3 |
| 1900 | DDP-2-3 | 62 | 19 | 53 | 3.8 | DDP-14-5 | 61 | 19 | 31 | 2.2 |
| 2000 | **DDP-3-3** | **61** | **18** | **55** | **3.7** | DDP-15-5 | 59 | 18 | 32 | 2.2 |
| 2100 | DDP-4-3 | 59 | 17 | 57 | 3.6 | DDP-16-5 | 58 | 16 | 34 | 2.1 |
| 2200 | DDP-5-3 | 58 | 16 | 59 | 3.6 | DDP-17-5 | 57 | 16 | 35 | 2.1 |
| 2300 | DDP-6-3 | 57 | 15 | 61 | 3.6 | DDP-18-5 | 56 | 15 | 36 | 2.1 |
| 2400 | DDP-7-3 | 56 | 14 | 63 | 3.5 | DDP-19-5 | 55 | 14 | 37 | 2.1 |
| 2500 | DDP-8-3 | 55 | 13 | 65 | 3.5 | DDP-20-5 | 54 | 13 | 38 | 2.0 |
| 2600 | DDP-9-3 | 54 | 13 | 67 | 3.5 | DDP-21-5 | 53 | 13 | 39 | 2.0 |
| 2700 | DDP-10-3 | 54 | 12 | 69 | 3.4 | DDP-22-5 | 52 | 12 | 40 | 2.0 |
| 2800 | DDP-11-3 | 53 | 12 | 71 | 3.4 | DDP-23-5 | 52 | 11 | 41 | 2.0 |
| 2900 | DDP-12-3 | 52 | 11 | 73 | 3.4 | DDP-24-5 | 51 | 11 | 43 | 2.0 |

The figures of DDP-3-3 represent system requirements for the SETV, meaning that the stage mass shall not exceed 25 t, the average power generation shall be at least 55 kW, the average specific impulse shall be at least 2000 s and the average thrust shall be at least 3.7 N. We will delve a bit deeper into the realisation options of DDP-3-3 utilising the nested HET, X3.

The need to throttle down the engine on the way to Ceres leads to a variation of both the specific impulse and the thrust. We have seen in Sect. 7.2.6 that this is in general achieved by reducing the beam current. Test data for X3 can be used to derive linear correlations for the $I_{sp}$ as function of the discharge power and the engine's rated discharge voltages [14]:

$$300\,\text{V:}\quad I_{sp} = 2.677\,\text{s/kW}\cdot P_d + 1848\,\text{s} \tag{10.5}$$

$$400\,\text{V:}\quad I_{sp} = 2.536\,\text{s/kW}\cdot P_d + 2108\,\text{s} \tag{10.6}$$

$$500\,\text{V:}\quad I_{sp} = 2.316\,\text{s/kW}\cdot P_d + 2411\,\text{s} \tag{10.7}$$

and for the thrust:

$$300\,\text{V:}\quad T = 49.6\,\text{mN/kW}\cdot P_d + 25.2\,\text{mN} \tag{10.8}$$

$$400\,\text{V:}\quad T = 55.1\,\text{mN/kW}\cdot P_d + 58.4\,\text{mN} \tag{10.9}$$

$$500\,\text{V:}\quad T = 60.9\,\text{mN/kW}\cdot P_d + 92.6\,\text{mN} \tag{10.10}$$

The total thruster efficiency, $\eta_T$, varies only slightly with discharge power, which can be ignored at this stage. TThese relations are of particular importance in determining the operating discharge voltage. It will be demonstrated that this is also of relevance

**Table 10.15** X3 performance for C-One mission

| $V_d$ | $\bar{I}_{sp}$ | $\bar{T}$ |
|---|---|---|
| V | s | N |
| 300 | 1998 | 2.8 |
| 400 | 2250 | 3.1 |
| 500 | 2541 | 3.5 |

to the PVA. Table 10.15 shows the X3 performance for a mission to Ceres and three different operating discharge voltages.

Operating X3 on 300 V would meet the average specific impulse requirement but not the thrust requirement (3.7 kN). This can be explained by the fact that the X3 thruster is limited to 100 kW but the initial demand is 145 kW. Operating the thruster at 500 V yields a higher performance but does still not match the thrust requirement as per EP model in order to achieve the mission within 3 years. As anticipated, deciding on the discharge voltage is delicate due to its various implications: firstly it drives the solar array voltage and thus its layout, it yields to large voltage conversion losses if the mismatch of consumer voltage demand and generated voltage is large. A quantitative analysis, including the solar array design, is therefore inevitable to find the optimal voltage operation point. However, this is beyond the scope of this introduction and is also not necessary for a preliminary design study. We select a thruster operating voltage of 300 V since this design offers greatest margin with respect to thrust and power generation.

Finally it shall be highlighted that the storage of 18 t xenon poses a challenge. We mentioned in Sect. 9.6.1 that the largest xenon tank in development, the L-XTA, has a capacity of 900 l, which corresponds to 1548 kg at a storage pressure of 177 bar. A cluster of 12 tanks with a total mass of 1020 kg will be needed to store all propellant. Whether less but larger tanks are advantageous from system perspective (less mass and less complex pressurisation assembly) and worth the development effort requires a dedicated trade-off at a later stage.

## *10.3.4 Thruster Configuration and Redundancy*

The engine trade-off analysis revealed that a nested HET should be the baseline for the SETV due to its compelling throttleability performance. The identified thruster, X3, is a 100 kW class thruster and therefore 45 kW short of the calculated requirement, 145 kW at BOL. This and the decision to operate the thruster at 300 V lead to a shortage in average thrust of 0.9 N. The power generation capacity of a 145 kW solar generator drops to 100 kW at about 1.25 AU. Additional propulsive support is required to bridge the gap in the first 0.25 AU but also to raise the average thrust

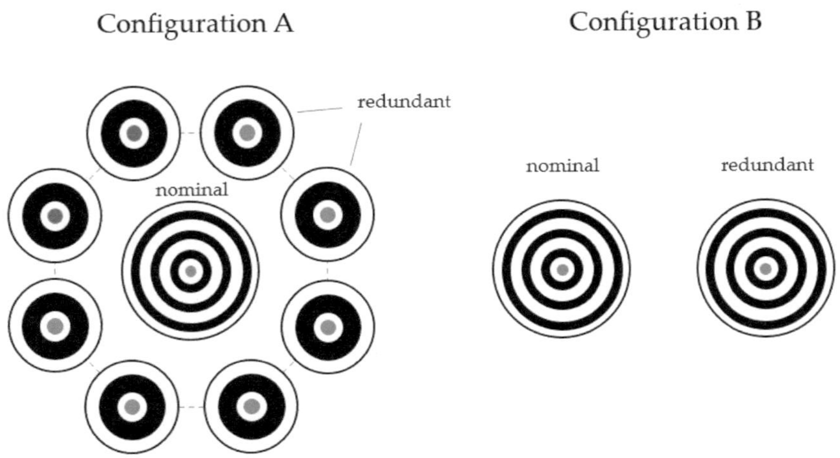


**Fig. 10.9** SEP-Engine-Configuration-A-B

magnitude. The solution to this problem should in addition be capable to serve as a *back-up* in case of a main engine failure. The following options exist:

configuration A: 1 × X3 and N × smaller thruster. The smaller thrusters support the main engine in the first 0.3 AU and serve thereafter as back-up engines in cold redundancy,

configuration B: 2 × X3, of which one is shut-off at 1.3 AU and serves thereafter as a back-up engine in cold redundancy.

Figure 10.9 depicts the two configurations. We start by discussing configuration A. Eight auxiliary engines surround the main engine equally spaced on a circle.

The exact geometry depends on several factors among them, a minimum distance to prevent electromagnetic interference and a save distance to gimbal the engines. The previously discussed HERMeS is a suitable auxiliary engine due to its high performance, Sect. 5.3.1. At 1 AU all eight engines are capable of providing 4.9 N at full throttle (i.e. 106 kW) [15]. The left side of Fig. 10.10 shows the thrust contributions and the resulting total thrust. A partition is assumed in which X3 receives 100 kW (i.e. X3 runs full throttle) and the auxiliary engines the remaining power. Note that the thrust magnitude of the auxiliary system reduces to zero due to this partition.

While focusing on average performances, it is important to keep in mind that the average is a result of a profile over distance. We are advised to introduce a further constraint from this profile, the initial thrust value, which is 7.7 N for the C-One mission. The combined thrust of configuration A is by 4% (0.3 N) lower than this requirement. A detailed comparison of both profiles, right hand side of Fig. 10.10 reveals that the difference vanishes quickly and reaches the required level at 1.2 AU.

It remains to verify that the auxiliary thrusters are capable to fulfil the back-up role in case of an X3 failure. Figure 10.10 (right hand side) depicts this *contingency*

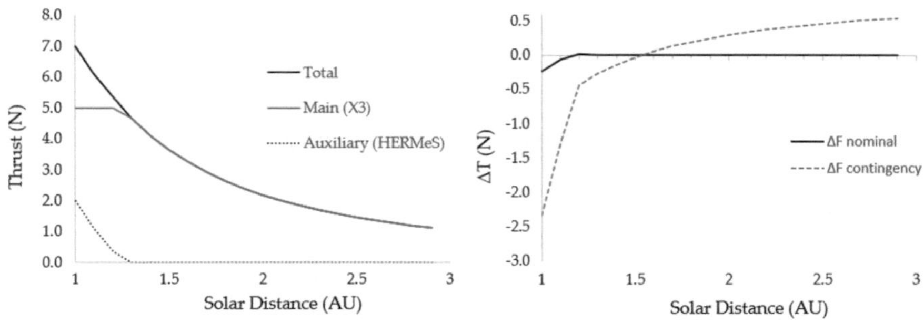


**Fig. 10.10** Thrust profiles over solar distance: combined thrust of main engine and auxiliary thruster (left), delta of required to available thrust for nominal and contingency operation

*case*. The auxiliary thrusters under-perform at the beginning, delivering 2.8 N less than nominally required. This deficit is recovered at about 1.6 AU, after which the delta becomes positive, so that overall the average thrust requirement is met. The under-performance in the first thrust arc and over-performance in the second do have consequences on the trajectory, specifically on flight duration, velocity demand and most relevant rendezvous requirements for the Mars swing-by and Ceres arrival. This demonstrates nicely the intricacy of mission design for EP missions.

In a quantitative trade-off, configuration B is the clear winner due to mass savings, higher mass-to-power ratio and cost efficiency—assuming sufficient space heritage for the nested HET technology has been accumulated. The mass of a single X3 thruster for instance is 230 kg while a single HERMeS has a mass of 50 kg [13]. Configuration A has a total thruster mass of 680 kg while B has only 460 kg. This does not yet account for the more complex and mass-intensive propellant management system of configuration A, which would further increase its mass disadvantage.

In this case, however, it is a qualitative factor that does make the difference, *versatility*. It is a veritable possibility that the same engines could suffer from the same failure mode, rendering redundancy useless. Contrary, it is very unlikely that physically different technologies experience simultaneously a failure. Not to speak of the same failure, which is almost impossible. To rely on *different* systems or technologies for the *same* function as back-up is called versatility.[28] An aspect that cannot be quantified in a straightforward way but rather depends on the judgment of the engineering team.

[28] Versatility plays a role in almost all engineering disciplines. Terrestrial traffic control combines sensor suits that apply different physical means, e.g. optical, RADAR, LIDAR and ultrasonic. An other common expression is *dissimilar redundancy*.

### 10.3.5 Solar Array Architecture

In Sect. 8.3.5, we identified that Rollout Solar Arrays (ROSA) will become state of the art technologies for very high power demands in the foreseeable future. Thanks to their flexible substrate, they offer very high specific power (kg/kW) and stowing efficiency (kW/m$^3$) values. These are reasons why all high power space systems, like the Chinese Tiangong, 100 kW, and the NASA's PPE for the lunar gateway, 70 kW (BOL), employ this technology. The first interplanetary mission to use ROSA was NASA's DART (Double Asteroid Redirection Test) mission with 6.6 kW. It is therefore a natural choice to baseline this technology for C-One. Prior to any discussion of the shape and location of the solar array, it is essential to establish a preliminary sizing of the array.

Solar Array Sizing
Starting point is always a power budget table, which considers both BOL (Earth departure) and EOL (Ceres arrival) requirements. A preliminary budget is provided in Table 10.16.

The budget considers the two main power consumers in the spaceship: the High Voltage Bus (HVB), consisting of the propulsion system and the Low Voltage Bus, consisting of the Low Voltage Bus (LVB). Unsurprisingly, the power demand is dominated by the electric propulsion system. For the avionics, an arbitrary provision is foreseen to be verified at a later design stage. Harness losses and voltage conversion losses have been considered with 2 and 5%, respectively. In addition, it is standard to consider a system margin of 20%, which is high but necessary in early design phases. The result is a total demand of almost 190 kW at BOL and 30 kW at EOL. We have seen previously that the design of a solar power generator is in general dominated EOL requirements.

To dimension the solar power generator, we apply the algorithm provided in Sect. 8.3.4 and worked-out in Annex D. The solar cell characteristics are listed in Table 10.17. The thermal impact and solar cell degradation, represented by efficiency factors at EOL, are considered. The result is a solar array with a total surface of 791 m$^2$. Its seize and capacity are provided in Table 10.18.

**Table 10.16** Preliminary power budget in kW, HBV (High Voltage Bus), LVB (Low Voltage Bus)

| Consumer | BOL | | EOL | |
|---|---|---|---|---|
| | HVB | LVB | HVB | LVB |
| Propulsion | 145.00 | | 22.00 | |
| Avionic | | 1.00 | | 1.00 |
| Harness losses (2 %) | 2.90 | 0.02 | 0.44 | 0.02 |
| Conversion losses (5 %) | 7.25 | 0.05 | 1.10 | 0.05 |
| Subtotal | 155.15 | 1.07 | 23.54 | 1.07 |
| System margin (20 %) | 33.03 | 0.21 | 4.71 | 0.214 |
| Total | 187 kW | | 30 kW | |

**Table 10.17** Baseline solar cell characteristics of IMM-$\alpha$ for C-One mission

| | | BOL | EOL |
|---|---|---|---|
| | | 1 AU (80 °C) | 3 AU (–106 °C) |
| Average open circuit | $V_{oc}$ | 4.01 V | 5.72 V |
| Average short circuit | $I_{sc}$ | 0.339 A | 0.025 A |
| Voltage MPP | $V_{mp}$ | 3.49 V | 5.34 V |
| Current MPP | $I_{mp}$ | 0.320 A | 0.024 A |
| Cell area | $A_{cell}$ | 30.18 $cm^2$ | |
| Fill factor | FF | 0.9 | |
| Cell width | $W_{cell}$ | 81 mm | |
| Cell length | $L_{cell}$ | 40 mm | |

**Table 10.18** SPG seize and performance, architecture C

| | BOL | EOL |
|---|---|---|
| Array voltage | 331 V | 507 V |
| Array current | 774 A | 59 A |
| Array power | 256 kW | 30 kW |
| Total SPG area | 791 $m^2$ | |
| $N_{cr}$ | 95 | |
| $N_{s,tot}$ | 2482 | |
| $N_{c,tot}$ | 235790 | |

As expected the available power at BOL is over-dimensioned, it exceeds the demand by 69 kW (27 %). It is important to note that this overcapacity would remain even in a more elaborated analysis. The reason is that the EOL demand is the sizing case. In principle, we can make a virtue of this, as the combined discharge power of nominal and redundant thrusters is around 200 kW, by making full use of the available power and running all thrusters at full throttle from the start of the mission. This in turn would increase the average thrust and reduce the flight time. Although a logical idea, there is a caveat. The designed trajectory assumes a pre-determined thrust profile to meet Mars at the right location, trajectory angles and time for the swing-by manoeuvre. More thrust means that the spaceship arrives too early at its rendezvous and misses the swing-by or the outcome is not the desired one, which is essentially the crux with SEP missions. This is essentially equivalent to a new mission baseline, starting the discussion anew and creates a circle. Hence, overcapacity exists by design, for SEP missions, but cannot be used to improve the mission profile. It also demonstrates again the intricacy and therefore iterative nature of spaceship design and mission design when employing an SEP system. We will see further below in the context of a contingency analysis for phase 1, that this overcapacity in power and thrust is actually an inherent *growth margin* that could help to alleviate short comings in the previous phase.

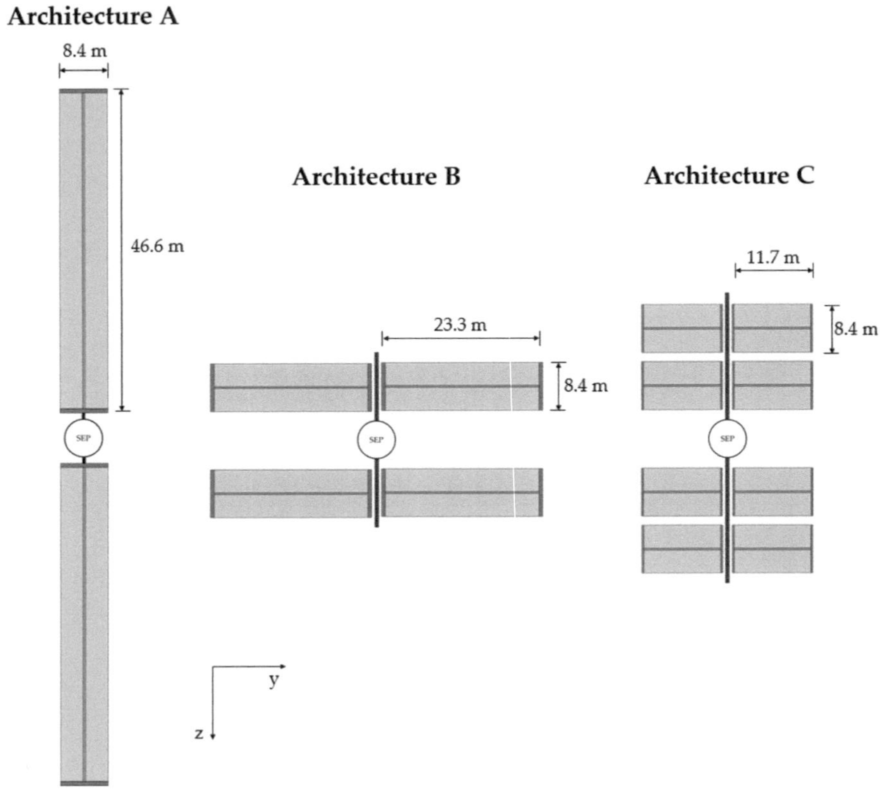


**Fig. 10.11** Three conceivable SPG architectures for C-One based on ROSA technology

We now turn our attention to the arrangement of the calculated photo-voltaic area to form an SPG. Three conceivable architectures are shown in Fig. 10.11 and detailed in Table 10.19.

Architecture A resembles the classical approach for satellites with two large wings consisting of one boom on each side and two blankets per boom, similar to that of the PPE design. Each boom can be rotated around its axis by the solar array drive mechanism (SADM). A single wing has a length of 46.6 m leading to a total wing span of about 100 m considering also an assumption for the SETV body and clearance to the array. Such a high aspect ratio (length/width) is prone to torsion and oscillations during thrust changes and other manoeuvres. It is uncertain whether this design will be sufficiently rigid—even with the addition of excessive stiffening mass. A detailed attitude control analysis coupled with a structural analysis is needed to judge its feasibility and work out a detailed design.

The other two architectures have smaller wing aspect ratios of 1.3 (B) and 0.3 (C), respectively. Both consist of two wings but in contrast to architecture A, they have more booms attached to a wing mast. Since the width of a blanket is determined

**Table 10.19** Geometric parameter of the three SPG architecture options A-C

| | | A | B | C |
|---|---|---|---|---|
| No. wings | – | 2 | 2 | 2 |
| No. booms per wing | – | 1 | 2 | 4 |
| No. blankets per blanket | – | 2 | 2 | 2 |
| Blanket length | m | 4.2 | 4.2 | 4.2 |
| Blanket width | m | 46.6 | 23.3 | 11.7 |
| Blanket aspect ratio | – | 11 | 5.5 | 2.8 |
| Blanket spacing | m | 0.5 | 0.5 | 0.5 |
| Boom width (total) | m | 8.9 | 8.9 | 8.9 |
| Boom length | m | 46.6 | 23.3 | 11.7 |
| Boom spacing | m | NA | 3 | 3 |
| Wing length | m | 8.9 | 18.4 | 37.3 |
| Wing width | m | 46.6 | 23.3 | 11.7 |
| Wing aspect ratio | – | 5.2 | 1.3 | 0.3 |
| Total SPG area | $m^2$ | 787 | 788 | 791 |
| Total SPG mass | kg | 1027 | 1029 | 1032 |

by the required voltage, all architectures have this geometric parameter in common (8.9 m). This leads to shorter blankets compared to (A). The disadvantage of architecture B compared to C is that in case of B the solar array protrudes into the plume of the thrusters leading to performance degradation by high energetic ion impingement from the thruster, depicted in Fig. 10.12. The plume of an electric thruster differs significantly from that of a thermal engine. It is considerably wider and the interactions of ions in the plume lead to back-scattering, see Sect. 7.2.2. Architecture B has therefore a clear disadvantage. During the course of its operation, the solar array must be maintained in a position that allows it to receive maximum sunlight, achieved by the SADM. An orientation in which the array points in direction to the vehicles y-axis, thus experiencing high loads of ion *impingement*, is therefore not avoidable. These qualitative considerations favour architecture C for C-One as baseline design.

### *10.3.6 Onboard Heat Dissipation and Thermal Radiator*

Once we have identified the power demand, we can size of the required thermal radiator and determine its mass. The power that is dissipated by the Power Management and Distribution (PMAD) system is already considered in the power budget, see Table 10.16, and amounts to 7.3 kW at BOL. Some of this heat is required to maintain operational temperature limits within the spaceship. For the design of the radiator we will however consider that the radiator has to reject the complete amount to derive an upper limit for its size and mass.

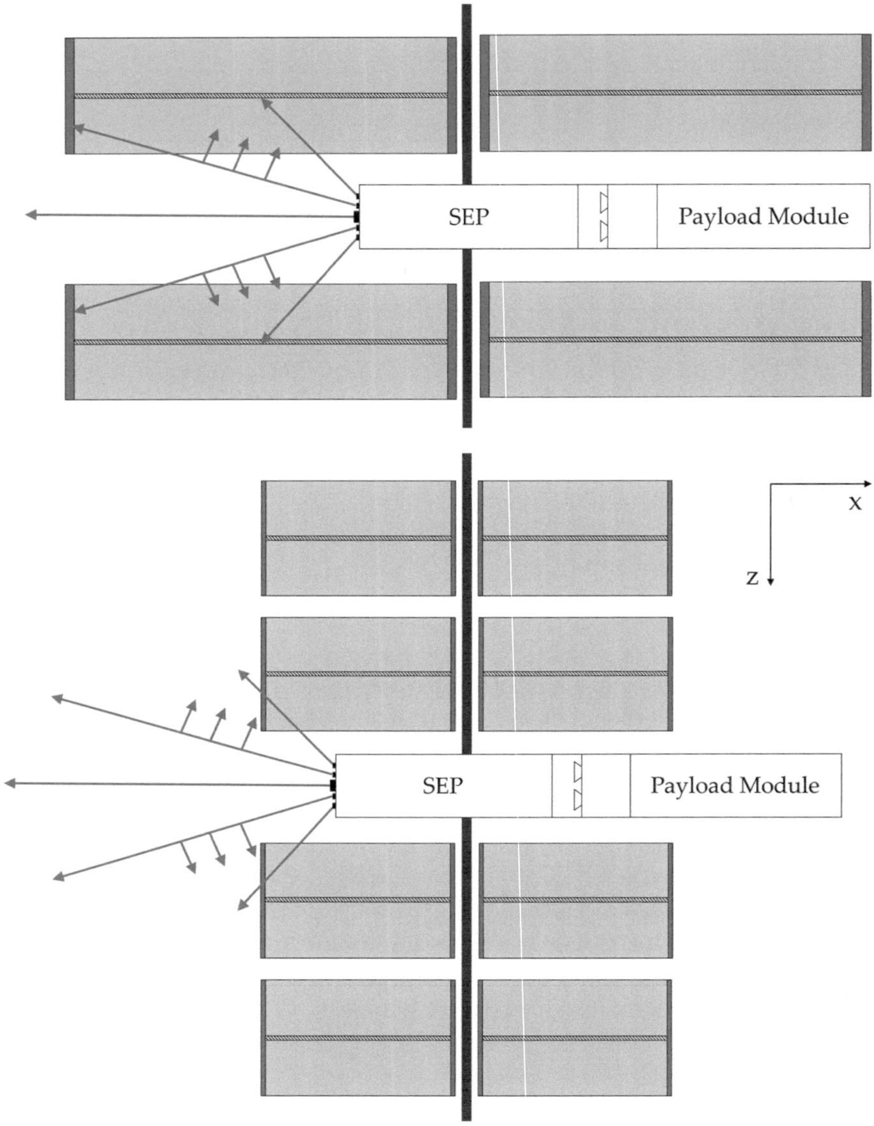


**Fig. 10.12** Plume impingement characteristics

For this we can use the values in Table 8.13 and scale them to $\dot{Q}_{\text{diss}} = 7,300\,\text{W}$. The incidence angle $\delta$ of the solar irradiance is 10 deg, which is the Ceres' orbital inclination. State of the art thermal radiator have a mass of 12 kg/m$^2$ [17]. Assuming an average radiator temperature of 30 °C, the specifications of the required thermal radiator can then be computed. Table 10.20 provides the preliminary top level specifications of the radiator.

**Table 10.20** SETV's thermal radiator

| | |
|---|---|
| $P_{diss}$ | 7,300 kW |
| $T_{rad}$ | 30 ° |
| $A_{rad}$ | 15 m$^2$ |
| $m_{rad}$ | 176 kg |

### *10.3.7 Preliminary Baseline Design SETV*

We summarise the outcome of the above analysis and the preliminary baseline design of the Solar Electric Transfer Vehicle (SETV). We start by a concise overview of the requirements and design decisions.

The function of the SETV follows two main requirements. They can be formulated in a partially-formal way as:

SETV-Func-REQ-1: The SETV shall transport the CLM from Earth's vicinity to a rendezvous with Ceres

SETV-Func-REQ-2: The SETV shall utilise a solar electric propulsion system

Figure 10.13 depicts the performed architecture relevant trade-offs, that are related to the propulsion system and the solar power generator and the path the discussion took.

Table 10.21 summarises all top-level requirements, starting with the mission and system requirements as well as the design choices and assumptions. Especially the latter need to be closely checked in follow-on studies—hence the remark TBC (to be confirmed).

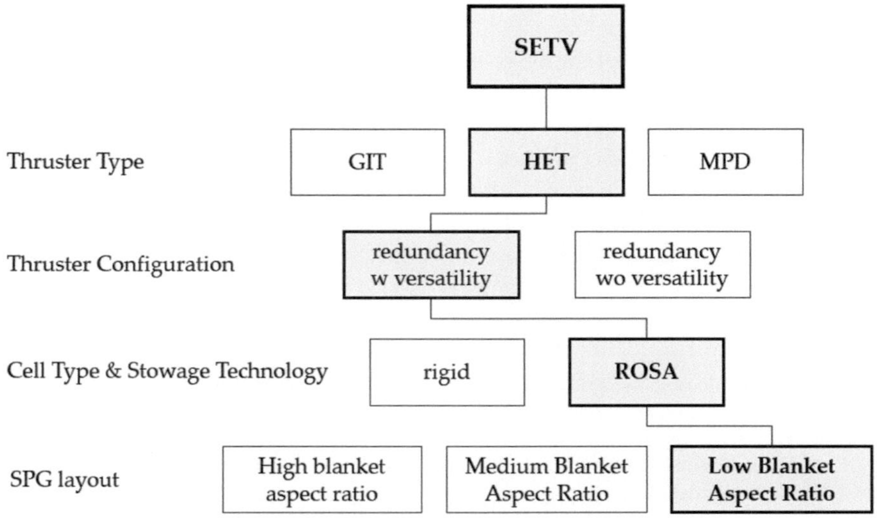


**Fig. 10.13** SETV propulsion system relevant architecture trade-off space

The predictions by the EP model are equivalent to top-down allocations. A brief bottom-up plausibility-check, considering the main mass contributors, is required. We start by examining the predicted mass for the entire electric and power subsystem, $m_{eps} = 1,670$ kg. We have have seen that the PVA mass, i.e. pure blanket mass, amounts to 1032 kg, thus leaving only a provision of 630 kg for the remaining elements: power processing units, harness and avionics. This seems marginal and requires more detailed modelling in the next design phase. On the other hand, the allocated mass for structure and the propulsion system, 5,428 kg, seems sufficient. We estimated the tank mass to 1020 kg and the thruster assembly to 680 kg, which leaves a total of 3728 kg for the carrying structure, the feed system and power processing units. A further cross-check is helpful: in NASA's NextSTEP programme a specific mass of less than 5 kg/kW is targeted. This figure includes a single X3 thruster and the complete feed system as well as the power processing units [14]. Based on this performance parameter, we arrive at 725 kg for a discharge power of 145 kW. Thruster configuration B ($2\times$X3) would require then 1450 kg, which can be considered as the lower limit for our baseline, configuration A. All in all, there is a good chance that any shortfall in the mass allocated to EPS could be made up by the mass not used by $m_{s,pms}$ (Table 10.21).[29]

Based on the above made design choices and derived requirements, we have established a baseline design, that serves as DDP for later detailed analyses (Table 10.22).

Other main contributors that have been missed out are the solar array drive mechanisms and the power management and distribution system. Both, however, do not pose any 'show-stopper' in terms of technical feasibility or mass contribution.

### *10.3.8 Launcher Compatibility Check SETV*

The above derived concept for the solar electric transfer vehicle (SETV) has a mass that is at the edge of heavy lift launch vehicle, 25 t. If this mass can be kept, there is a large number of launch vehicles that can support its launch. The crucial factor is the required volume of the SPG in launch configuration. All elements must be foldable on-ground and unfolded in space. A large amount of literature exists that all confirm the feasibility of this challenge [16] for large SPGs. A detailed volume check needs to be performed in a follow-on phase to identify techniques and their cost in terms of mass.

[29] Such swap in allocated mass between the subsystems are common. The systems engineer must keep a close eye on the maturity of the design and the accuracy of each subsystem's mass budget to assess reallocation. Mass budgets tend to change significantly, for example when design changes are made, and because small contributors are often ignored in early design phases that tend to show up at CDR (Critical Design Review), the nominal starting point for assembly and testing. To counteract this trend, space agency standards demand a system margin of 20% on the dry mass in early design phases—up to SRR (System Requirements Review).

**Table 10.21** Derived Requirements, design choices and assumptions for SETV

| *Mission and system requirements* | | | |
|---|---|---|---|
| Velocity demand | 6875 | m/s | Mission requirement |
| Flight time (target–max) | 3–5 | yrs | Mission requirement |
| Min. $F_{\odot}$ | 150 | W/m$^2$ | Mission requirement |
| Payload mass (CLM) | 36 | t | System requirement |
| *Overall architecture* | | | |
| Structural index, $\sigma$ | 0.30 | | Assumption (TBC) |
| Specific mass, $\alpha$ | 0.03 | kg/W | Assumption (TBC) |
| Max. structure & PMS Mass, $m_{\text{s,pms}}$ | 5,428 | kg | Design choice |
| Min. propellant mass, $m_{\text{p}}$ | 18,093 | kg | Design choice |
| Max. EPS mass, $m_{\text{eps}}$ | 1,670 | kg | Design choice |
| Max. total wet mass, $m_{\text{tot}}$ | 25,191 | kg | Design choice |
| *Electric propulsion system* | | | |
| Min. thruster efficiency, $\eta_{\text{T}}$ | 0.66 | – | Design choice |
| Avg. discharge power, $P_{\text{d}}$ | 55 | kW | Design choice |
| Average thrust, $F$ | 3,700 | N | Design choice |
| Min. $I_{\text{sp}}$ | 2000 | s | Design choice |
| Avg. discharge voltage, $V_{\text{d}}$ | 300 | V | Design choice |
| No. main thrusters | 1 | | Design choice |
| No. back-up thrusters | 8 | | Design choice |
| *Solar power generator* | | | |
| Min. power generation BOL | 145 | kW | Derived requirement |
| Min. power generation EOL | 30 | kW | Derived requirement |
| Min. array voltage BOL | 330 | kW | Derived requirement |
| Solar cell type | Flexible | | Design choice |

## 10.4 Phase 1 Earth Escape

The first phase consists of the Earth escape burn. This is certainly the most difficult part, as the payload mass, consisting of the scientific payload, landing module and solar electric transfer ship, must be accelerated out of the Earth's sphere of influence, Sect. 2.2.4, to a hyperbolic escape velocity. What makes this manoeuvre difficult is the fact that it starts deep in the gravity well. Escape is therefore achieved with help of chemical high thrust stages, called hereafter booster stages, or simply booster or stage.

**Table 10.22** Main elements of SETV baseline architecture

| *xenon propellant tank–single tank* | |
|---|---|
| Baseline tank | L-XTA (MT-A) |
| No. tanks | 12 |
| Tank volume | 900 l |
| Tank diameter | 1144 mm |
| Tank height | 1306 mm |
| Tank dry mass | 85 kg |
| *Main thruster–single thruster* | |
| Baseline main thruster | X3 (nested HET) |
| Mean efficiency, $\bar{\eta}_{\mathrm{T}}$ | 0.66 |
| Nominal thrust | 6490–1.950 mN |
| Nominal $I_{\mathrm{sp}}$ | 1930–2150 s |
| Mass | 230 kg |
| Nominal, $V_{\mathrm{d}}$ | 300 V |
| Throttle range in $P_{\mathrm{d}}$ | 100–30 kW |
| Diameter | 800 mm |
| *Auxiliary and back-up thruster–single thruster* | |
| Vernier Thruster | HERMeS (HET) |
| Mean Efficiency, $\bar{\eta}_{\mathrm{T}}$ | 0.675 |
| Vernier Thruster Thrust | 612 mN |
| Vernier thruster $I_{\mathrm{sp}}$ | 2826 s |
| Nominal, $V_{\mathrm{d}}$ | 600 V |
| Throttle Range in $P_{\mathrm{d}}$ | 13.1–2.7 kW |
| Mass | 50 kg |
| *Solar power generator* | |
| Cell technology | IMM-$\alpha$ (SolAero-RocketLab) |
| No. wings | 2 |
| No. Booms per wing | 4 |
| No. Blankets per Blanket | 2 |
| Blanket length | 4.2 m |
| Blanket width | 11.7 m |
| Boom length (total) | 8.9 m |
| Wing length | 37.3 m |
| Total SPG area | 791 m$^2$ |
| Total SPG mass | 1032 kg |

The impulsive velocity increase to reach Psyche's escape velocity can be computed with help of $C3$, see Sect. 2.2.3 starting from an initial orbit altitude of 400 km:

$$C3 = v_\infty^2 = 34\,\text{km}^2/\text{s}^2, \tag{10.11}$$

$$v_{\text{esc}} = \sqrt{2\frac{\mu}{r}} = \sqrt{2}v_{\text{circ}} = 10.85\,\text{km/s}, \tag{10.12}$$

$$v_{\text{inj}} = \sqrt{v_{\text{esc}}^2 + v_\infty^2} = 12.32\,\text{km/s}. \tag{10.13}$$

and for the impulse:

$$\Delta v_{\text{esc,id}} = v_{\text{inj}} - v_{\text{circ}} = 4.65\,\text{km/s}. \tag{10.14}$$

We are advised to consider an uncertainty margin on this value: 10% is typical for deterministic $\Delta v$, which leads to $\Delta v_{\text{esc,id}} = 5.11$ km/s. The second mission requirement is the payload mass. We calculated above a CLM mass of 36 t and an SETV mass of 25 t. Given the early design stage and the complexity of the SETV, we are advised to anticipate a potential mass increase by considering a 10% on the SETV mass.[30] The two main mission requirements for the Earth escape manoeuvre are:

- accelerate a payload mass of 63.5 t,
- reach a $C3$ of 34 km$^2$/s$^2$, equivalent to an ideal $\Delta v_{\text{esc,id}}$ of 5.11 km/s.

For comparison a launcher has to have a $\Delta v$ capacity of about 9 km/s. The high figure of 5.11 km/s demonstrate the significant challenge that this task presents. Consequently, the required space transportation systems will be large. We therefore concentrate on high energetic cryogenic propellant combinations, methalox and hydrolox, due to their superior specific impulse. Other combinations, with lower $I_{\text{sp}}$, will lead to lower performances resulting in prohibitively large system masses—which cannot be launched in the worst case.

To determine the design departure point (DDP) for the boosters, we perform first a staging analysis for both propellant combinations assuming for simplicity a constant dry mass index of $\sigma = 0.16$. The primary objective is to facilitate and illustrate more effectively the comparison between the propellant combinations. A more detailed analysis should take the propellant mass into account as shown in Table 6.2. Our simplified approach is however rather conservative—meaning that the real dry mass

[30] There are standards set by the space agencies to deal with margins, called the margin philosophy. It applies to masses from equipment to system level, their power consumption, $\Delta v$ requirement, onboard memory storage, data rates, etc. The values depend on the design maturity and decrease as the project matures. As the name suggests, it is indeed a philosophy and there are different approaches. For example, the European Space Agency (ESA) has a mass margin of 5% for equipment, even if it has been built and has flight heritage and is therefore fully known. Roscosmos does not do that. This leads to misunderstandings in international projects. A senior systems engineer once said that 20% of his time was spent fighting for margin—either with the customer to reduce margin requirements, or with colleagues to ensure that the agreed margin was met.

**Table 10.23** Methalox performance for $\Delta v_{\text{esc,id}} = 5.11\ \text{km s}^{-1}$, $\sigma = 0.16$, $I_{\text{sp}} = 360\,\text{s}$ for 1S: single booster, 2S: two booster and 3S: three booster configuration

| | $\Delta v_{\text{i,id}}$ (m/s) | | | Wet mass (t) | | | Propellant mass (t) | | |
|---|---|---|---|---|---|---|---|---|---|
| | 1S | 2S | 3S | 1S | 2S | 3S | 1S | 2S | 3S |
| 1st stage | 5110 | 1599 | 966 | 499 | 173 | 105 | 430 | 149 | 91 |
| 2nd stage | | 3511 | 1420 | | 173 | 105 | | 149 | 91 |
| 3rd stage | | | 2725 | | | 105 | | | 91 |
| Total | 5110 | 5110 | 5110 | 499 | 346 | 315 | 430 | 298 | 273 |

**Table 10.24** Hydrolox booster stages, $\Delta v_{\text{esc,id}} = 5.11\ \text{km s}^{-1}, \sigma = 0.16$, $I_{\text{sp}} = 460\,\text{s}$ for 1S: single booster, 2S: two booster and 3S: three booster configuration

| | $\Delta v_i$ (m/s) | | | $m_{0,i}$ (t) | | | $m_{p,i}$ (t) | | |
|---|---|---|---|---|---|---|---|---|---|
| | 1S | 2S | 3S | 1S | 2S | 3S | 1S | 2S | 3S |
| 1st stage | 5110 | 1775 | 1088 | 233 | 98 | 62 | 201 | 84 | 54 |
| 2nd stage | | 3335 | 1515 | | 98 | 62 | | 84 | 54 |
| 3rd stage | | | 2507 | | | 62 | | | 54 |
| Total | 5110 | 5110 | 5110 | 233 | 195 | 186 | 201 | 168 | 161 |

index is expected to be even lower—and is therefore acceptable at this stage.[31] There is another implication to a constant dry mass index within a multi-booster configuration, *commonality*. In a multi-booster assembly it is reasonable to keep the design as similar as possible. Making use of commonality in the design helps to reduce the development effort and ultimately saves cost.

Table 10.23 shows the results for methalox based booster configurations: a single booster (1S), a two booster (2S) and a three booster configuration (3S). Note that due to the requested commonality, all boosters within a configuration have the same wet mass and propellant mass. A three booster staging configuration leads to a mass of 105 t per stage. As expected, a single booster configuration exceeds the three-stage configuration by 184 t (58%!). Among the three configurations only the first can be launched with near term super heavy lift launch vehicles, like Starship or LM9.

The same exercise is repeated for the propellant combination hydrolox, Table 10.24. The trend is the same, the best performance in terms of overall booster system mass is achieved with a three-stage booster configuration.

In contrast to methalox, the masses are much lower. A super heavy lift launch vehicle is capable of lifting a booster of a two-stage configuration into orbit. The comparison of both tables, demonstrates nicely the impact of the specific impulse

[31] The dry mass index is also a function of the propellant density, see end of Sect. 6.1.1. Since methalox is a denser propellant combination than hydrolox, it will also lead to a lower $\sigma$. For the sake of a lucid discussion, we will ignore this correlation in the current discussion.

on stage mass and eventually the entire booster stack. Even a single stage hydrolox configuration is more mass efficient than a three-stage methalox configuration.

In view of this compelling performance advantage we select hydrolox as baseline propellant combination. It remains to select a staging configuration. At this point we need to perform a qualitative launcher compatibility check to support the down-selection. The aforementioned super heavy lift launch vehicle, LM9 and Starship will be capable to launch a single stage of a two-stage configuration, pending a volume check. However, as we have seen in Sect. 8.2.2 the average density of hydrolox (355 kg/m$^3$ for $r = 5.8$) is the lowest of all propellant combinations leading to large overall dimensions. In fact, despite the mass savings, a hydrolox stage has a larger tank volume than a comparable methalox stage due to the propellant density ratio of both stages. Methalox is 2.35 times denser than hydrolox.[32] Volume compatibility is a critical subject and we are advised to baseline a three-stage configuration.

The overall *booster stack mass* sums up to 186.5 t and in combination with the Ceres Landing Module and the Solar Electric Transfer Vehicle, the C-One spaceship has a mass of 250 t.

Finally a remark on the concept of operation (ConOps) should be made. It is important to note that while the booster stages function similarly to a launcher, there are distinct differences. In both cases a stage is jettisoned once depleted. However, in case of C-One the ignition of the subsequent stage is delayed until shortly before perigee is reached to maximise the achievable $\Delta v$, because of the Oberth-Effect. This means that there are coasting phases in between the burns in which the stack experiences weightlessness. A partially filled tank could suffer from a tank pressure drop, but this is not the case in the ConOps baseline, but needs to be accounted for in case of off-nominal conditions.

### *10.4.1 Thrust Requirement and Engine Configuration*

Unlike launchers, spacecraft do not generally require high acceleration levels and consequentially high thrust propulsion systems. This is due to the fact that they do not need to lift the system against the weight force, as they are already in an orbital state and the manoeuvre is tangential to the gravity vector. However, we need to account for the fact that the above derived *total stack mass* for C-One is 250 t and the velocity to be reached is indeed high. Furthermore, the journey starts deep in Earth's gravity well. All three factors render the ideal computation carried out at the beginning of this section for the $\Delta v_{\mathrm{esc,id}}$ insufficient, and a closer look into the dynamics of the escape manoeuvre is required.

The allocations of $\Delta v_{\mathrm{i,id}}$ in Table 10.24 are based on ideal values that assume an impulsive manoeuvres. We have seen previously in Sect. 2.2.3 that an impulsive manoeuvre is equivalent to a thrust impulse that is infinitely short-duration. However, in our case, this is not a viable assumption for the escape manoeuvres. A deviation

[32] Methalox has a density of 833 kg/m$^3$ for a mixture ratio of $r = 3.6$.

from the ideal case is equivalent to losses, expressed in an increased velocity increment and finally additional propellant. To compute the real propellant consumption in a finite burn, it is necessary to solve the equations of motion introduced in Sect. 2.2.3. To illustrate the effect, we chose the expander cycle engine RL10B-2. Based on the results a graph can be generated that illustrates the interdependence of three masses:

$\Delta m_{\mathrm{p,gl}}$: additional propellant needed to achieve the required orbital energy, $m_{\mathrm{p,gl}} = m_{\mathrm{p,real}} - m_{\mathrm{p,id}}$, hereafter referred to as gravity losses

$m_{\mathrm{eng,i}}$: the engine mass needed to achieve the thrust magnitude, i denotes the number of engines

$m_{\mathrm{sum,gl}}$: sum of $\Delta m_{\mathrm{p,gl}}$ and $m_{\mathrm{eng,i}}$

Figure 10.14 depicts the opposing trends for the first booster stage of C-One—the entire stack is accelerated. The additional propellant mass, $\Delta m_{\mathrm{p}}$, due to the finite manoeuvre decreases with increasing thrust magnitude. This improvement requires an invest in terms of engine mass. For a specific thrust-to-weight ratio, the engine mass of a hypothetical engine can be computed, which gives a linear slope. The sum of both masses has a minimum that in theory depends only on the engine's thrust-to-weight ratio and the specific impulse, dashed line. The assumed hypothetical engine has the feature that it can be *continuously* scaled for fixed engine parameters, T/W and $I_{\mathrm{sp}}$. This helps to establish the curves and to shed light into the overall behaviour. A real engine configuration cannot be scaled continuously, though. We deliberately chose a hypothetical engine with parameters that are identical to the specifications of the RL-10B-2 engine. Mapping real engine configurations (single, double, triple etc.) onto the smooth curves leads to discrete points depicted as dots. The results show that the utilisation of a two-engine configuration of RL-10B-2 s significantly surpasses the performance of a single engine. The propellant mass saving is 1777 kg

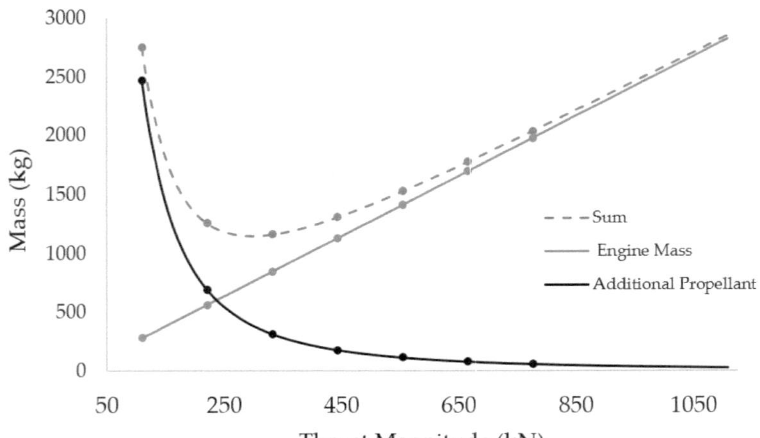


**Fig. 10.14** Computation of gravity losses and invested engine mass based on RL10B-2 characteristics

**Table 10.25** Optimum engine configurations of first stage for first manoeuvre

| Engine | No. | Thrust | $(a_0/g_0)$ | T/W | $m_{p,gl}$ | $m_{eng}$ | $m_{sum}$ |
|---|---|---|---|---|---|---|---|
| | (–) | (kN) | (–) | (–) | (kg) | (kg) | (kg) |
| LE-5B-2 | 2 | 289 | 0.12 | 51 | 422 | 579 | 1,002 |
| RL10-B2 | 3 | 332 | 0.14 | 40 | 316 | 848 | 1,164 |
| RS25 | 1 | 2279 | 0.93 | 73.1 | 7 | 3,179 | 3,186 |

while the invested mass in terms of an additional engine is 283 kg. Hence the overall mass saving by adding an engine is 1500 kg. Despite being closer to the theoretical optimum, a three-engine configuration only yields a marginal improvement of 96 kg. Adding a fourth engine does not justify the investment in terms of mass and the additional engine mass exceeds even the propellant mass savings.

In order to study the effect with real rocket engines, we repeat the above discussion for 3 further engines. The results are provided in Table 10.25. To ensure an objective comparison, the optimum configuration was chosen for each engine type. The two engine configuration of the Japanese LE-5B-2 s has the best overall performance in terms of $m_{sum}$. Note that it also has the highest gravity losses, $m_{p,gl}$, but again the lowest invest in terms of $m_{eng}$. The powerful RS25 that propelled NASA's space shuttle and now the SLS has practically no gravity losses owing to its high absolute thrust. However, its weight renders it the heaviest configuration.

In this context, it is instructive to analyse the initial acceleration level measured in $g_0 = 9.81$ m/s. A single RS25 is capable to accelerate the 250 t of the C-One spaceship with almost 1-$g_0$. This value increases as the propellant of the first booster stage is consumed. The lowest acceleration occurs with the most mass efficient configuration: $2 \times$ LE-5B-2.

The discussion reveals that there is no single parameter that can be used to define an optimum configuration. Instead, a graph like the one shown in Fig. 10.14 should be created for each engine configuration.

We discussed before the effects of gravity losses for the first manoeuvre, which is performed with the first booster stage. The losses are different for the second and the third manoeuvres in a sense that the optimum shifts towards higher thrust levels. The effect is exemplified by considering the same engine type, RL10-B2, for all boosters, Table 10.26. If all stages were propelled by a single RL10-B2, the additional propellant mass required to compensate for the finite thrust losses increases from 2472 kg for the first stage to 8732 kg for the third stage. In the latter case, this corresponds to 16.4% of the ideally required propellant, 53140 kg, for the last manoeuvre.

The optimal number of engines for each stage is not uniform. The number of engines increases by one for each booster stage, starting with three for the first booster and increasing to five for the third booster. This increase in engines is not exclusive to the RL10-B2, but also applies to other thrust rocket motors listed above. RS25 is an exception, a single engine remains the optimum configuration for all booster

**Table 10.26** Gravity losses effect for all three-stage with RL10-B2, respective optimum configuration in bold

| No. | Thrust (kN) | 1st Stage | | 2nd Stage | | 3rd Stage | |
|---|---|---|---|---|---|---|---|
| | | $m_{p,gl}$ | $m_{sum}$ | $m_{p,gl}$ | $m_{sum}$ | $m_{p,gl}$ | $m_{sum}$ |
| | | (kg) | (kg) | (kg) | (kg) | (kg) | (kg) |
| 1 | 110 | 2,472 | 2,755 | 8,970 | 9,251 | 8,732 | 9,014 |
| 2 | 221 | 695 | 1,260 | 2,758 | 3,319 | 3,547 | 4,112 |
| **3** | 332 | 316 | **1,164** | 1,293 | 2,134 | 1,840 | 2,688 |
| **4** | 443 | 179 | 1,310 | 742 | **1,864** | 1,106 | 2,237 |
| **5** | 554 | 115 | 1,529 | 480 | 1,882 | 732 | **2,146** |
| 6 | 665 | 80 | 1,776 | 335 | 2,018 | 518 | 2,215 |
| 7 | 776 | 59 | 2,038 | 247 | 2,210 | 385 | 2,364 |

**Table 10.27** Optimum engine configurations for each stage and resulting optimum mass

| | 1st Stage | | 2nd Stage | | 3rd Stage | |
|---|---|---|---|---|---|---|
| | No. | $m_{opt}$ | No. | $m_{opt}$ | No. | $m_{opt}$ |
| | | (kg) | | (kg) | | (kg) |
| LE-5B-2 | 2 | 1,002 | 4 | 1,597 | 4 | 1,835 |
| RL10-B2 | 3 | 1,164 | 4 | 1,882 | 5 | 2,146 |
| Vinci | 2 | 1,384 | 3 | 2,174 | 3 | 2,439 |
| RS25 | 1 | 3,186 | 1 | 3,208 | 1 | 3,225 |

stages. Table 10.27 shows the optimum engine configuration and the corresponding mass optimum for all four investigated engines per stage.

Note that the single $\Delta$ for each manoeuvre is different which leads to longer burn times for similar propellant loading. A longer burn time means a larger deviation from the ideal impulse. Further, the fact that the third stage has less initial mass to accelerate than the second stage is more than compensated by the larger velocity demand, 2507 m/s to 1515 m/s, see Table 10.24.

Contingency Analysis—Engine String Failure

In order to select a baseline, it is necessary to consider a number of additional factors, with the most significant being the *engine-out* failure. This is a *contingency case* that is always a topic of heated debate in spaceflight. As mentioned above, the crewed Apollo missions, for instance, had no back-up engine, neither for the Command and Service Module (CSM) nor for the Lunar Module (LM). Compared to the bold 60 s, the emphasis on safety in crewed missions has undergone a significant shift. In case for a robotic mission engine redundancy is in principle still debatable. We have seen (footnote 15 on "Separated and Unified Propulsion Systems") that the Cassini

Spacecraft to Saturn had two engines so did the Japanese robotic lunar lander SLIM.[33] Reliability comes at a price: the larger the mission the greater the implications of a full redundant propulsion system mainly in terms of mass, test effort and cost. One may argue that a chemical rocket motor is less prone to failures compared to an electric propulsion systems—mainly due to thorough testing on ground and its short operation time.[34] Furthermore, the shorter the operation time, the fewer things can go wrong, correct? A rocket motor from the legendary RL-10 family is arguably the most reliable high thrust engine that exists. On the other hand, a mission like C-One is not only prestigious but also cost intensive and will not be carried out by a single space fairing nation—failure is not an option.[35]

It is almost certain, that full redundancy will be a system requirements applicable to all subsystems, including the propulsion system. We therefore adopt a multi-engine configuration. In order to decide on the optimum number of engines, we will discuss the implications an engine-out failure has on the performance and the different options to arrange the engines on the basis of the Japanese expander cycle engine, LE-5B-2.

According to the commonality requirement stated above, we are seeking a global optimum configuration, i.e. for the entire stack and not a local one for the single booster stage. This implies that all stages shall share the same engine configuration. Figure 10.15 shows potential multi-engine configurations.

Thermal engines that operate in vacuum have high expansion ratios, $\epsilon$, to achieve a high specific impulse. This leads to large nozzle diameters, in case of LE-5B-2 the nozzle exit diameter is 1.7 m. This will inevitably become a driver of the stage diameter or at least an additional constraint. Furthermore, thrust vector control requires (TVC) *stay-out zones* to avoid nozzle collision, which drives the diameter for the aligned three-engine configuration (C3-A).

The LE-5B engine in combination with configuration C2 offers a good compromise between the technical needs with respect to gravity losses, contingency and cost saving. We adopt *C2 as baseline* for all three booster stages.

We briefly analyse the consequences of an engine-out failure in the first and the third stage. Instead of propellant mass, we will this time consider the impact on $\Delta v$ on stage and stack level. Table 10.28 shows the performance degradation in $\Delta v_{\mathrm{i}}$ and $\Delta v_{\mathrm{esc}}$ for C2 and C3-B.

In a C2 configuration an engine-out failure in the first stage would reduce the stage's achievable velocity increment, $\Delta v_1$, by merely 20 m/s and the impact on the escape velocity would be reduced by 47 m/s. This value, in the order of 1% of the requirement, is in fact low. The same failure in the third stage is more severe. It leads to a deficit of 533 m/s, which is 10% of the requirement. It became evident that the

[33] The redundant main engine saved partially SLIM's mission, see Sect. 9.2.4, as one engine was literally lost during descent.

[34] It is best practice that thermo-chemical engines are hot fire tested before and after assembly. The latter is referred to as wet rehearsal and necessary to identify potential workmanship errors.

[35] This famous phrase in spaceflight dates back to Gene Kranz, the flight director at NASA during the Apollo missions. It is also the title of his autobiography "Failure Is Not an Option" [23].

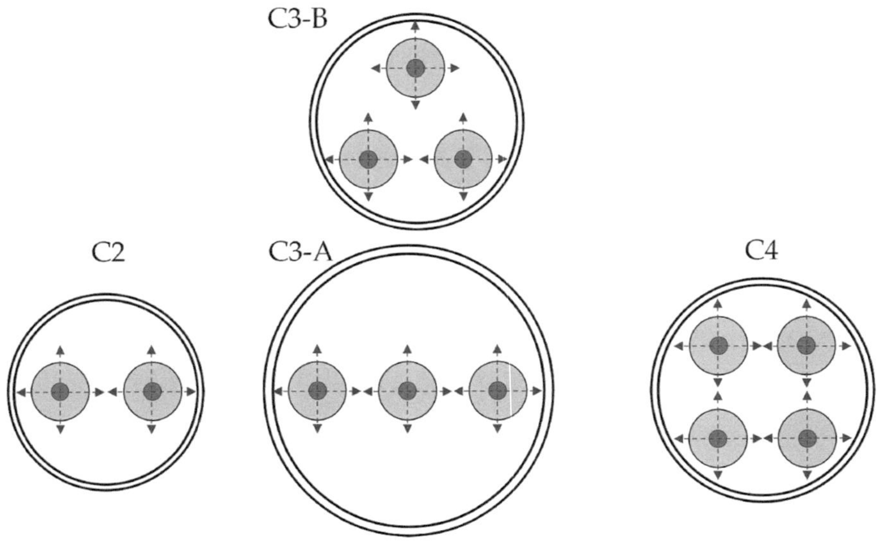


**Fig. 10.15** Multiple potential engine configurations with LE-5B-2

**Table 10.28** Performance degradation in an engine-out failure scenario for the 1st stage or the 3rd stage

| | C2 | | C3-B | |
|---|---|---|---|---|
| | 1st stage | 3rd stage | 1st stage | 3rd stage |
| $\Delta(\Delta v_i)$ | –20 m/s | –221 m/s | –15 m/s | –112 m/s |
| $\Delta(\Delta v_{esc})$ | –47 m/s | –533 m/s | –35 m/s | –265 m/s |

location of the engine-out failure within the three boosters has a significant impact on the overall outcome. Furthermore, the analysis indicates that a failure at the stage level can result in a substantially larger system failure. The same analysis for the C3-B configuration leads to significantly lower deficits in $\Delta v$. The reason for the better performance of C3-B is obvious: that after the contingency, there are still two engines left, and a two engine configuration has a very good performance in terms of gravity losses as discussed above. In case of C2, the stage is left with the relatively poor performance of a single engine configuration.

Even if not the decisive factor, the results in the table consider in addition thrust reduction due to steering losses by engine gimbaling. This is necessary in order to ensure thrust vector point through the stack's centre of mass. A consistent analysis would also require knowledge of the movement of the centre of mass during tank depletion. In order to facilitate the analysis, it is assumed that the remaining operational engines can be gimbaled. The resulting *steering losses* are for all engines the same. Assuming a typical gimbal angle of 15° leads to a thrust reduction of 3% which corresponds to 5 kN for C2 and 10 kN for C3-B. This force represents a lateral load case and must be structurally analysed.

The here discussed energetic degradation, i.e. reduction of $\Delta v_{esc}$, in case of a contingency needs to be compensated by the solar electric transfer vehicle (SETV). As previously identified, space vessels with electric propulsion systems have more control on their trajectory (powered flight) compared to thermo-chemical systems whose trajectory is purely *ballistic* (unpowered flight). However, the SETV is scheduled to perform a gravity-assist manoeuvre with Mars and Ceres. Failure to achieve these manoeuvres would have significant implications for the mission, unless a suitable backup transfer trajectory exists. The latter would entail a longer transfer time. Consequently, it is essential to extend the contingency analysis and assess the SETV's ability to cope with an engine-out failure in the third stage. The fact that the SETV has excess power and thrust available, as discussed above, helps to mitigate for phase 1 short comes.

A probabilistic-based consideration leads us to the question of the likelihood of (a) an engine failing, and (b) the engine in the third stage failing. This question is in the realm of RAMS engineers, which stands for Reliability, Availability, Maintainability and Safety. The result of this analysis will be a quantitative likelihood value that takes the complete propulsion string into account, i.e. PMS, DHS (HW and SW) as well as ground operation, if involved. These analyses are typically not subject of early concept phases. We continue with the *working hypothesis* that the dual engine configuration C2 is sufficiently reliable and that the SETV is capable to compensate the altered initial conditions of the heliocentric transfer trajectory due to its inherent growth margin.

Next we will determine the required propellant mass, which is different for each stage and not possible to deduce a priori. The first stage, for instance, has the lowest $\Delta v$ requirement but needs to push the complete stack (250,000 kg). The third stage has to push only 125,650 t but needs to achieve the highest $\Delta v$. In addition, the gravity losses are not the same for all stages. Table 10.29 provides an overview of all stages and reveals that the third stage dominates the propellant need because of the gravity losses. The highest amount of propellant must be selected and, in accordance with the requirement for commonality, must be imposed on all stages.

The propellant budget is not yet complete. Additional inevitable losses have to be considered. The three main contributors are:

*Residuals* occur in all types of propellant systems and refer to propellant mass that cannot be extracted from the tanks. It stays in the tank in different forms: as vapour mixed with the pressurant gas, as liquid that adheres to the walls or in form of bubbles floating in the tank. Typically, 2% of the totally loaded propellant is allocated for residuals.

**Table 10.29** Propellant demand per stage, C2 with LE-5B-2

| Propellant mass | $m_{p,id}$ | $m_{p,gl}$ | $m_{p,tot}$ |
|---|---|---|---|
| 1st stage | 54,949 | 422 | 55,371 |
| 2nd stage | 54,886 | 1,701 | 56,587 |
| 3rd stage | 54,727 | 2,355 | 57,082 |

*Chilldown losses* refer to propellant mass that is only needed for propulsion systems that run with cryogenic propellant. The propellants are stored at extremely low temperatures and have very low boiling points: 90 K (–183 °) for LOX and 20 K (–253 °) for LH2, see Sect. 8.2.2. In contrast the feed lines and the engine are relatively hot, which causes the propellant to evaporate immediately when it comes in contact with it. This causes thermo-mechanical stresses and distortions [19]. To prevent this, it is important to bring the feed system to working temperature before operation. This preconditioning is called chilldown and it foresees to flush the feed system with the cryogenic propellant. This process is purely passive, propellant is not burned. We allocate 500 kg for this operation based on experience for upper stage engines.

*Boil-off losses* refer the propellant mass that is lost due to parasitic heat transfer. Tanks of cryogenic propellants cannot be hermetically insulated from external heat sources (e.g. irradiance, Earth albedo, heat dissipation from avionics). The heat entering the tanks is called parasitic heat and causes the propellant to boil. To prevent overpressure in the tank, this vapour has to be vented to space and is lost for the manoeuvre. The longer the stage remains in space, the greater the boil-off losses. For what concerns the boil-off losses during the assembly phase of C-One we expect compensation by in-orbit refuelling. Given the current state of this technology and business case, it is prudent to assume that this technology will be mature and available as a commercial service in ten years' time.[36] However, the time between refuelling completion and Main Engine Cut Off (MECO) needs still to be calculated for the propellant budget. Boil-off losses of a cryogenic propellant combination depend on several design and environmental factors as well as the mission scenario since the latter determines the time in orbit and the stage's orientation relative to the heat sources. In the context of our preliminary design a ballpark value is sufficient to understand the order of magnitude. Data of the Centaur upper stage for long duration LEO missions have shown that the boil-off can be limited to 2% per day of the loaded propellant mass with pure passive means [18]. The team argues that this value can be brought to 1% if more advanced insulation means are used. As a working hypothesis it is justified to adopt this value since it is based on in-flight measurements of a very similar vehicle. To determine overall mission duration, we need to expand the analysis to the timetable of the escape manoeuvres. For this we need to apply the astrodynamical formula introduced in Sect. 2.2.2 for the simplified 2 dimensional case. Table 10.30 shows the change of the orbital parameters and Fig. 10.16 depicts the resulting orbits in a simplified 2 dimensional scenario.[37]

The initial orbit is circular and has an altitude of 400 km. The manoeuvre of the first stage changes the shape of the orbit and the circular orbit becomes an ellipse with and altitude of 5,556 km. The second manoeuvre has a larger impact and raises the

[36] In fact, the Starship user guide offers in-orbit refuelling alongside classical orbit delivery service. In addition to SpaceX there are a number of commercial companies developing this technology and plan to offer refuelling service, too. Nonetheless, the total boil-off losses need to be determined to place the right order for the vendor.

[37] The line of apsides rotates slightly after each manoeuvre. For the sake of clarity, this effect is ignored in the depiction. The focus is on the change of its length.

**Table 10.30** Sequence of orbital changes during from first to third manoeuvre, phase 1

| | | Inital orbit | Intermediate orbit 1 | Intermediate orbit 2 | Final orbit |
|---|---|---|---|---|---|
| $h_p$ | km | 400 | 400 | 400 | 400 |
| $h_a$ | km | 400 | 5,556 | 53,031 | $\infty$ |
| $a$ | km | 6,778 | 9,356 | 33,094 | $\infty$ |
| $T$ | h | 1.5 | 2.5 | 16.6 | $\infty$ |
| $t_{burn}$ | min | 29 | 29 | 29 | – |
| $t_{burn}/T$ | % | 31 | 19 | 3 | – |

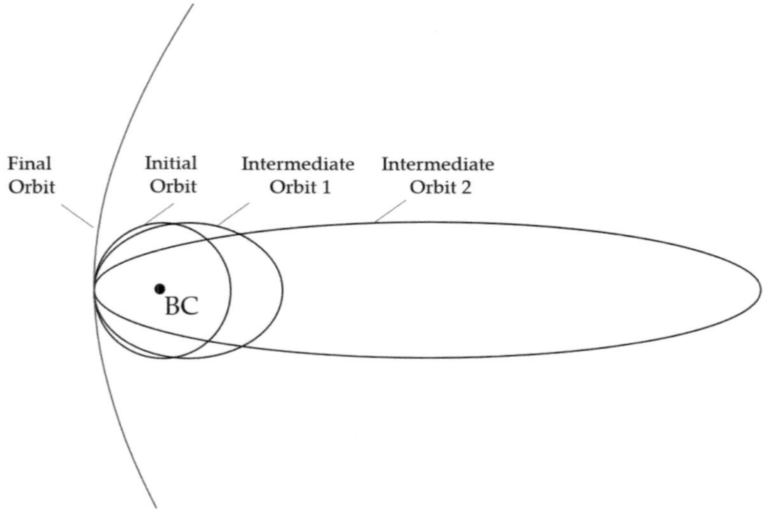


**Fig. 10.16** Simplified depiction of phase 1 escape manoeuvres by the three booster stages

apogee to an altitude of 53,031 km, which is above GEO. Finally, the last manoeuvre reaches the desired escape hyperbola.

Interestingly, the table also shows the manoeuvre times in terms of engine burn time, $t_{burn}$. The LE-5B-2 consumes propellant on a rate of 32.87 kg/s. Since the available and usable propellant mass is the same for all stages, 57,082 kg, the burn time is also the same. However, note that the orbital fraction $t_{burn}/T$ is 31% for the first manoeuvre and reduces down to 3% for the last burn. Boil-off losses for the third stages are largest since the overall time spent in space is longest. It remains the question how much time will pass from final refuelling till the first manoeuvre. It is reasonable to take a provision of six hours at this stage until a more elaborated mission plan is established. This time should be sufficient to establish a safe clearance between the C-One and the propellant tanker. Hence, a total mission duration of slightly less than one day results for the third stage. Assuming 2% losses boil-off per day yields about 1,000 kg for the third stage (Table 10.31).

**Table 10.31** Booster propellant budget

| | |
|---|---|
| Ideal impulse | 54,727 kg |
| Gravity losses | 2,355 kg |
| Chill-down losses | 500 kg |
| Residuals | 1,152 kg |
| Boil-off losses[a] | 1,000 kg |
| *Total propellant mass* | **59,734 kg** |
| Propellant mass LOX | 49,778 kg |
| Propellant mass LH2 | 9,956 kg |

[a]Commercial in-orbit refueling service assumed

## *10.4.2 Thermo-Mechanical Architecture*

We have seen in Sect. 9.6.4 that for vehicles with large amount of propellant, i.e. low dry mass index, an *integral tank design* is the best option from a mass efficiency point of view. The mechanical design of large structures requires detailed structural analysis,[38] which is clearly beyond the scope of this introductory book. We therefore restrict ourselves for illustrative reasons to a geometry based discussion, which in fact is always the starting point before conducting quantitative analyses for verification.

There are three different types of stage design, that have proven their worth. Figure 10.17 shows the three options for a booster stage with the propellant load specified above and an ullage volume of 5%. For the sake of comparison it is assumed that all stages have the same diameter. We will see later that length and diameter are actually determined by the available payload envelope of the fairing.

A *common bulkhead* design leads to the shortest architecture. Compact designs are in general more lightweight and in our geometry-based consideration, this feature is a direct consequence of the smaller surface area. The price that needs to be paid for this compactness is a heat conductive bridge formed between fuel and oxidiser via the bulkhead. Heat flows from the 'hot' LOX (90 K) tank into the cold fuel LH2 (20 K) tank, causing the fuel to boil. Fortunately, the enthalpy of evaporation, $h_{evp}$, of LH2 is large and it takes a considerable amount of heat to let 1 kg boil-off, refer to Table 8.8, which makes LH2 a good heat sink. On the other hand, the loss of heat from the LOX tank bears the veritable risk of freezing.[39] To mitigate this problem a foam insulation is generally used. Overall the boil-off and freezing issue remains significant especially if it comes to long journeys with cryogenic propellant, for instance into cislunar space.

The *separated tank* design mitigates the coupled boil-off-freezing problem by pure physical separation. In addition the tank domes are covered by MLI (Multi Layer

[38] Using finite element method (FEM).

[39] The situation is reversed and even more severe in case of methalox. Heat flows from the hot liquid methane (110 K) into the LOX tank and causes LOX to boil. The situation is more severe because of a) enthalpy of evaporation of LOX is only half of that of LH2 and b) liquid methane is closer to its freezing point than liquid LOX [20].

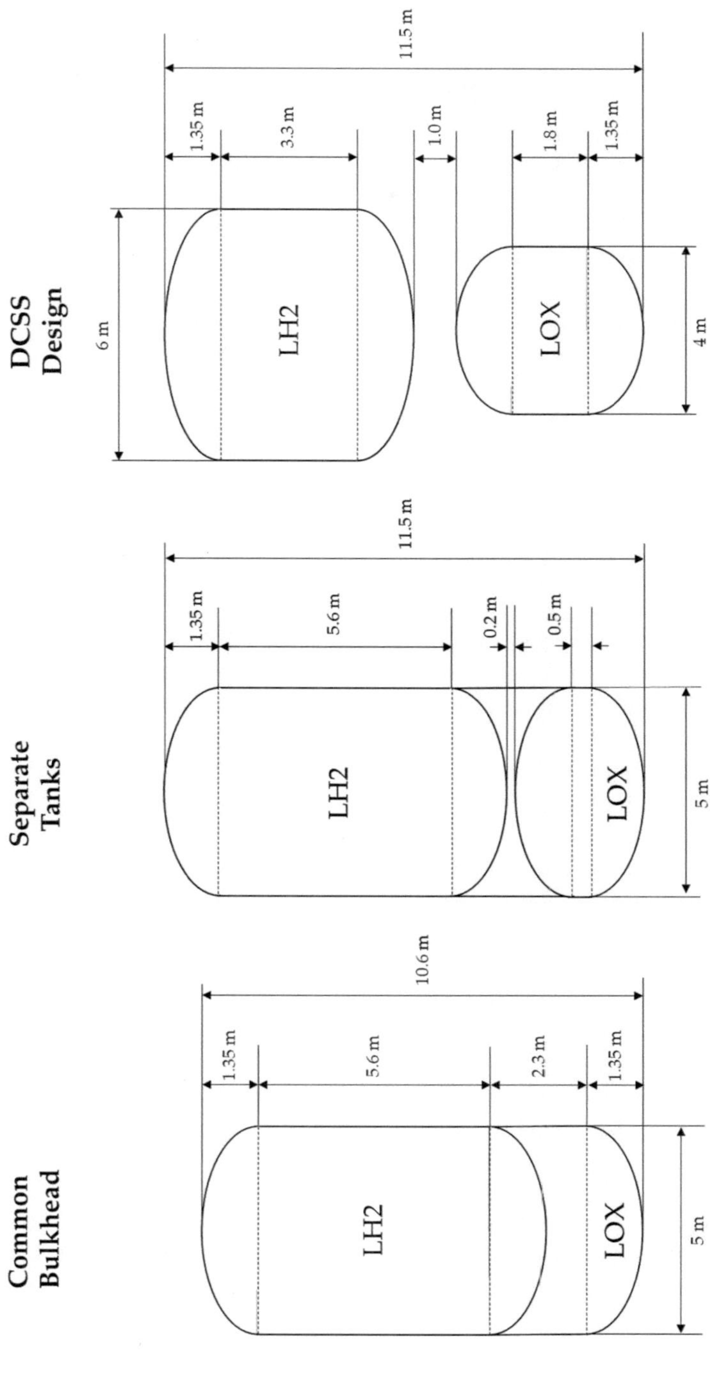


**Fig. 10.17** Comparison of mechanical architecture options for a booster stage

Insulation) blankets, a technology that limits radiative heat transfer. Conductive heat transfer between the two propellants still takes place through the inter-tank structure which encloses the two opposing domes. This separation is the reason, why this architecture is about 1 m larger than that of the common bulkhead and will be more massive. The mass increased related to this architecture is also due to an additional tank dome. From a geometric point of view, there is a constraint that must be regarded by all three architectures, but it is the separate tank design that is most constrained. The constraint applies to the oxidiser tank diameter. For structural reason, the two bulkheads are not allowed to get too close and a minimum distance must be kept for the cylindrical part of the tank. A reasonable limit is 0.5 m, which is achieved for a stage diameter of 5 m. An attempt to further reduce the overall tank height, 11.5 m, by increasing the diameter is therefore not possible with this architecture. The structural reason for this, is that two domes should not be attached to each other and a cylindrical segment is necessary to separate the two elements.

The *DCSS type* design is by definition also a separate tank architecture but with the decisive difference that both tank diameters are decoupled. This design offers several advantages firstly, by decoupling the two geometries, both tanks, for LH2 and LOX, are allowed to be designed according to their geometric and structural needs and secondly, thermal decoupling between the two tanks is even better than for the simple separated tank design. The launcher interfaces mechanically the stage at the LH2 cylindrical tank segment, which is long and able to carry the main load path. During launch, the upper part of the stage drags the LOX tank behind it, which then experiences tensile forces. This situation is reversed after separation from the launcher when the vehicle is in space but at much lower acceleration forces.[40] Furthermore, by decoupling the two tanks it is possible to accommodate significantly more propellant while limiting the overall height of the stage. A good example is the Exploration Upper Stage of SLS Block 1B, see Sect. 9.6.4. This is because the DCSS type stage attaches the LOX tank to the LH2 tank by means of struts instead of a shell. Not only are these struts lighter, they can be made of low conductive materials. Besides these advantages there are similar disadvantages like for the separate tank design: the overall surface is larger than for the common bulkhead architecture leading to a higher dry mass index. The fact that the launch adapter needs to be stretched up to the LH2 tank increases the structural mass if it stays attached. If it remains with the launcher, it reduces the launcher's payload capacity.

In addition, to these geometric considerations, structural considerations must also be taken into account. We shall stress here two subjects: the tank dome height and the height over diameter ratio. The first, the tank dome height cannot be too flat as this causes a phenomenon called *compression hoop stresses*. The assumed value of 1.35 m is a working hypothesis based on best engineering guess (BEG) and needs later verification. The second is related to the minimum stiffness requirement imposed by the launcher on the payload. This stiffness requirement is more difficult to meet, the larger the height over diameter (H/D) ratio. This is intuitive, since stiffness is the

[40] This is how the load case for the stage looks like if a classical launch is assumed. If the stage is inside a payload bay like of Starship, the situation may be very different.

resistance of a structure to deform (e.g. bend) and a slim stage with a large H/D ratio is easier to bend than a small and wide stage. To increase the stiffness of slim stages, it is necessary to increase the wall thickness. Again, this is intuitive: a heavier cylinder is harder to bend than a lighter one.

The discussion so far enables us to select a baseline architecture for the C-One booster stages. Given the above made assumption that there will be a servicing mission to refuel the booster stages prior to mission start and given the short mission time of each stage, the need to reduce boil-off losses by stage design is not essential. Dimension and overall mass become then the main parameters. The common bulkhead design excels in both. Its diameter can be increased to 9 m before the cylindrical part of the LH2 tank hits the threshold of 0.5 m—thereby reducing the overall tank assembly height to a mere 3.9 m. This flexibility could proof helpful when analysing the fairing's static envelope, see Fig. 10.2. The common bulkhead excels also in terms of tank mass as discussed above and we will see further below that this advantage will become even more pronounced when considering the remaining structure not yet discussed, e.g. thrust frame, upper and lower inter-stage structure. We therefore, baseline the common bulkhead architecture for the three booster stages.

Launch and Stacked Configurations

In our analytical approach, we have restricted the discussion above to the tank only. We complete the picture in this paragraph by considering the launch and stacked configuration of the stages. Figure 10.18 shows the launch configuration of a 5 m diameter common bulkhead stage. The total height increases by 5.5 m because of the upper compartment and the lower assembly.

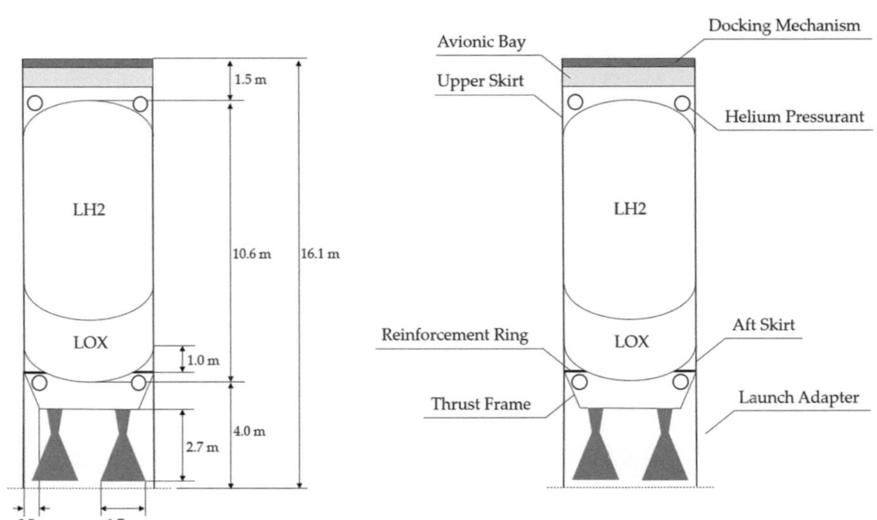


**Fig. 10.18** Booster architecture for a common-bulkhead configuration with an outer diameter of 5 m

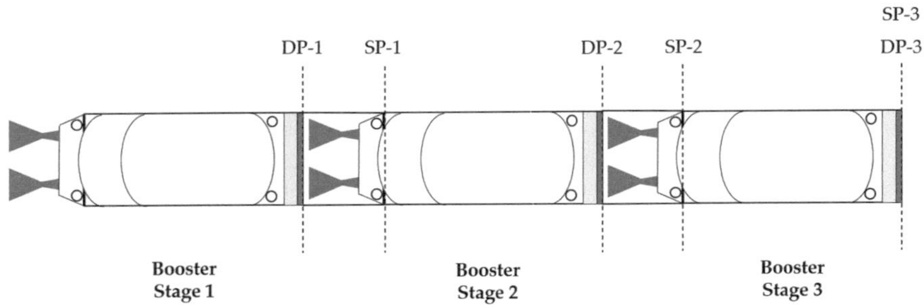


**Fig. 10.19** Stacked booster configuration with separation planes (SP), docking planes (DP)

The first contains the avionic bay (e.g. DHS, TTC, batteries), propellant for attitude control and the docking mechanism. A height provision of 1.5 m is foreseen—a value that needs to be verified with a more detailed accommodation analysis. The lower assembly contains the propellant management system, helium pressurant vessels, a thrust frame to which the engines are mounted and the launch adapter. The length of the lower assembly is determined by the thrust frame and engine length. It could be reduced, if the engine offers the feature of a deployable nozzle pending the available stage diameter. As this is not the case for the selected baseline, LE-5B, we will not assume this beneficial feature. The width of the launch adapter depends among other factors on the launcher interface provision. All in all the allocated height of the lower assembly is 4 m leading to a total stage height of 16.1 m for a 5 m diameter configuration.

In the next step, we consider the stacked booster stages configuration without SETV and CLM, Fig. 10.19. The basic depiction shows that like for launchers, inter-stages are needed to connect the single booster stages to each other. These inter-stages need to be longer than the engines to ensure clearance during booster stage separation. Separation should occur at the side of the firing stage to reduce the mass of the remaining stack, like done for launchers during ascent. In total there are three separation planes (SP), two among the booster stages and the third between booster stage 3 and SETV.

### 10.4.3 Launcher Compatibility Analysis

Like for the CLM an SETV a launcher compatibility check needs to be performed. In contrast to the others, we do have all information to perform payload envelope check besides the pure mass check. The dry mass index of 16% leads to a total wet mass of 69 t for a single booster, of which 9.5 t is dry mass. Hence, a fully loaded booster stage can only be launched by a super heavy lift launch vehicle, see Table 10.1. Since we have assumed earlier in-orbit refuelling as a service, it is only logical to consider the option of launching an under-loaded booster and refuelling more in orbit. This

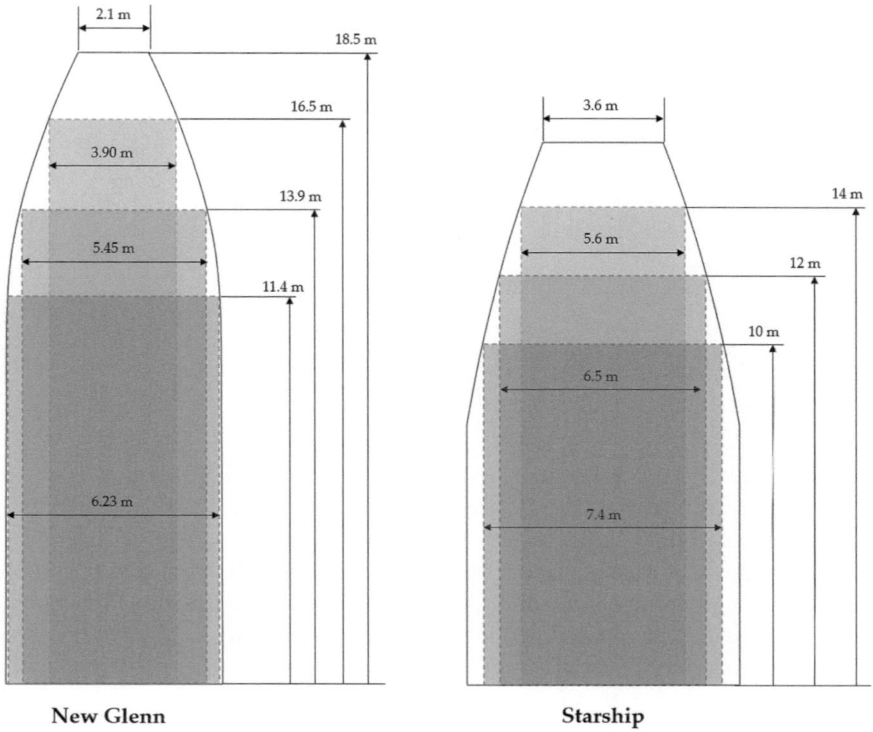


**Fig. 10.20** Admissible height-to-diameter ratios of NewGlenn's [3] and Starship's [4] payload envelope

'trick' increases the number of suitable launchers, like New Glenn with a capability of 45 t into LEO, and Falcon 9 Heavy with 63 t. The latter has a too small payload envelope provision and needs to be discarded from the trade-off [21]. We restrict the volume compatibility analysis to New Glenn and Starship.

Figure 10.20 depicts the available volume for a rectangular shaped payload such as the booster inside both payload fairings. Table 10.32 compares the fairing's admissible height-to-diameter ratios with the stage's geometric manifold.

The result is that a hydrolox based booster stage fits marginally into the Starship payload bay and does marginally *not* fit into the fairing of New Glenn. Marginal compliance at this early design stage could go either way. On the one hand we have assumed that the upper compartment requires a height of 1.5 m which is rather conservative and height saving is conceivable. We further assumed in Table 10.32 that this value does not change with the diameter of the stage, i.e. the available volume increases with the diameter of the stage. However, the equipment housed in this volume (avionics, etc.) is not volume dependent. It is safe to assume that a detailed accommodation analysis will result in a lower height for the upper compartment.

**Table 10.32** Payload envelope compliance check, non compliant (NC), marginally compliant (MC), marginal non compliant (MNC)

| Diameter | 5.0 | **6.0** | 6.5 | 7.0 |
|---|---|---|---|---|
| Upper compartment | 1.5 | 1.5 | 1.5 | 1.5 |
| Overall tank height | 10.6 | 7.6 | 6.6 | 5.8 |
| Lower assembly | 4.00 | 4.00 | 4.00 | 4.00 |
| Total stage height | 16.1 | **13.1** | 12.1 | 11.3 |
| New Glenn | 14.7<br>NC | 12.1<br>NC | –<br>NC | –<br>NC |
| Starship | 15<br>NC | **13.2**<br>**MC** | 12<br>MNC | 11<br>MNC |

Height optimisation is also possible for the lower assembly leading to a shorter thrust frame, potentially at the expense of additional structural mass.

This discussion highlights two crucial aspects that require further consideration. The first is the fact that super heavy lift launch vehicle tend to have a relatively low payload volume provisions which is equivalent to high payload densities. The second reveals a general problem with hydrolox as a rocket propellant—its low mean density is its tender spot. Launch service providers are known to be conservative in the user guide, promising rather less than what is actually possible. They explicitly ask the customer to consult the service provider for technical iteration to find a solution for marginal non compliance. Fairings have been adapted in past missions. Chances to achieve compliance in a joint effort are therefore likely.

### *10.4.4 Preliminary Baseline Design Booster Stage*

We summarise the outcome of the above discussion and analyses for the booster stage. This summary forms the DDP for the next design iteration step. We start by formulating the main functional requirement:

*BSK-Func-REQ-1*: The Booster Stack (BSK) shall accelerate the payload (SETV & CLM) to escape velocity starting from LEO

Figure 10.21 shows the performed architecture relevant trade-offs related to the propulsion system and the thermo-mechanical layout of the booster stage and the path we took in the discussion.

Table 10.33 lists the main mission and system requirements as well as the design choices and derived requirements. These requirements were used to develop a baseline design, which is outlined in Table 10.34.

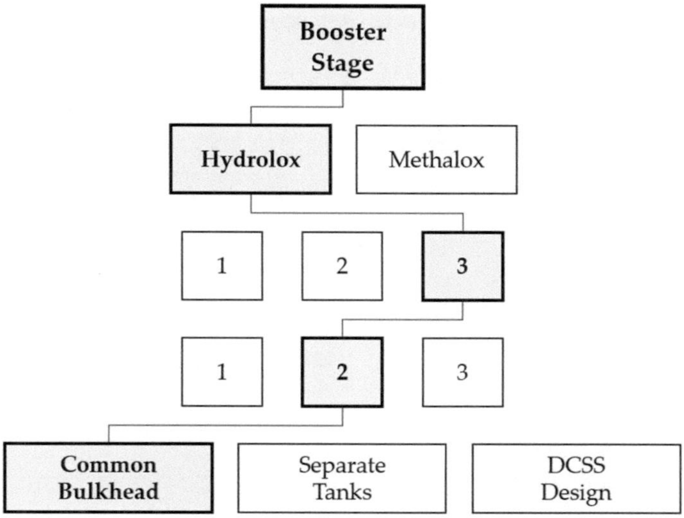


**Fig. 10.21** Booster architecture trade-off space

**Table 10.33** Derived requirements, design choices and assumptions for the booster stages

| *Mission and system requirements* | | | |
|---|---|---|---|
| Escape energy, $C3$ | 34 | km/s$^2$ | Mission requirement |
| Initial orbit altitude | 400 | km | Mission requirement |
| Payload Mass (CLM) | 63.5 | t | System requirement |
| *Overall booster architecture* | | | |
| No. stages, | 3 | | Design choice |
| Tank architecture | Common bulkhead | | Design choice |
| Max. dry mass | 9557 | kg | Derived requirement |
| Max. Tot. wet mass, $m_{tot}$ | 69,292 | kg | Derived requirement |
| Max. stage diameter $D_{st}$ | 6 | m | Design choice (TBC) |
| Max. stage height $H_{st}$ | 13.1 | m | Derived requirement |
| *Booster propulsion system* | | | |
| Propellant combination | Hydrolox | | Design choice |
| Min. Tot. propellant mass, $m_p$ | 59,734 | kg | Derived requirement |
| Mixture Ratio | 5 | | Design choice |
| No. Engines | 2 | | Design choice |
| Min. Tot. Thrust, $F_{tot}$ | 288 | kN | Design choice |
| Min. $I_{sp}$ | 445 | s | Design choice |
| Thrust vector control | Engine Gimballing, 15 | ° | Design choice (TBC) |

**Table 10.34** Main elements of SETV baseline architecture

| *LH2 propellant tank* | |
|---|---|
| Overall tank height | 6.0 m |
| Cylindrical tank segment | 3.3 m |
| Dome height | 1.35 m |
| Dome type | Elliptical |
| *LOX propellant tank* | |
| Overall tank height | 2.9 m |
| Cylindrical tank segment | 1.6 m |
| Dome height | 1.35 m |
| Dome type | Elliptical |
| *Main engine* | |
| Cycle type | Expander bleed cycle |
| Baseline main engine | LE-5B-2 |
| Nominal thrust | 144 kN |
| Nominal $I_{sp}$ | 447 s |
| Engine mass | 290 kg |
| Nozzle exit diameter | 1.7 m |

## 10.5 Phase 0 Launch and In-Orbit Assembly

After we have gained an understanding of the overall C-One architecture in terms of mass and partially of the layout in terms of dimension. It is possible to proceed and discuss a preliminary launch and assembly scenario with focus on the sequence of both. In total a minimum of five elements need to be launched:

*Ceres Landing Module*: the CLM consisting of the Ceres Payload Module (CPM) and the Ceres Landing Propulsion Module (CLPM) with a combined mass of $\simeq$36 t,
*Solar Electric Transfer Vehicle*: the SETV with a mass of $\simeq$25 t,
*Booster Stages*: the three hydrolox booster stages with a wet-mass of $\simeq$70 t each.

Minimum, because the CLM elements, CPM and CLPM, could be launched separately pending detailed structural analysis. If co-manifested its launch requires a heavy lift launch vehicle, like NewGlenn or Falcon 9 heavy but presumably with an enlarged fairing.

The SETV is at the edge of the capability of a medium launch vehicle, like Vulcan. The packing volume of its solar power generator (SPG) will be the primary driver for the launch. While a single solar array can be compactly stowed in a mandrel, the entire assembly, including the long wing masts, will require significant space. Special folding techniques are currently being investigated to reduce the overall accommodation volume thereby ensuring sufficient stiffness to withstand launch loads [22]. The cryogenic booster stages can be launched fully loaded by a super heavy lift launch vehicle. Alternatively, they can be launched by a heavy lift launch vehicle but with a lower filling ratio, but this should not be the preferred launch option. The fifth launch is needed to for re-fuelling,

In-Orbit Assembly

The assembly of vehicles in space can be performed either by *docking* or *birthing*. We will discuss in the following the two options for C-One without drawing a baseline the intention is rather to provide insight in the dependencies and criticality of this task.

Docking refers to the ability of a spacecraft to mate with another vehicle, either autonomously or under the guidance of a crew member, if one is present. Birthing denotes a mating procedure in which the approaching free-flying vessel is captured with a mechanism, in general a robotic arm. The international space station (ISS)[41] has been assembled using both methods. Supply vessels such as the former ATV (Autonomous Transfer Vehicle) and the current Dragon capsule from SpaceX dock autonomously with the ISS. Northrop Grumman's Cygnus and Jaxa's HTV, on the other hand, are birthed to the ISS by a robotic arm that is capable to wander over the space station like a caterpillar. The first group is equipped with a full sensor suit and a fine control propulsion system, while the second group is rather passive. Whether a robotic arm is capable to handle such large masses must be analysed and a detailed trade-off of both options is needed. The selected technology is relevant for the final assembly plan, among other factors. The cryogenic booster stages, for instance, represent the only time-critical element of the spaceship due to boil-off losses. While in-orbit refuelling is assumed as a service, minimizing boil-off losses is still essential to reduce the propellant required to the extent that only a single servicing mission is necessary. It is, therefore, reasonable to demand that the boosters are launched last, which reduces their orbital dwell time and consequentially propellant losses.

Given this premise, launch and mating of the CLM and the SETV will take place first. The three boosters will be mated to them in consecutive order, such that the spaceship grows from the 'head to the tail', i.e. from the CLM to booster stage 1. If a robotic arm is used the dynamics of the mating process are strongly dependent on the mass ratio of the involved objects. The berth place is ideally considerably larger than the free-flying object, as this would reduce the complexity of the stack dynamics. This is difficult to achieve for C-One. The mass of CLM, 36 t, is not overwhelmingly lighter than that of SETV, 25 t. Together they have a mass of 61 t, which is in the order of the fully loaded booster stage, 70 t, leading again to a ratio of about one. The situation changes for the remaining two booster stages, the ratio of stack mass to booster mass improves successively to $\sim$2 and then $\sim$3. A detailed dynamic analysis is required to assess the feasibility of a robotic arm for C-One assembly and a mixed approach might be the outcome.

[41] The name is in fact misleading as China is explicitly excluded from participation which prompted the nation to design its own space station, Tiangong (heavenly palace).

Concept of Operation—ConOps

An important element of the systems engineering discipline is the *Concept of Operation*, short ConOps. In essence it is a synopsis of how a system or entire mission works and often presented in form of a depiction. The mission ConOps for C-One is depicted in Fig. 10.22.

The mission commences with the execution of the first manoeuvre (M1) using booster 1, which propels the spaceship to the first intermediate orbit (IMO-1). The depleted booster is then separated and the spaceship is aligned by the RCS so that it can perform the second manoeuvre. The orbital period of IMO-1 is 150 min of which 29 min is the burn duration of M1. Hence, the remaining time for safe separation and re-orientation is lower, about 120 min. In contrast to launchers, where gravity helps to gain clearance between the depleted and the active stage, active means need to be taken to achieve safe distance. This can be achieved by retro-thruster located at the depleted stage and/or by a mechanical separation system. The important question for operations is again the time needed to fulfil both tasks. Detailed GNC and multi-body analyses are required to determine the available time margin.

There are two open questions that need to be addressed in this context: firstly, what happens to the booster stages after separation and secondly, could the solar power generator of the SETV be deployed earlier? The first and second booster will stay in bound orbits with very low perigees and due to the ballistic factor of the stages, also with a very high orbital lifetime of thousands of years. The apogee of IOM-1 where booster 1 is flying is relatively close, 5,556 km. The remaining mass of the booster, dry mass plus residuals, is about 10 t. A salvage mission is doable and probably worth it. The second stage presents a significantly greater challenge due to its apogee of 53,031 km, which is beyond GEO. A salvage mission is complex and requires thorough assessment, and an electric tug is likely the best option to recover this asset and refurbish it for a future mission. The third booster stage, however, is beyond reach and thus lost in a heliocentric orbit.

It would be ideal to deploy the SETV's solar power generator (SPG) in LEO prior to mission start where potential issues could be solved by a servicing mission. This confidence measure has several drawbacks: berthing with a fully deployed SPG is extremely challenging due the high moments of inertia (MoI). Furthermore, the risk of space debris impacts is exacerbated due to the increased cross section of C-One. A further challenge is the escape phase: the spaceship experiences a linear acceleration during the single manoeuvre (M1-M3), which increases after each manoeuvre but stays relatively low ($\simeq 0.2 \cdot g_0$). However, engine ignition and shut-down will cause the arrays to oscillate around several axes, which give rise to dynamic loads. The second challenge is reorientation after each manoeuvre. We have seen that the available time is small between M1 and M2 and in view of the increased MoIs this will take either more orbits in IM-1 or larger RCTs, which again leads to the problem of array oscillations and dynamic loads (not to speak of boil-off losses). Given these challenges and associated risks it is advisable to baseline SPG deployment after M3 and booster stage 3 separation—pending detailed analysis.

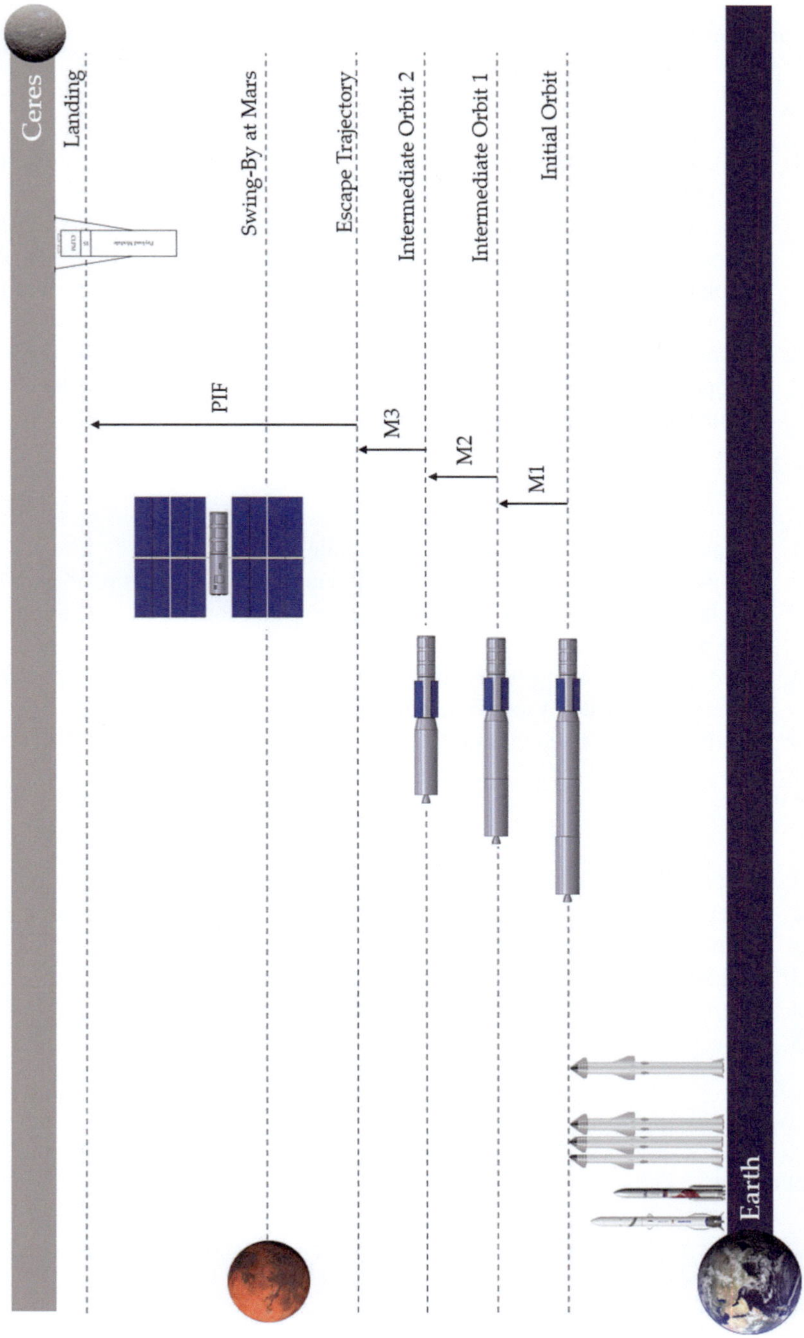


**Fig. 10.22** Concept of operation (ConOps) for C-One

Assembly of C-One and refuelling take place in a low Earth orbit (LEO). After final system check out, the mission is initiated. We decided to break down the mission in distinct phases, the nominal single steps per phase are:

*Phase 1*:

Step 1: Initiate the first manoeuvre by firing booster stage 1 until main engine cut off (MECO).
Step 2: The second manoeuvre (M2) takes place around perigee of IOM-1 using booster 2, which brings C-One into the intermediate orbit 2 (IOM-2). The orbital period is now 16.6 h, which grants sufficient time to execute the above discussed tasks
Step 3: Firing booster 3 around perigee of IOM-2 is the final manoeuvre (M3). It propels the spaceship to the final orbit: a hyperbolic escape trajectory with course to Mars. This concludes phase 1.

*Phase 2*:

Step 1: Separation of booster 3 from the remaining spacecraft.
Step 2: Deployment of the solar power generator and system power-up follow. This includes powering the thrusters, reorienting the spacecraft, and initiating the powered interplanetary flight toward Mars.
Step 3: Perform Mars swing-by, reorienting the spacecraft and continue powered interplanetary flight toward Ceres rendezvous.
Step 4: Arrival at Ceres. Align trajectory with Ceres and maintain safe distance.

*Phase 3*:

Step 1: Separation of CLM from SETV and initiation of orbit insertion.
Step 2: Prepare Ceres landing manoeuvre: CLM reorientation and propulsion system pressurisation.
Step 3: Initiate descent by retrograde burn, Fig. 10.4.
Step 4: Throttle engines to match anticipated flight profile
Step 5: Touch-down on Ceres.

## References

1. Gamgami, F., Bos, R., & Zandbergen, B. T. C. (2013). *Large solar electric transfer stages for lunar exploration*
2. United Launch Alliance. (2023). Vulcan Launch Systems User's Guide
3. New Glenn Payload User's Guide (2018)
4. Starship User's Guide (2020)

5. Wilhite, A., Wagner, J., Tolson, R., & Moen, M. (2008). *Lunar module descent mission design*. https://doi.org/10.2514/6.2008-6939
6. Russell, C., Capaccioni, F., Coradini, A., De Sanctis, M. C., Feldmann, W., Jaumann, R., Keller, H., Mccord, T., Mcfadden, L., Mottola, S., Pieters, C., Prettyman, T., Raymond, C., Sykes, M., Smtih, D., & Zuber, M. (2007). Dawn mission to vesta and ceres. *Earth, Moon and Planets, 101*, 65–91 (2007). https://doi.org/10.1007/s11038-007-9151-9
7. Liu, J., Xu, B., Li, C., & Li, M. (2022). Lifetime extension of ultra low-altitude lunar spacecraft with low-thrust propulsion system. *Aerospace, 9*(6). https://doi.org/10.3390/aerospace9060305
8. Klem, M. D., Smith, T. D., Wadel, M. F., Meyer, M. L., Free, J. M., & III, H. A. (2011). Liquid oxygen/liquid methane propulsion and cryogenic advanced development. In *62nd international astronautical congress 2011* (pp. 6256–6267). IAC.
9. Hammock, C. E., & Fisher, A. (1973). Apollo experience report: Descent propulsion system.
10. Kalia, P., Evans, J. W., Menzel, M., & Kilic, H. A. (2023). Managing risk for the james webb space telescope deployment mechanisms: Enabling first light. In *2023 annual reliability and maintainability symposium (RAMS)* (pp. 1–6). https://api.semanticscholar.org/CorpusID:257958875
11. Vanthuyne, T. (2009). An electrical thrust vector control system for the vega launcher. https://api.semanticscholar.org/CorpusID:164208979
12. Tsuda, Y., Kato, T., Shiraishi, M., & Matsuoka, M. (2015) Trajectory navigation and guidance operation toward earth swing-by of asteroid sample return mission "Hayabusa2".
13. Jovel, D., Walker, M., & Herman, D. (2022). Review of high-power electrostatic and electrothermal electric propulsion. *Journal of Propulsion and Power*. https://doi.org/10.2514/1.B38594
14. Jackson, J., Allen, M. P., Myers, R. M., Hoskins, A., Soendker, E. H., Welander, B., Tolentino, A., Hablitzel, S., Hall, S. J., Gallimore, A. D., Jorns, B. A., Hofer, R. R., & Goebel, D. M. (2017). 100 kW nested hall thruster system development IEPC-2017-219. https://api.semanticscholar.org/CorpusID:53706554
15. Hofer, R., Lobbia, R., Chaplin, V., Ortega, A. L., Mikellides, I., Polk, J., Kamhawi, H., Frieman, J., Huang, W., Peterson, P., & Herman, D. (2019) Completing the development of the 12.5 kW hall effect rocket with magnetic shielding (HERMeS). In *36th international electric propulsion conference*. IEPC Paper 2019-193.
16. Straubel, M., Hillebrandt, M., & Hühne, C. (2016). Evaluation of different architectural concepts for huge deployable solar arrays for electric propelled space crafts.
17. Mason, L. (2001). A comparison of brayton and stirling space nuclear power systems for power levels from 1 Kilowatt to 10 Megawatts. https://doi.org/10.1063/1.1358045
18. Kruif, J. S. D., & Kutter, B. (2007). Centaur upperstage applicability for several -day mission durations with minor insulation modifications. https://api.semanticscholar.org/CorpusID:112831667
19. Moreau, G.-M., Le Thanh, Kc., Bachelet, C-H., & Duri. D. (2015). Toward the chill-down modeling of cryogenic upper-stage engines under microgravity conditions using the thermal-hydraulic code COMETE. In *EU-CASS 2015—6th European conference for aeronautics and space sciences*. Cracovie. cea-02500837.
20. Gamgami, F., Rohrbeck, M., Perczynski P., Schonenborg, R., Beaurain, A., Lassoudiere, F., Tomassini, A., (2016). The choice of high trust liquid propulsion stages in human exploration of the solar system. In *Space propulsion conference, SP2016-3125077*
21. Space Exploration Technologies Corp. (2021). Falcon User's Guide.
22. Straubel, M., Hillebrandt, M., & Hühne, C. (2016). Evaluation of different architectural concepts for huge deployable solar arrays for electric propelled space crafts.
23. Kranz, G. (2000). *Failure is not an option*. Berkley Publishing. ISBN 0-425-17987-7.

# Part V
# The Near Future: Nuclear-Based Space Propulsion

# Chapter 11
# Nuclear Propulsion Technology and Systems

**Abstract** The intention of this chapter is to conclude the book by presenting nuclear-based propulsion technology in its two forms: nuclear thermal propulsion (NTP) and nuclear electric propulsion (NEP). Rather than providing a synopsis, as done in previous chapters, the focus will be on the basic working principles and system implications. Achievements made so far will be provided as illustrative examples.

## 11.1 Nuclear Propulsion Then and Now

Nuclear propulsion technology has long been identified as a key enabler for space travel, due to its high power density and longevity. Compared to a chemical propulsion system a nuclear reactor heats the propellant by transferring the released nuclear energy in the fission process, thereby replacing the combustion process and eliminating the need for an oxidizer. This type of propulsion is called nuclear thermal propulsion (NTP), and the engine itself is often referred to as a nuclear thermal rocket (NTR) in US and nuclear rocket engine (NRE) in Russian literature. In the context of electric propulsion, the reactor replaces the solar power generator, removing the bothersome distance dependence of the power output. This type of propulsion is referred to as nuclear electric propulsion (NEP).

The US Rover/NERVA projects from 1955 till 1973 made great strives in understanding the processes and material sciences required to safely operate a space nuclear reactor. In total 22 reactor types have been studied with thermal power outputs from 44 to 4200 $MW_{th}$. Table 11.1 provides a selection of reactors constructed and tested. In the end stood the Nuclear Engine for Rocket Vehicle Application (NERVA) a full fledged nuclear thermal rocket, depicted in Fig. 11.1. In the meantime the Soviet Union worked on its nuclear thermal engine dubbed RD-0410 from 1965 till the 1980s [5].

Interest is once again high. For instance, the two famous US agencies, NASA and DARPA have joined forces to develop DRACO (Demonstration Rocket for Agile

 
F. Gamgami, *Space Propulsion and Spaceship Design*, Springer Praxis Books,
https://doi.org/10.1007/978-3-031-71336-1_11

**Table 11.1** Selection of reactor versions and their performances within project ROVER [3]

| Date | Reactor version | Maximum power ($MW_{th}$) | Time at maximum power (min) |
|---|---|---|---|
| Jul.1959 | KIWI-A | 70 | 5 |
| Sept. 1964 | KIWI-B4E | 900 | 8 |
| May 1965 | NRX-A3 | 1,122 | 13 |
| Feb. 1967 | Phoebus 1B | 1,500 | 30 |
| Dec. 1967 | NRX-A6 | 1,600 | 62 |
| Jun. 1968 | Phoebus 2A | 4,200 | 12 |
| Dec. 1968 | Pewee | 514 | 40 |
| Jul. 1972 | Nuclear furnace | 44 | 109 |

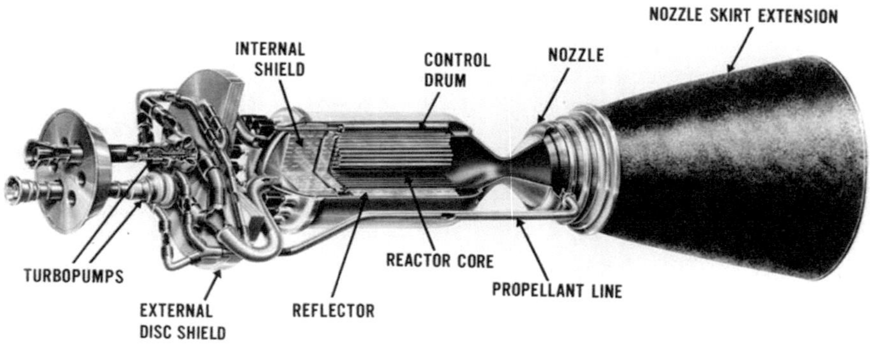


**Fig. 11.1** Nuclear Engine for Rocket Vehicle Application (NERVA) layout. *Credit* Atomic Energy Commission/NASA

Cislunar Operations) a nuclear thermal rocket.[1] With a budget of \$499 Million the primary objective is to demonstrate the reactor with state of the art technology.[2] Therefore, propulsion performance in terms of $I_{sp}$ is not the main objective of DRACO, which explains the target of merely 700 s. One of the key reasons for conducting in-space testing is that, in contrast to the times during the project Rover, environmental regulations have become considerably more rigorous. This results in high testing costs to prevent radioactive exhaust gases from entering the atmosphere.

[1] While NASA's interest is of scientific nature, DARPA—being a defence agency—is rather interested in its military use.

[2] NASA will manage the development of the nuclear reactor build by the private company BWXT, while DARPA is responsible for regulatory approval and overseeing the overall spacecraft, which is build by defence giant Lockheed Martin and dubbed X-NTRV. The mission will last several months and is only limited by the loaded hydrogen propellant. The launch will be provided by Space Force and is planned for 2027 into a low Erth orbit between 700 and 2,000 km. Foust, J. NASA and DARPA to partner on nuclear thermal propulsion demonstration. SpaceNews (January 24, 2023).

It is important to note that the high specific impulse of 800–1000 s is only possible with hydrogen as propellant due to its low molecular mass of about 2, refer to Eq. 5.7.

The discussion commences with an examination of the principles underlying an NTP, after which the NEP will be addressed. The technology will be made tangible by applying both to the C-One mission.

## 11.2 Nuclear Thermal Propulsion NTP

An NTP shares the major building blocks of a chemical propulsion system: the propellant management system with the turbomachinery and the need to store cryogenic propellant, the expansion nozzle, the need for engine cooling. This becomes obvious when examining the NERVA engine, which was an expander bleed cycle. In essence, the combustion chamber is replaced by a *radioactive core*. It is therefore the radioactive core that is at the centre of interest.

### *11.2.1 Nuclear Reactor Technology for Space*

During the almost two decades long Rover/NERVA projects, a vast body of knowledge was created. The current overview will focus on these achievements. These reactors used highly enriched uranium (HEU) but current regulations restrict its use—mainly to contain proliferation. The alternative is high-assay low-enriched uranium (HALEU) with a uranium-235 enrichment between 5 and 20% maximum. The term enrichment refers to the relative amount of uranium-235 to uranium-238.[3] The first isotope is fissionable and the latter not. In contrast, HEU has a uranium-235 concentration above 90%. HALEU based reactors have not been tested yet and this is also the reason why this technology is the prime focus of DRACO.[4] The nuclear fission process is depicted in Fig. 11.2. A uranium-235 atom is composed of 92 protons and 143 neutrons. When a single neutron is added, it becomes unstable and fissions into two lighter atoms, such as barium (Ba), krypton (Kr), strontium (Sr), cesium (Cs), iodine (I), and xenon (Xe), releasing a substantial amount of energy. In addition, 2–3 neutrons are released, with the exact number depending on the products. These neutrons go on to fission other uranium-235 (short U-235) atoms, thereby causing a chain reaction.

$$^{235}_{92}\mathrm{U} + ^{1}_{0}\mathrm{n} \rightarrow ^{141}_{56}\mathrm{Ba} + ^{92}_{36}\mathrm{Kr} + 3^{1}_{0}\mathrm{n} + \Delta E \tag{11.1}$$

[3] Naturally occurring uranium has a ratio of less than 1 %. Weapons grade enrichment is above 90%.

[4] To put it in the words of DRACO's project manager; 'Collecting data on the HALEU reactor will define mission success.' A HALEU based reactor is safer to handle and reduces expensive regulatory and security measures during development and ground operation.

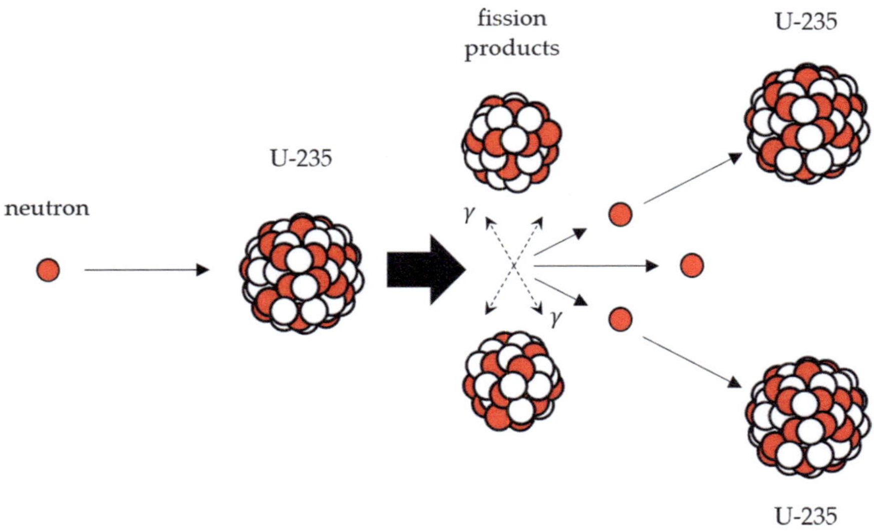


**Fig. 11.2** Depiction of Uranium-235 nuclear fission process and chain reaction

The energy $\Delta E$ released is about 200 MeV of which about 160 MeV is kinetic energy of the atoms and the rest is carried away by the released particles: protons, neutrons, antineutrino $\bar{\nu}_e$ and $\gamma$ rays. The lighter products are in general unstable isotopes themselves, which makes them radioactive. They decay by a process referred to as *minus-beta* decay: simply speaking, a neutron inside the atom of a lighter product transforms into a proton thereby emitting an electron $e^-$, which for historic reasons got the symbol $\beta^{-1}$, hence the name minus-beta decay. Krypton-92, for instance, decays into Rubidium-92, ${}^{92}_{36}\mathrm{Kr} \rightarrow {}^{92}_{37}\mathrm{Rb} + \beta^{-1} + \bar{\nu}_e$, which is again unstable and the process continuous until the stable isotope Zirconium-92 is reached:

$$\text{Krypton-92} \rightarrow \text{Rubidium-92} \rightarrow \text{Strontium-92} \rightarrow \text{Yttrium-92} \rightarrow \text{Zirconium-92}$$

The entire decay process takes statistically ~63 h and releases additionally 13.7 MeV, which is a significant amount that needs to be considered in the design and operation of the reactor.

Since uranium depletes in this process, it is called fissile fuel, nuclear fuel or simply fuel. The engineering challenge of a nuclear reactor for a NTP is to pack the fuel densely and to maximise the heat transfer to the propellant. This domain is called fuel architecture.

Fuel Architecture

Due to its low melting temperature of 1405 K (1132 °C) U-235 is not employed in pure form but rather in form of nitride (UN), oxide ($UO_2$) or carbide ($UC_2$) compounds, called fuel beads, suspended in a graphite matrix [4].

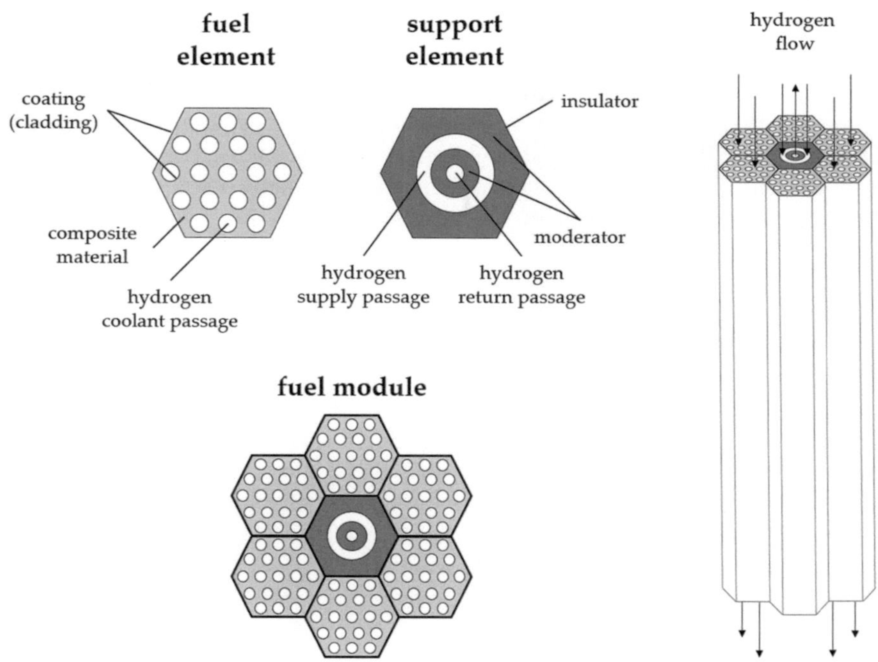


**Fig. 11.3** Schematic of nuclear fuel module used in the Rover/NERVA project [4]

Contact between hydrogen and graphite must be prevented as this causes massive erosion, since hydrogen at high temperature (>200 K) is very aggressive and washes out the graphite. Protection of the graphite was achieved with help of a coating made of NbC or ZrC. Figure 11.3 shows the cross section of a *fuel module* which consists of 6 *fuel element* with 19 coolant channels each and a *support element* in the middle that is also called *tie tube*. They have a characteristic hexagonal shape. The hydrogen flows through the coolant channels to the nozzle, except in the tie tube, where it is diverted at the end and flows back upwards. This flow joins the hydrogen used to cool the nozzle to then drive the turbopump before entering the fuel elements. The support element consists of a *moderator*, which is needed to slow down the neutrons in order to enable the nuclear chain reaction discussed above. Without a moderator, the neutron's velocity would be too high, causing the cross section for interaction to be too low and thus making a hit with a uranium atom unlikely.[5]

In the Soviet Union, engineers and scientists took another approach. Instead of suspending uranium compounds in a graphite matrix, they created a carbide solid solution, which simply speaking is a homogeneous material made of (UZr)C. The fuel elements resembled twisted ribbon and ZrH was used as moderator [5]. This approach enabled much higher propellant exit temperatures of 3300–3500 K—a direction that

[5] Figuratively speaking, the situation is akin to a very fast golf ball skimming over the hole without falling into it.

is currently investigated again. This highlights a further important difference of NTP compared to a chemical propulsion systems. While the latter is limited by the delta in bonding energy of initial and final products, an NTP is only limited by the ability of the core material to withstand the temperatures and efficient cooling techniques. This is why NTP technology is strongly driven by material sciences. Interestingly, the operating temperature within a nuclear reactor is actually below the combustion temperature in high thrust engines, see Table 8.5.

Reactor Core

The task of the reactor core is to sustain the above described chain reaction. In nuclear engineering terminology, the reactor is called *subcritical*, if the neutron population decreases due to leakage, for instance. Consequentially there are not sufficient neutrons to maintain the chain reaction, which decreases then over time. The reactor is called *supercritical*, if the neutrons population increases and the chain reaction intensifies. This is a dangerous state, that could lead to a reactor runaway in which the power output is increases in an uncontrolled manner. A good reactor design seeks a *critical* state, which describes a self-sustaining stable chain reaction. We will see in the following how this is achieved.

The nuclear reactor consists of fuel modules placed within an internal pressure vessel surrounded by a layer of neutron reflective material, called *reflector* and *control drums*. These elements are shrouded in an external pressure shell. Figure 11.4 shows the cross section of the core.

The fission process takes place in the fuel elements, see discussion above. The reflector, as the name suggests, is needed to reflect the neutrons leaving the core back into the core in order to maintain the chain reaction. Hence, the reflector prevents a

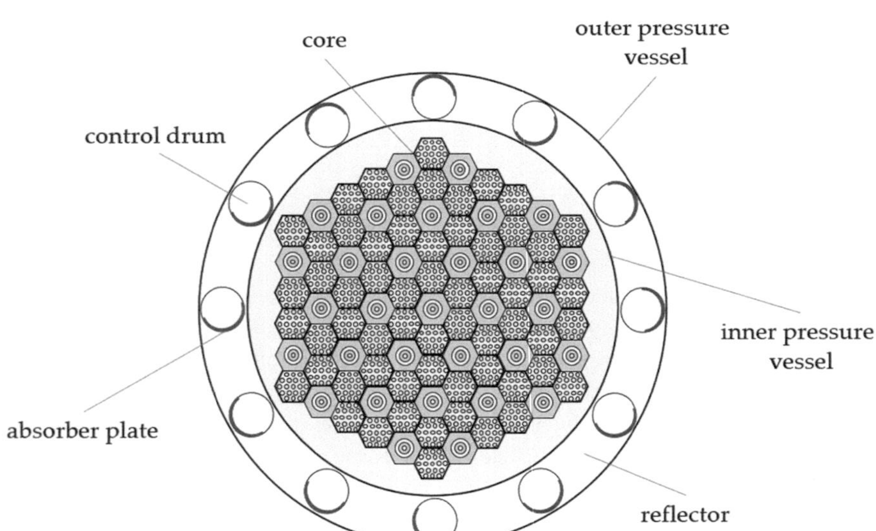


**Fig. 11.4** Schematic of the cross section of a nuclear reactor

subcritical chain reaction. NERVA, for instance, used beryllium (Be) as a reflective material but also BeO is possible. The temperature limit of Be is 1000 K, which only can be maintained, if the reflector is also cooled by convection, i.e. the propellant flow. Temperature spikes must be prevented as they cause thermo-mechanical stress, expressed in cracks [4].

The control drums are also made of neutron-reflecting material except on one side, which is able to absorb neutrons to prevent a supercritical chain reaction. This Janus-like nature of the drums allows neutron population modulation meaning that the reactor can be switched ‘on’ or ‘off’ by increasing or decreasing the number of neutrons reflected, depending on which side is facing the reactor core. Neutron absorption is achieved with Boron carbide $B_4C$ during for the Rover reactors. The control drums must therefore be rotatable.

Out of the 22 tested reactor types, it was Pewee 1 that stood out. With 508 $MW_{th}$, it did not have the highest thermal power output, but it boasted the highest average power density, 2,340 $MW/m^3$, and highest average exit-gas temperature of 2,539 K. This performance corresponds to an ideal vacuum $I_{sp}$ of 901 s. Its core consisted of 402 fuel elements within an overall core diameter of just 0.53 m, containing 36.4 kg of highly enriched uranium $>90\%$. Postmortem inspection did not show excessive damage [2].

A further crucial element of both, NTP and NEP systems, is radiation shielding. this time the source is not from out side the spacecraft as discussed in Sect. 3.2, but from within. The shield is placed directly on top of the radiator and requires cooling as well. The material of the shield must protect the remaining spacecraft from both, the neutron flux and the gamma rays. Neutron shielding is achieved by low atomic number elements like beryllium, lithium or boron.[6] Gamma rays can be shielded by high atomic number elements like tungsten [12].

NTP for C-One

In order to analyse the impact of a nuclear thermal propulsion system in C-One, we need to select a specific engine configuration. The Rover program focused on reactor technology with NERVA being the only designed full scale propulsion system, which is not necessarily suitable. The Rover/NERVA programme provided the opportunity for the acquisition of knowledge, which enabled the formulation of several detailed NTP concepts. We will rely on these concepts, for the C-One system analyses. Table 11.2 shows the results of numerical performance analysis for two engines based on NERVA reactor technology [6]. One with a small thrust level of about 33 kN dubbed ‘small NERVA’ and a second with a larger thrust level, 113.3 kN, in the range of RL-10B-2 dubbed ‘large NERVA’. Unlike their predecessor, NERVA, a closed expander cycle has been considered in the flow simulations for both engines.

Besides the high specific impulse, which was expected, it is above all the thrust-to-weight ratio that stands out. The ratio is only one tenth of a comparable chemical engine, which is due to the high reactor mass.

---

[6] Hydrogen is also a highly effective shielding element, available in the form of a propellant. However, since it is a consumable, it cannot be considered in the shielding design.

**Table 11.2** Numerical Simulation of NTP engine performance and properties based on a NERVA type reactor core [6]

| | | Small nerva | Large nerva |
|---|---|---|---|
| *Engine* | | | |
| Expansion ratio | – | 330 | 300 |
| Specific impulse | s | 899.6 | 911.5 |
| Thrust | kN | 33.2 | 113.3 |
| Chamber inlet pressure | bar | 48.1 | 83.8 |
| Chamber pressure | bar | 41.4 | 68.9 |
| Thrust to weight | – | 1.96 | 3.49 |
| Tot. engine mass | kg | 1,730 | 3,305 |
| *Reactor only* | | | |
| Diameter | cm | 87.7 | 98.5 |
| Fuel length | cm | 88.9 | 132 |
| U-235 mass | kg | 27.5 | 36.8 |
| Total power | MW | 161.6 | 555 |
| Reactor mass | kg | 1,435 | 2,645 |

We have seen in Sect. 10.4 that two parameters are essential for a staging analysis: specific impulse $I_{sp}$ and dry mass ratio $\sigma$. Repeating the $\sigma$-constrained optimisation with the same dry mass index of 16%, however, leads to unrealistic results. The dry mass for a three stage NTP configuration would be then 2.8 t, which is lower than the engine mass itself of 3.3 t. We, therefore, need to increase the dry mass index. We have seen in Sect. 6.1.1 that $\sigma$ is essentially a statistical value that compensates for the lack of a detailed mass budget. The issue described above shows clearly the lack of a statistical data base for NTP propelled stages. In order to proceed we assume a conservative value of 35% but must be aware of a potential inconsistency. Table 11.3 shows the $\sigma$-constrained staging optimisation results for an NTP based stack a large NERVA engine.

**Table 11.3** NTP stages with hydrolox, $\sigma = 0.35$, $I_{sp} = 911.5$ s, 1S: single booster, 2S: two booster and 3S: three booster configuration

| | $\Delta v_i$ (m/s) | | | $m_{0,i}$ (t) | | | $m_{p,i}$ (t) | | |
|---|---|---|---|---|---|---|---|---|---|
| | 1S | 2S | 3S | 1S | 2S | 3S | 1S | 2S | 3S |
| 1st stage | 5110 | 2076 | 1308 | 90 | 40 | 26 | 67 | 30 | 19 |
| 2nd stage | | 3034 | 1633 | | 40 | 26 | | 30 | 19 |
| 3rd stage | | | 2169 | | | 26 | | | 19 |
| Total | 5110 | 5110 | 5110 | 90 | 81 | 78 | 67 | 60 | 58 |

The dry mass for a three stage configuration (3S) is about 6 t. Considering that the large NERVA engine has a mass of 3.3 t this leaves only 2.7 t for remaining dry mass (structure, PMS, avionics and docking mechanism). This is clearly not enough. A two stage configuration (2S) has an absolute dry mass of about 10 t, leaving a comfortable 6.7 t for the dry mass minus the engine mass. This is challenging but not unrealistic. Finally, a single stage configuration (1S) leaves 20 t dry mass for all other elements besides the engine. This on the other hand seems a lot. Instead of prematurely revising the dry mass index it is worth to complete the analysis and to consider the gravity losses.

Applying the same analysis as we did for the hydrolox case in Sect. 10.4.1 yields Fig. 11.5. It shows the additional propellant needed to compensate for the gravity losses in case of a singe booster stage configuration.

The additional propellant needed for single engine configuration is a staggering 16 t which reduces to 7.6 t for a dual engine configuration. The sum of this mass and the two nuclear engines yields a total mass of 14.2 t compared to 19.2 t in a single engine configuration. Hence, the invest of a second engine saves 5.1 t of system mass, which corresponds to 6% of the total stage mass. It is questionable that this saving justifies the financial invest and technical risk of developing two engines for a single booster stage.

The results show that the high specific impulse in combination with the low thrust-to-weight ratio of nuclear thermal propulsion systems make the system behaviour very different to a classical chemical-based system. From an orbit mechanical point of view an NTP system will exhibit a long burn duration, twice of that of a chemical-based system. As a result the manoeuvre is even less impulsive, which is an additional explanation of the large gravity losses.

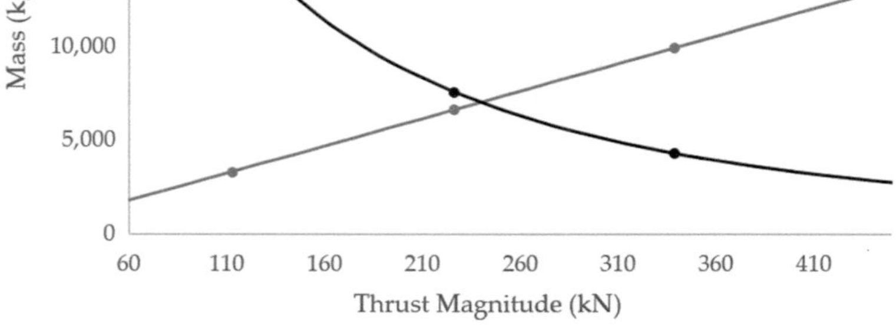


**Fig. 11.5** Computation of gravity losses and invested engine mass based on 'large NERVA' engine characteristics [6]

Finally, a programmatic aspect shall be highlighted. A nuclear thermal stage represents a high asset vehicle that will not be discarded in space mainly for two reasons. Firstly, the risk of an uncontrolled re-entry in Earth's atmosphere is too high. This could even occur if it leaves Earth's sphere of influence and enters a heliocentric orbit. Secondly, the development and production of a nuclear reactor is expensive and its lifetime is very long. Hence, from a sustainability point of view, the vehicle should be salvaged and re-used. These consideration will have a consequences for the mission planning leading to a mission and system architecture which the NTP stage enters an orbit from which it can be salvage or from where it can return by itself.

## 11.3 Nuclear Electric Propulsion–NEP

The reactor design of an NEP system resembles terrestrial reactor designs albeit at lower power output levels. The targeted power level is in the range of 10–1000 $kW_e$ [8]. This is two to three orders of magnitude less than NTP reactors. Also the operational temperature is much lower, 1350 K. The reactor of an NEP is in principle agnostic of the attached load. It can be used to power an electric engine or any other power consumer in orbit or on a planetary surface.[7] An NEP system consists of several critical subsystems:

- Reactor Core
- Power Conversion
- Shielding
- Heat Rejection: Cooling System and Thermal Radiator

We will limit the discussion in this section to the working principle and system design of NEPs as well as its impact on the overall spaceship design.

The genuine energy form a reactor creates is heat, i.e. kinetic energy. In case of an NTP this heat is then transferred directly to the propellant, a process that is highly efficient. Applying Eq. 5.29 to the above introduced large NERVA engine, yields an $\eta_T$ of 92%. Energy conversion techniques for NEP reactors are by far not as efficient. Firstly because of the power conversion mechanism that is needed to convert the released heat in the reactor into electricity and secondly because of the efficiency factor of the electric thruster itself. Table 11.4 shows two conceivable groups, static and dynamic, for power conversion. Static conversion systems have an efficiency of merely 5–10%. They are used as power sources on spacecraft beyond the Jupiter system, where the solar flux is too low to be effectively utilised by a solar power generator without resulting in excessive dimensions. Dynamic power conversion systems apply thermodynamic cycles that are known since the 19th century.[8] They

[7] The energy intense resource mining and its processing via ISRU on Moon and Mars would greatly benefit from nuclear reactors.

[8] It is somehow ironic that the technology of the future follows the same principles as the first steam machines.

**Table 11.4** Power conversion technology for NEP, based on [1]

| Conversion method | Conversion type |
|---|---|
| Thermoelectric | Static |
| Thermionic | Static |
| Brayton cycle | Dynamic |
| Stirling cycle | Dynamic |
| Rankine cycle | Dynamic |

are twice as efficient as static systems, i.e. 10–20% [8]. The dynamic conversion methods produce an alternating current (AC), which has to be rectified into direct current for the electric propulsion system and the other spaceship consumers. This process reduces further the overall efficiency of an NEP system.

An NEP system, therefore, generates high amounts of waste heat, either in the reactor or in the power conversion system. This heat must be rejected efficiently into space, which is again achieved by radiators. There are in principle two options to transport the large amounts of heat to the radiators. Either by actively pumping a fluid from the reactor to the radiator or by heat pipes. The first method applies liquid metals, like lithium, sodium or potassium, or a combination of sodium and potassium, NaK.

The thermodynamic cycles have very different operation characteristics and scaling capabilities depending on the power regime, and applied medium. A detailed trade-off is needed on a case by case basis.

Like for NTP systems, the Soviet Union and later Russia worked too on NEP technology based on both *thermoelectric* and *thermionic* power conversion. The programme name of the latter is TOPAZ, which stands for 'Thermionic Experiment with Conversion in Active Zone'. After the fall of the Soviet Union, the US purchased two units from the Russian Federation. TOPAZ II had a nominal thermal power output of 115 $kW_{th}$ generated by 37 thermionic fuel elements (TFE), which were fueled with $UO_2$ pellets at 96% enrichment.

The total mass of U-235 in the reactor was 27 kg. The reactor was cooled by sodium-potassium NaK [9]. The conversion efficiency was 5% which translates to 5.75 $kW_e$ for a reactor with a mass of about 1000 kg [10]. This example demonstrates the small yield and the high invest that comes with NEP systems. Dynamic conversion promises a higher yield. The US SP-100 programme aimed at a scalable space reactor with a power provision of 10–1000 $kW_e$ and a specific power of 26 $kW_e$/kg. The projected total system mass including the power converter, the radiator panels, the pump system and related structure was 4,575 kg [8]. Figure 11.6 depicts the layout of the system. The project was cancelled in 1994 and never developed into a flight configuration.

System Design Impact and C-One Application

A spaceship with a nuclear reactor as its power source has a distinctive layout: Firstly, the reactor is placed on a boom in flight direction to create distance between

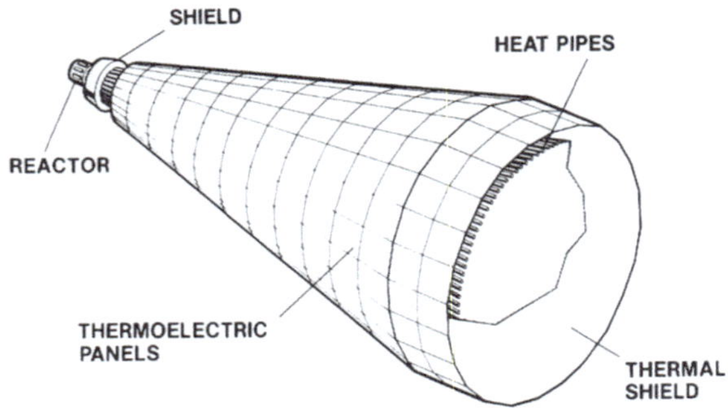


**Fig. 11.6** SP-100 concept for NEP [7]. *Credit* Los-Alamos-National-Library

the radiation source and the electronic equipment and payload. Secondly, it has a massive thermal radiator to dissipate the massive excess heat. Furthermore, like in case of an NTP, there is dedicated shield behind the reactor to protected the rest of spaceship from the harmful radiation.

The power conversion system is placed right behind the shield to reduce thermal losses. This closeness requires so called *radiation hardened* electronics, which are expensive and have a long development time. This is clearly visible in Fig. 11.7. It depicts a concept study of a 1.6 MWe class NEP system conducted by NASA [12]. Table 11.5 provides NEP relevant specifications of this concept study. The arguably most visible part is the radiator, which exhibits an X-shape. The configuration allows for a high heat rejection capability by an overall optimum view factor into space. Like

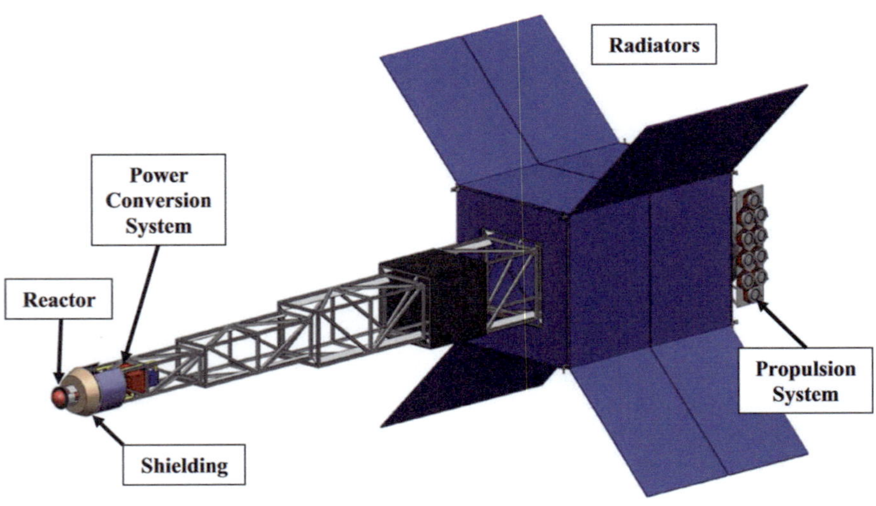


**Fig. 11.7** Concept for a spaceship with nuclear electric propulsion for interplanetary transfer [11]. *Credit* NASA

**Table 11.5** System and propulsion characteristics of NEP NASA concept study for a crewed mission to Mars [12]

| | |
|---|---|
| Reactor temperature | 1200 K |
| Thermal power | 5.8 MWth |
| Electric power | 1.6 MWe |
| EP thruster, HET[a] (15+1) | 16×100 kWe |
| Power conversion | Brayton cycle |
| Radiator area (500 K) | 2500 m$^2$ |
| System mass | 22.2 t |

[a]NASA-457 with nominal $I_{sp} = 2600$ s

for an SPG based propulsion system, the final design is limited by the available fairing volume and packing density of the radiator during launch.[9] The concept foresees $H_2O$ based heat pipes.The achieved specific power is 68.5 We/kg. This value is three times higher than that of the SP-100 programme. Given the fact that 26 We/kg (at 100 kWe) was already an ambitious projection and the fact that the specific power improves significantly with the power output, we are well advised to adopt a smaller value for C-One. A value of 20 We/kg should be sufficiently conservative for a preliminary analysis. This corresponds to an $\alpha$ of 0.05 W/kg, which is slightly larger than the value adopted for a solar power generator system, 0.03 W/kg. While comparing the two values, it should be noted that the thermal radiator mass is in general not included in the EPS mass of SEP systems. This is different for NEP systems in which the thermal radiator is regarded an important and integral part. Consequentially it is part of the specific power value together with the Power Management and Distribution (PMAD) system. It shall be further noted that the thruster do not align with the symmetry axis. This is because the depiction of the spaceship is not complete. The missing part is the chemical stage and the habitation module. The chemical stage is needed for Mars orbit insertion, which as discussed earlier is not recommended to conduct with electric propulsion.

We apply this to the second phase vehicle, which due to the change in power source we now call 'Nuclear Electric Transfer Vehicle' (NETV). Table 11.6 shows the system level requirements for two different transfer times: 3 years according to the requirement and a fast transfer of 1.7 years (20 months). The NETV mass for the fast transfer matches the mass of the SETV computed before. Hence, according to the EP model parameters an NEP system can reduce the travel time by 16 months (43%) for the same wet mass. Note that the electric power demand increases by almost the same factor (48%).

[9] In orbit assembly of the radiator, could alleviate the issue and enable larger radiators. This necessitates a space-shipyard.

**Table 11.6** NETV system requirements according the EP model, $I_{sp} = 2600\,\text{s}$, $\eta_T = 0.66$, $\tau = 3\,\text{yrs}$

| | $\tau = 3\,\text{yrs}$ | $\tau = 1.7\,\text{yrs}$ |
|---|---|---|
| Wet mass, $m_0$ | 21 t | 25 t |
| Reactor system mass | 3.5 t | 6.6 |
| Propellant mass | 13 t | 14 t |
| Electrical power | 69 kWe | 132 kWe |
| Thrust | 3.6 N | 6.8 N |

All in all the results are not very different from those previously calculated for the SETV but the interpretation is different now. Since the reactor's output power is not dependent on distance like the solar power generator, the system does not need to be scaled up. In fact, the reactor thermal output does experience a degradation as the nuclear fuel depletes and more detailed models are need to account for this effect in a later phase.

## References


1. National Academies of Sciences, Engineering, Medicine, on Engineering, D., Sciences, P., Aeronautics, Board, S. E., & Committee, S. N. P. T. (2021). *Space nuclear propulsion for human mars exploration*. National Academies Press.
2. Finseth, J. L., (1991). Overview of rover engine tests-final report. NAS 8-37814
3. GUNN, S. (n.d.). Development of nuclear rocket engine technology. In *25th joint propulsion conference*. https://doi.org/10.2514/6.1989-2386
4. Walton, J. (n.d.). An overview of tested and analyzed NTP concepts. In *Conference on advanced SEI technologies*. https://doi.org/10.2514/6.1991-3503
5. Vadim, Z., & Vladimir, P. (2007). Russian nuclear rocket engine design for mars exploration. *Tsinghua Science and Technology, 12*(3), 256–260. https://doi.org/10.1016/S1007-0214(07)70038-X
6. Belair, M. L., Sarmiento, C. J., & Lavelle, T. L. (2013). Nuclear thermal rocket simulation in NPSS. https://api.semanticscholar.org/CorpusID:109402533
7. International Atomic Energy Agency (2005). *The role of nuclear power and nuclear propulsion in the peaceful exploration of space*. https://www-pub.iaea.org/MTCD/publications/PDF/Pub1197_web.pdf
8. Bennett, G. L., Hemler, R. J., & Schock, A. (1996). Status report on the U.S. space nuclear program. *Acta Astronautica, 38*(4), 551–560. https://doi.org/10.1016/0094-5765(96)00038-0
9. Voss, S., & Reynolds, E. L. (1994). An overview of the nuclear electric propulsion space test program (NEPSTP) satellite. https://doi.org/10.2514/6.1994-3818
10. Lao, X., Lin, Y., Dai, C., Liu, C., & Lyu, W. (2019). Core physics calculation study of miniature lead-bismuth cooled nuclear reactor. *IOP Conference Series: Earth and Environmental Science, 354*, 012028. https://doi.org/10.1088/1755-1315/354/1/012028

11. Bragg-Sitton, S., Werner, J., Johnson, S., Houts, M., Palac, D., Mason, L., Poston, D., & Qualls, A. (2011). *Ongoing space nuclear systems development in the United States*.
12. Oleson, S., Turnbull, E., Burke, L., Packard, T., Mason, L., Fittje, J., Colozza, A., Schmitz, P., McCarty, S., Yim, J., Klefman, B., Gyekenyesi, J., Faller, B., Smith, D., Tian, L., Austin, C., Simon, W., Heldman, C., Theofylaktos, O., & Weckesser, N. (2022). *1.9MWe nuclear electric propulsion-chemical propulsion piloted mars opposition vehicle* (pp. 240–248). https://doi.org/10.13182/NETS22-38628

# Appendix A
# Planetary Parameter

*All's well that ends well*

See Table A.1.



F. Gamgami, *Space Propulsion and Spaceship Design*, Springer Praxis Books,
https://doi.org/10.1007/978-3-031-71336-1

**Table A.1** Planetary parameter of the solar system [4]

| Property | Mercury | Venus | Earth | Moon | Mars | Jupiter | Saturn | Uranus | Neptune | Pluto |
|---|---|---|---|---|---|---|---|---|---|---|
| Mass ($10^{24}$ kg) | 0.33 | 4.87 | 5.97 | 0.073 | 0.642 | 1898 | 568 | 86.8 | 102 | 0.013 |
| Diameter (km) | 4879 | 12,104 | 12,756 | 3475 | 6792 | 142,984 | 120,536 | 51,118 | 49,528 | 2376 |
| Density (kg/m$^3$) | 5429 | 5243 | 5514 | 3340 | 3934 | 1326 | 687 | 1270 | 1638 | 1850 |
| Gravity (m/s$^2$) | 3.7 | 8.9 | 9.8 | 1.6 | 3.7 | 23.1 | 9 | 8.7 | 11 | 0.7 |
| Escape velocity (km/s) | 4.3 | 10.4 | 11.2 | 2.4 | 5 | 59.5 | 35.5 | 21.3 | 23.5 | 1.3 |
| Rotation period (hours) | 1407.6 | –5832.5 | 23.9 | 655.7 | 24.6 | 9.9 | 10.7 | –17.2 | 16.1 | –153.3 |
| Length of day (hours) | 4222.6 | 2802 | 24 | 708.7 | 24.7 | 9.9 | 10.7 | 17.2 | 16.1 | 153.3 |
| Distance from sun ($10^6$ km) | 57.9 | 108.2 | 149.6 | 0.384 | 228 | 778.5 | 1432 | 2867 | 4515 | 5906.4 |
| Perihelion ($10^6$ km) | 46 | 107.5 | 147.1 | 0.363 | 206.7 | 740.6 | 1357.6 | 2732.7 | 4471.1 | 4436.8 |
| Aphelion ($10^6$ km) | 69.8 | 108.9 | 152.1 | 0.406 | 249.3 | 816.4 | 1506.5 | 3001.4 | 4558.9 | 7375.9 |
| Orbital period (days) | 88 | 224.7 | 365.2 | 27.3 | 687 | 4331 | 10,747 | 30,589 | 59,800 | 90,560 |
| Orbital velocity (km/s) | 47.4 | 35 | 29.8 | 1.0 | 24.1 | 13.1 | 9.7 | 6.8 | 5.4 | 4.7 |
| Orbital inclination (deg) | 7 | 3.4 | 0 | 5.1 | 1.8 | 1.3 | 2.5 | 0.8 | 1.8 | 17.2 |
| Orbital eccentricity | 0.206 | 0.007 | 0.017 | 0.055 | 0.094 | 0.049 | 0.052 | 0.047 | 0.01 | 0.244 |
| Obliquity to orbit (deg) | 0.034 | 177.4 | 23.4 | 6.7 | 25.2 | 3.1 | 26.7 | 97.8 | 28.3 | 119.5 |
| Mean temperature (°C) | 167 | 464 | 15 | –20 | –65 | –110 | –140 | –195 | –200 | –225 |
| Surface pressure (bar) | 0 | 92 | 1 | 0 | 0.01 | Unknown | Unknown | Unknown | Unknown | 0.00001 |
| Number of moons | 0 | 0 | 1 | 0 | 2 | 95 | 146 | 28 | 16 | 5 |
| Ring system? | No | No | No | No | No | Yes | Yes | Yes | Yes | No |
| Global magnetic field? | Yes | No | Yes | No | No | Yes | Yes | Yes | Yes | Unknown |

*Credit* NASA Goddard Space Flight Center

# Appendix B
# Hohmann Transfer within the Coplanar Approximation

## B.1 Introduction

This analysis has been kindly contributed by Dr. Davide Amato[1] and Anton Sabin-Viorel,[2] and considers the design of an interplanetary trajectory from the Earth to Ceres using chemical propulsion, i.e. an impulsive transfer. The levels of sophistication will increase:

- Interplanetary Hohmann transfer,
- Patched conics,
- Lambert's problem.

The usage of the B-plane will also be examined.

## B.2 Hohmann Transfer within the Coplanar Approximation

At a first level of approximation, we can model the two-way trajectory to Ceres with two coplanar Hohmann transfers. We assume that the Earth and Ceres are on circular orbits of semi-major axes $a_1 = 1.007$ au, $a_2 = 2.7663$ au, respectively. The gravitational parameter of the Sun is $\mu = 39.476926\,\mathrm{au}^3/\mathrm{yr}^2 = 1.327\,1244 \times 10^{11}\,\mathrm{km}^3\,\mathrm{s}^{-2}$.

The spaceship is injected onto the outbound Hohmann transfer through a $\Delta v$ that increases its velocity from the Earth's circular velocity $v_1$ to the velocity at perihelion of the Hohmann transfer, $v_{\mathrm{H,D}}$ (where the subscript $D$ stands for "departure"). Once the spaceship arrives at Ceres, a second $\Delta v$ is necessary to increase its velocity from the velocity at aphelion of the Hohmann transfer, $v_{\mathrm{H,A}}$ (where the subscript $A$ stands

[1] Lecturer in Spacecraft Engineering at Imperial College London.

[2] Ph.D. Student in the Astrodynamics and Space Missions Department, Delft University of Technology.



F. Gamgami, *Space Propulsion and Spaceship Design*, Springer Praxis Books,
https://doi.org/10.1007/978-3-031-71336-1

for "arrival") to the circular velocity of Ceres, $v_2$. The circular velocities of the Earth and Ceres are

$$v_1 = \sqrt{\frac{\mu}{a_1}} = 6.2809 \text{ au yr}^{-1} \tag{B.1}$$

$$v_2 = \sqrt{\frac{\mu}{a_2}} = 3.7777 \text{ au yr}^{-1}. \tag{B.2}$$

The departure and arrival velocities of the Hohmann transfer can be obtained by considering that

$$v_{H,D} = \frac{h}{a_1} \tag{B.3}$$

$$v_{H,A} = \frac{h}{a_2}, \tag{B.4}$$

where $h$ is the specific angular momentum of the transfer orbit, which can be shown to be

$$h = \sqrt{2\mu \frac{a_1 a_2}{a_1 + a_2}}. \tag{B.5}$$

Plugging B.5 into Eqs. B.3 and B.4 we obtain

$$v_{H,D} = \sqrt{2\mu \frac{a_2}{a_1(a_1 + a_2)}} \tag{B.6}$$

$$v_{H,A} = \sqrt{2\mu \frac{a_1}{a_2(a_1 + a_2)}}. \tag{B.7}$$

Considering the above equations, we compute the departure, arrival, and total $\Delta v$'s for the outbound transfer as

$$\Delta v_D = v_{H,D} - v_1 = 1.3309 \text{ au yr}^{-1} = 6.3092 \text{ km s}^{-1} \tag{B.8}$$

$$\Delta v_A = v_2 - v_{H,A} = 1.0241 \text{ au yr}^{-1} = 4.8548 \text{ km s}^{-1} \tag{B.9}$$

$$\Delta v = \Delta v_D + \Delta v_A = 2.3550 \text{ au yr}^{-1} = 11.1640 \text{ km s}^{-1}. \tag{B.10}$$

By symmetry, the total $\Delta v$ for the return transfer is equal in magnitude to the outbound transfer.

### *B.2.1 Mission Timing*

We consider here a launch window from 1st January 2030 to 31st December 2035. The heliocentric orbital elements of the Earth and Ceres on 1st January 2030 in the International Celestial Reference Frame of epoch J2000 are listed in Table B.1. From Table B.1, we see that the phase angle $\phi = \theta_2 - \theta_1$ at $t_0$, corresponding to 1st January 2030 00:00 TDB, is

$$\phi(t_0) = \theta_2 - \theta_1 = 244.59^\circ. \tag{B.11}$$

We also note that the rate of change of the phase angle is

$$\dot{\phi} = n_2 - n_1 = -281.37^\circ \, \mathrm{yr}^{-1}, \tag{B.12}$$

where $n_1, n_2$ are the mean motions of Earth and Ceres, respectively. For the spaceship to rendezvous with Ceres on the outbound transfer, both bodies must be at the same position when the spaceship reaches the apohelion of the transfer. This imposes a constraint on the phase angle at the start of the outbound transfer, which must satisfy

$$\phi(t_{\mathrm{O,D}}) = 180^\circ - n_2 \Delta t, \tag{B.13}$$

where $t_{\mathrm{O,D}}$ is the time of departure for the outbound transfer, and $\Delta t$ is the transfer time on the outbound Hohmann transfer orbit. The latter is half the orbital period of the transfer orbit $T$, which has semi-major axis

$$a = \frac{a_1 + a_2}{2}, \tag{B.14}$$

therefore

$$\Delta t = \frac{T}{2} = \frac{\pi}{\sqrt{\mu}} \left( \frac{a_1 + a_2}{2} \right)^{3/2} = 1.2925 \, \mathrm{yr} \tag{B.15}$$

**Table B.1** Orbital elements of the Earth and Ceres in the ICRF on 1st January 2030 00:00 TDB (Temps Dynamique Barycentrique/Barycentric Dynamical Time). The true longitude is $\theta = \Omega + \omega + \nu$

| | Earth | Ceres | |
|---|---|---|---|
| $a$ | 1.0007 | 2.7663 | au |
| $e$ | $1.7345 \times 10^{-2}$ | 7.9411e–2 | |
| $i$ | $3.8560 \times 10^{-3}$ | 10.584 | ° |
| $\Omega$ | 206.06 | 80.174 | ° |
| $\omega$ | 255.85 | 72.295 | ° |
| $\nu$ | 358.27 | 192.30 | ° |
| $\theta$ | 100.18 | 344.77 | ° |

*Source* Horizons system [1]

and $\phi(t_{\mathrm{O,D}}) = 78.87°$ according to Eq. B.13. The time required for the phase angle to go from $\phi(t_0)$ to $\phi(t_{\mathrm{O,D}})$ is

$$\Delta t_0 = \frac{\phi(t_{\mathrm{O,D}}) - \phi(t_0)}{\dot{\phi}} = 0.589 \text{ yr}. \tag{B.16}$$

Thus, the first Hohmann transfer opportunity occurs on 4th August 2030 02:40 TDB. However, this is not the only transfer opportunity in the launch window. The phase angle is periodic in the synodic period

$$T_{\mathrm{syn}} = \frac{360°}{|\dot{\phi}|} = 1.2794 \text{ yr}, \tag{B.17}$$

therefore, the potential departure epochs (at which the phase angle takes the value in Eq. B.13) take place at intervals of $T_{\mathrm{syn}}$,

$$t_{O,D}^{(k)} = t_{O,D} + kT_{\mathrm{syn}}. \tag{B.18}$$

The departure epochs within the 2030–2039 launch window are listed in Table B.2. In the following, we will consider the mission timeline corresponding to the first transfer opportunity of 4th August 2030 02:40. Figure B.1 a shows the geometry for this transfer; note that the Earth completes more than one revolution during the transfer because $\Delta t > 1$ yr. Also, note that the phase angle at arrival of the outbound transfer is

$$\phi(t_{O,A}) = 180° - n_1 \Delta t = -284.80°. \tag{B.19}$$

The return transfer consists of another Hohmann transfer orbit, this time with apohelion at the departure point (at Ceres) and perihelion at the arrival point (at the Earth). For the Earth to be at the same position as the spacecraft at perihelion of the transfer orbit, the phase angle at the start of the return Hohmann transfer must satisfy

$$\phi(t_{R,D}) = n_1 \Delta t - 180° = -\phi(t_{O,A}). \tag{B.20}$$

**Table B.2** Possible departure epochs for the Earth-Ceres Hohmann transfer. All times are TDB

| Epoch |
|---|
| 4th August 2030 02:40 |
| 14th November 2031 10:02 |
| 23rd February 2033 17:24 |
| 6th June 2034 00:46 |
| 16th September 2035 08:07 |

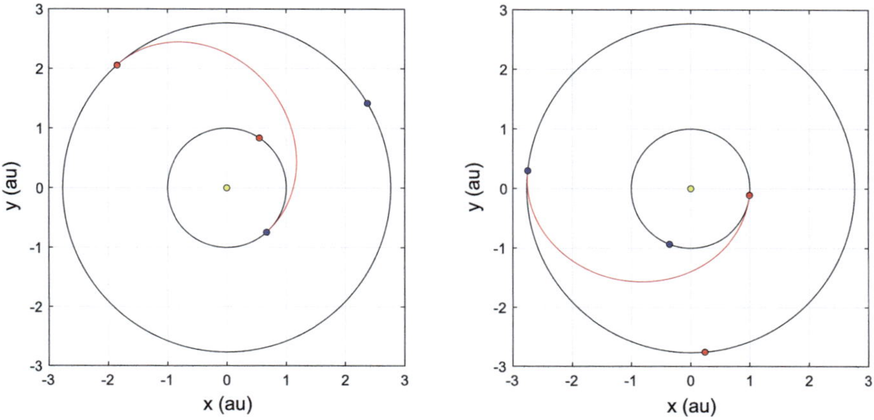


(a) Outbound. (b) Return.

**Fig. B.1** Outbound and return transfer geometries in the ICRF for the 4th August 2030 02:40 departure date. The initial and final positions of Earth and Ceres are shown in blue and red, respectively. The Hohmann transfer orbit is also shown in red

After arriving at Ceres, we need to wait for the phase angle to go from the value $\phi(t_{O,A})$ to

$$\phi(t_{R,D}) = -\phi(t_{O,A}) + k360^\circ \tag{B.21}$$

in order to start the return Hohmann transfer. In Eq. B.21, the departure phase angle at for the outbound transfer $\phi(t_{R,D})$ is defined up to integer multiples of 360°; in this case $k = -2$. The wait time spent at the asteroid is thus

$$\Delta t_{\text{wait}} = \frac{-2\phi(t_{O,A}) + k360^\circ}{n_1 - n_2} = 0.5345 \text{ yr.} \tag{B.22}$$

This is the minimum time that we need to wait for the Hohmann transfer geometry to be satisfied for the return transfer. If longer times at the asteroid are desired, we will need to wait for an integer number of synodic periods in addition to the minimum wait time. Finally, the arrival phase angle in the outbound transfer is

$$\phi(t_{R,A}) = \phi(t_{R,D}) + \dot{\phi}\Delta t = -78.87^\circ. \tag{B.23}$$

The mission opportunities corresponding to the launch epochs in Table B.2 are shown in Fig. B.2, with each opportunity having a total mission duration of $2\Delta t + \Delta t_{\text{wait}} =$ 3.1195 yr.

Note that, if a non-zero eccentricity is considered for the orbits of Earth and Ceres, Eq. B.34 is not valid because the rate of change of the phase angle is not equal to the difference of the mean motions. If non-zero eccentricity is considered, then

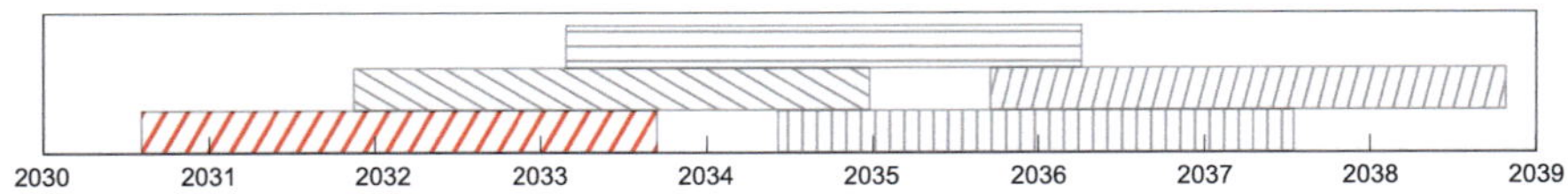


**Fig. B.2** Potential mission opportunities within the 2030–2039 launch window. Each of the opportunities corresponds to a mission lasting 3.12 years, departing on one of the launch dates in Table B.2

$$\dot{\phi} = \dot{\theta}_2 - \dot{\theta}_1 = \frac{h_2}{r_2^2} - \frac{h_1}{r_1^2} \tag{B.24}$$

where $h_2, h_1$ are the angular momenta of Ceres and Earth, respectively. The effect of non-zero eccentricity will be considered in the following section, along with that of the inclination.

## B.3 Patched Conics Analysis

The simplified analysis in the previous section is useful to obtain an estimate for the minimum-$\Delta v$ transfer between Earth and Ceres. However, several simplifying assumptions have been made:

- The orbits of Earth and Ceres are circular and coplanar.
- The transfer takes place at one specific geometry, namely, the Earth at departure (arrival) and Ceres at arrival (departure) are in conjunction for the outbound (return) transfer.

The first assumption neglects out-of-plane components of the $\Delta v$ that must be assigned to match the out-of-plane position component at the end of the transfer. This has important consequences for the total $\Delta v$, as we will see. The second assumption constrains the orbit transfer to one specific geometry which repeats every synodic period, which results in a severe constraint on a launch window.

We can remove both of these assumptions by considering the interplanetary transfer as a solution to *Lambert's problem*, consisting in finding the solution to the two-point boundary value problem

$$\begin{cases} \ddot{\boldsymbol{r}} = -\frac{\mu}{r^3}\boldsymbol{r} \\ \boldsymbol{r}(t_0) = \boldsymbol{r}_0 \\ \boldsymbol{r}(t_1) = \boldsymbol{r}_1 \end{cases} \tag{B.25}$$

where $\boldsymbol{r}_0, \boldsymbol{r}_1$ are the positions at departure and arrival of the transfer, respectively. The transfer $\Delta t = t_1 - t_0$ is given, and corresponds to the difference between the departure and arrival times. The Lambert solution specifies the velocities on the transfer orbit at departure and arrival, $\boldsymbol{v}_T(t_1), \boldsymbol{v}_T(t_2)$. To inject the spacecraft from the departure to the transfer orbit, and to match the position and velocity of Ceres, the total required $\Delta v$ is

$$\Delta v = \Delta v_D + \Delta v_A, \tag{B.26}$$

where the departure and arrival $\Delta v$'s are, respectively,

$$\Delta v_D = \|\boldsymbol{v}_T(t_1) - \boldsymbol{v}_E(t_1)\| \tag{B.27}$$

$$\Delta v_A = \|\boldsymbol{v}_C(t_2) - \boldsymbol{v}_T(t_2)\|, \tag{B.28}$$

and $\boldsymbol{v}_E$, $\boldsymbol{v}_C$ are the velocities of Earth and Ceres, respectively.

For an extensive qualitative analysis of Lambert's problem and a thorough exposition of its methods of solution, we refer the reader to Chap. 6 of [2] and Chap. 4 of [3].

Figure B.3 shows the porkchop plot for the outbound transfer from Earth to Ceres. Departure dates are between February 2030 and February 2031, corresponding to the first synodic period in the 2030–2040 window. The black lines in the porkchop plot are contours of total $\Delta v$ as defined in Eq. B.26, with the numerical values expressed in km s$^{-1}$. The plot also shows the departure and arrival date for the first Hohmann transfer in the 2030–2040 window, i.e., the first row of Table B.2 corresponding to a departure date on 4th August 2030. The effect of the non-zero inclination of Ceres is to create a steep "ridge" in the $\Delta v$ surface corresponding to transfers with a transfer angle $\Delta\theta$ close to 180°. The geometry of these transfers dictates large out-of-plane $\Delta v$'s that are necessary to provide the out-of-plane component of position at arrival. The Hohmann transfer sits in the middle of the ridge, as it takes place for $\Delta\theta = 180°$. Below the ridge we have Type I transfers, which take place for $0° \leq \Delta\theta \leq 180°$ whereas above the ridge we have Type II transfers taking place for $180° < \Delta\theta \leq 360°$. Within this analysis, the minimum-$\Delta v$ transfers entail a total $\Delta v$ of about 16 km s$^{-1}$. The minimum-$\Delta v$ Type I transfer departs close to the Hohmann departure date, but it has a shorter time of flight of 330 d. The minimum-$\Delta v$ Type II transfer arrives close to the Hohmann arrival date, with a longer time of flight of 460d. Figures B.4 and B.5 are porkchop plots for the outbound and return transfers in the entire window between 2030 and 2035.

The return transfers start from March 2031, corresponding to the earliest feasible arrival dates at Ceres during the departure transfer. Transfers with $\Delta v$ greater than 30 km s$^{-1}$ have not been plotted as they would be extremely difficult achieve with future propulsion technology. The topology of the transfer alternates between consequent synodic periods. For example, it is clear from Fig. B.4 that departure dates between April and December 2030 with minimum $\Delta v$'s around 15 km s$^{-1}$, are less favourable than those between July 2031 and April 2032 with minimum $\Delta v$'s around 12 km s$^{-1}$. The topology of the transfers differs between outbound and return: outbound porkchops are more elongated vertically. As the Earth travels faster than Ceres on its orbit, the timing of the transfer is more sensitive to the departure date for outbound transfers, and to the arrival date for return transfers. Analysis of these porkchop plots allows choosing appropriate mission schedules. It seems that the 2031 launch window presents the most favourable opportunities.

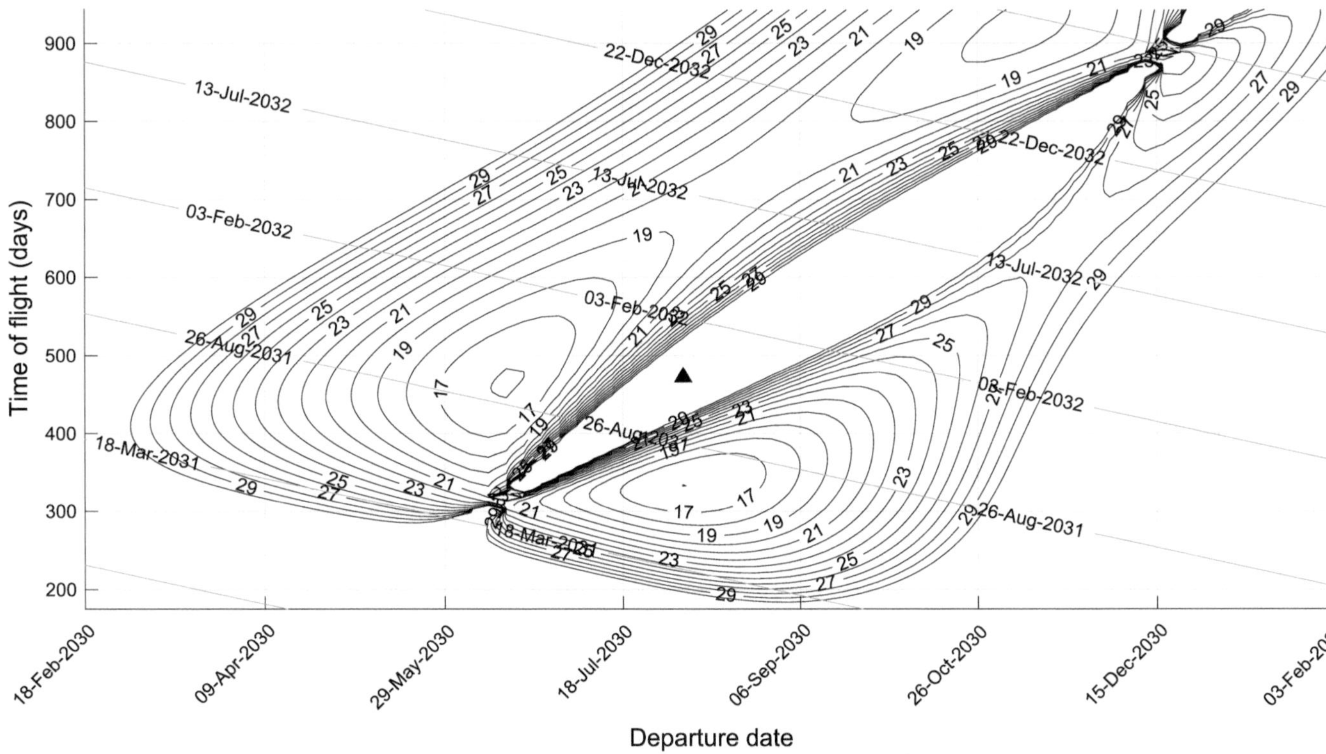


**Fig. B.3** Porkchop plot for the outbound transfer to Ceres, with departure dates between February 2030 and February 2031. The Hohmann transfer with departure in August 2030 is marked with a black triangle. Grey lines represent constant arrival times. Contour labels represent the total $\Delta v$ in km s$^{-1}$

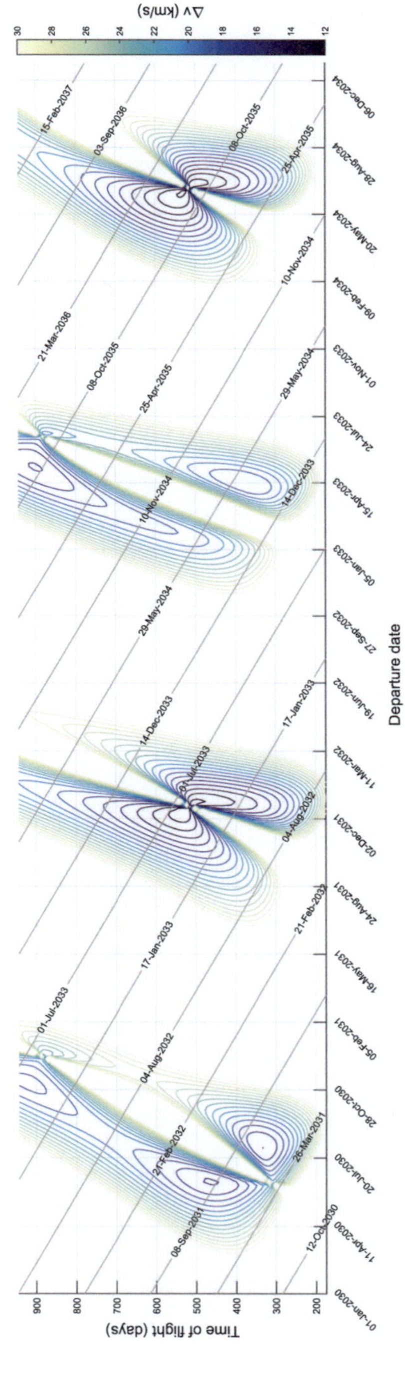


**Fig. B.4** Outbound porkchop plot for all mission opportunities between 2030 and 2035

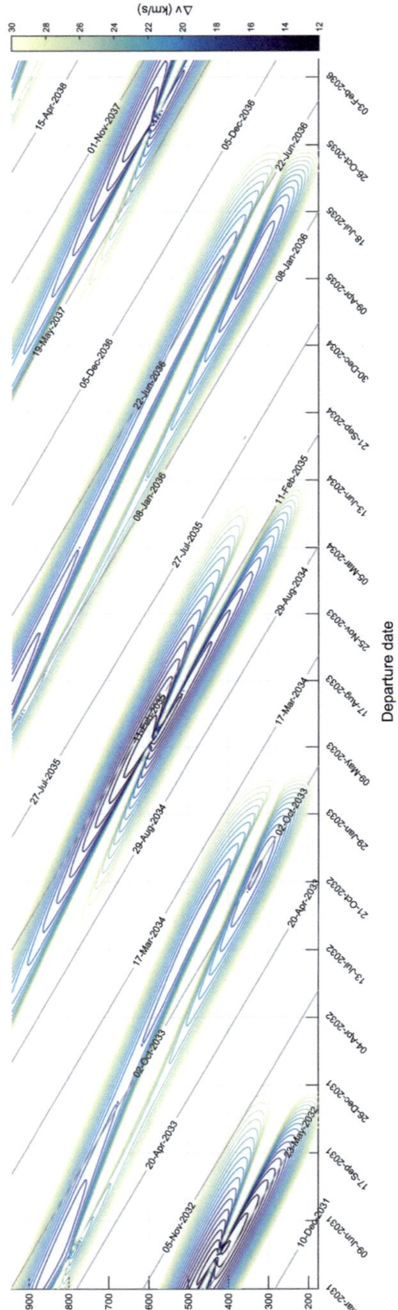


**Fig. B.5** Return porkchop plot for all mission opportunities between 2030 and 2035

## B.4 Mars-Ceres Mission Analysis

The previous sections have analysed at various levels of sophistication a potential interplanetary mission to Ceres with Earth as a starting point. It was shown in Sect. B.3, under the assumption of Keplerian orbits, that a minimum outbound $\Delta v$ of 16 km $s^{-1}$ would be required for the outbound transfer, whereas a $\Delta v$ of 12 km $s^{-1}$ is estimated for the return trip, leading to a rather expensive total $\Delta v$ of 28 km $s^{-1}$. This section, therefore, explores the option of a Ceres mission with Mars as a starting point, and performs the same $\Delta v$ budget and timing analyses with various degrees of accuracy. Section B.4.1 gives a lower-bound estimation of the $\Delta v$ required for such a mission using Hohmann transfers, followed by an estimation of the time windows in which such transfers would be possible in Sect. B.4.2. Consequently, a patched conic analysis of the outbound and inbound transfers for such a mission is given in Sect. B.4.3, for the time period between 2030 and 2040.

### *B.4.1 Hohmann Transfer Analysis*

The first-level approximation for a Mars-Ceres mission follows a similar procedure to that in Sect. B.2, with the difference that the Mars and Ceres orbits are not assumed co-planar. Instead, they are assumed to be circular orbits, with a radius equal to the semi-major axis given in the Horizons catalog [1] for the date January 1st, 2030 at 00:00:00. The ephemerides of the two bodies at this date are given in Table B.3:

Under the assumption of a Hohmann transfer, the semi-major axes of both the outbound and return transfer orbits are given as:

$$a = \frac{a_M + a_C}{2} = 2.14 \text{ au.} \tag{B.29}$$

**Table B.3** Orbital elements of Mars and Ceres in the ICRF on 1st January 2030 00:00 TDB. The true longitude is $\theta = \Omega + \omega + \nu$

| Parameter | Mars | Ceres | Unit |
|---|---|---|---|
| $a$ | 1.5236 | 2.7663 | au |
| $e$ | $9.3352 \times 10^{-2}$ | $7.9411 \times 10^{-2}$ | |
| $i$ | 1.84736 | 10.584 | ° |
| $\Omega$ | 49.472 | 80.174 | ° |
| $\omega$ | 286.78 | 72.295 | ° |
| $\nu$ | 358.27 | 192.30 | ° |
| $\theta$ | 334.52 | 344.77 | ° |

*Source* Horizons system [1]

where $a_M = r_M$ is the semi-major axis of Mars' orbit, while $a_C = r_C$ is the semi-major axis of Ceres. Then, the velocities of the spacecraft before and after the first impulse of the outbound transfer are given by:

$$v_{1_s} = \sqrt{\frac{\mu}{r_M}} = 4.64 \text{ au yr}^{-1}, \qquad v_{1_e} = \sqrt{\mu\left(\frac{2}{r_M} - \frac{1}{a}\right)} = 5.38 \text{ au yr}^{-1}. \quad \text{(B.30)}$$

Similarly, the initial and final in-plane velocities of the spacecraft for the second impulse of the transfer are given as follows:

$$v_{2_s} = \sqrt{\mu\left(\frac{2}{r_C} - \frac{1}{a}\right)} = 2.83 \text{ au yr}^{-1}, \qquad v_{2_e} = \sqrt{\frac{\mu}{r_C}} = 3.49 \text{ au yr}^{-1}. \quad \text{(B.31)}$$

For an in-plane transfer, the total $\Delta v$ budget would be computed straightforwardly, as $\Delta v_{tot_o} = v_{2_e} - v_{2_s} + v_{1_e} - v_{1_s} = 1.4$ au yr$^{-1}$. However, for this analysis, a plane-change maneuver is also considered. For a minimum $\Delta v$ estimation, it is assumed that this maneuver takes place simultaneously with the second impulse, yielding a $\Delta v$ for this impulse of:

$$\Delta v_2 = \sqrt{v_{2_s}^2 + v_{2_e}^2 - 2v_{2_s}v_{2_e}\cos(\Delta i)} = 0.81 \text{ au yr}^{-1} \quad \text{(B.32)}$$

This leads to the following total $\Delta v$ for the outbound Hohmann transfer: $\Delta v_{tot_o} = 1.56$ au yr$^{-1}$ $= 7.39$ km s$^{-1}$. By symmetry, the inbound total $\Delta v$ would have the same value as the outbound one ($\Delta v_{tot_i} = \Delta v_{tot_o}$), resulting in a total $\Delta v$ for the entire mission of $\Delta v_{tot} = 14.8$ km s$^{-1}$.

### *B.4.2 Hohmann Transfer Timing*

Similar to the analysis for the Earth-Ceres mission shown in Sect. B.2.1, a slightly extended launch window between the 1st of January 2029 and the 31st of December 2036 is considered. The reason for the 2-year extension is that the synodic period between Mars and Ceres is larger than that between Earth and Ceres, resulting in scarcer Hohmann transfer opportunities. Making use of the orbital elements for Mars and Ceres given in Table B.3, retrieved from [1] for the 1st of January 2030, the phase angle between the two planets at time $t_0$ is

$$\phi(t_0) = \theta_2 - \theta_1 = 10.25°, \quad \text{(B.33)}$$

where the subscripts 1, 2 stand for Mars and Ceres, respectively. At the same time, the rate of change of the phase angle is

$$\dot{\phi} = n_2 - n_1 = -113.173^\circ\,\text{yr}^{-1}, \tag{B.34}$$

The apoapsis constraint for the spacecraft is also given as

$$\phi(t_{\text{O,D}}) = 180^\circ - n_2 \Delta t, \tag{B.35}$$

where $t_{\text{O,D}}$ is the time of departure for the outbound transfer, and $\Delta t$ is the transfer time on the outbound Hohmann transfer orbit. The latter is half the orbital period of the transfer orbit $T$, which has semi-major axis $a$ as previously computed. Hence, the period is:

$$\Delta t = \frac{T}{2} = \frac{\pi}{\sqrt{\mu}}(a)^{3/2} = 1.571\ \text{yr} \tag{B.36}$$

with $\phi(t_{\text{O,D}}) = 57.1^\circ$ according to Eq. B.35. The time required for the phase angle to go from $\phi(t_0)$ to $\phi(t_{\text{O,D}})$ is

$$\Delta t_0 = \frac{\phi(t_{\text{O,D}}) - \phi(t_0)}{\dot{\phi}} = -0.443\ \text{yr}. \tag{B.37}$$

Hence, the first Hohmann transfer opportunity occurs on 23rd July 2029 04:16 TDB. Additional transfer opportunities are obtained by adding integer multiples of the synodic period as:

$$T_{\text{syn}} = \frac{360^\circ}{|\dot{\phi}|} = 3.181\ \text{yr}, \qquad t_{O,D}^{(k)} = t_{O,D} + kT_{\text{syn}}. \tag{B.38}$$

The negative $t_{O,D}^{(k)}$ justifies the extension of the launch window to January 2029. The possible departure dates are summarised in Table B.4. The return transfers are considered assuming a minimum wait time, computed as in Sect. B.2.1, with the value $\Delta t_{wait} = 1.048$ yr. If one considers only the first mission opportunity, the Hohmann transfers as well as the positions of Mars and Ceres are given in Fig. B.6a, b. The remaining mission windows are shown in Fig. B.7. As for the Earth-Ceres

**Table B.4** Possible departure epochs for the Mars-Ceres Hohmann transfer. All times are TDB

| Epoch |
|---|
| 23rd July 2029 04:16 |
| 27th September 2032 00:39 |
| 2nd December 2035 21:02 |
| 6th February 2039 17:25 |

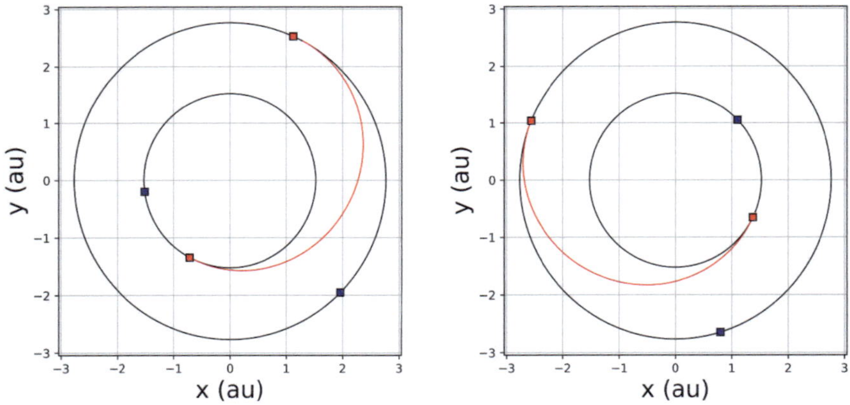


(a) Outbound.

(b) Return.

**Fig. B.6** Outbound and return transfer geometries in the ICRF for the 23rd July 2029 04:16 departure date. The initial and final positions of Mars and Ceres are shown in blue and red, respectively. The Hohmann transfer orbit is also shown in red

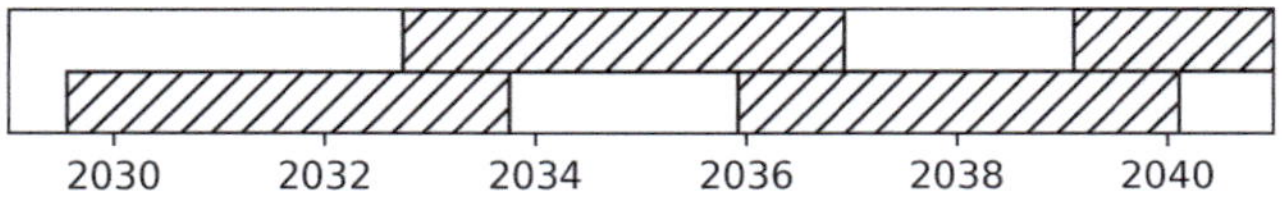


**Fig. B.7** Potential mission opportunities within the 2029–2041 launch window. Each of the opportunities corresponds to a mission lasting 4.19 years, departing on one of the launch dates in Table B.4

Hohmann analysis, assuming a non-zero eccentricity for the orbits invalidates the above results. The next subsection shall consider general Keplerian orbits in a patched conics analysis of the potential mission opportunities and $\Delta v$ budgets.

## *B.4.3 Patched Conics Analysis*

A similar patched conics analysis to that from Sect. B.3 is performed here for the Mars-Ceres mission. Figure B.8 shows the porkchop plot for the outbound transfer from Mars to Ceres, with departure dates between February 2029 and December 2031. As for the departure plot for the Earth-Ceres mission, two regions can be distinguished, that are separated by a ridge, corresponding to a $\Delta\theta = 180°$, where large out-of-plane $\Delta v$'s are required for a transfer to occur. To the right of the ridge, Type I Lambert solutions are found, whereas to the left, Type II solutions can be seen. The region with minimum $\Delta v$ can be seen for a Type II transfer, around July 2029, with an estimated $\Delta v_{o_{min}} = 8$ km s$^{-1}$. Note that this value is close to the Hohmann analysis estimation of 7.39 km s$^{-1}$. Figures B.9 and B.10 constitute the porkchop

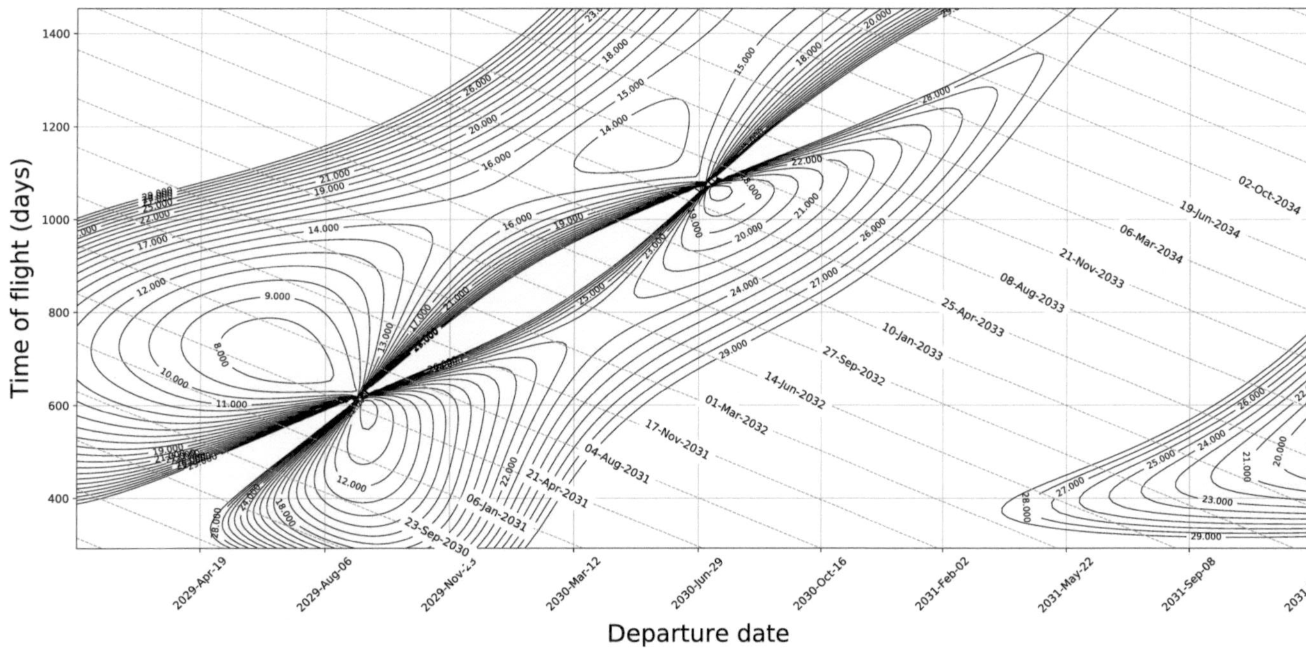


**Fig. B.8** Porkchop plot for the outbound transfer to Ceres, with departure dates between February 2029 and December 2031. Grey lines represent constant arrival times. Contour labels represent the total $\Delta v$ in km $s^{-1}$

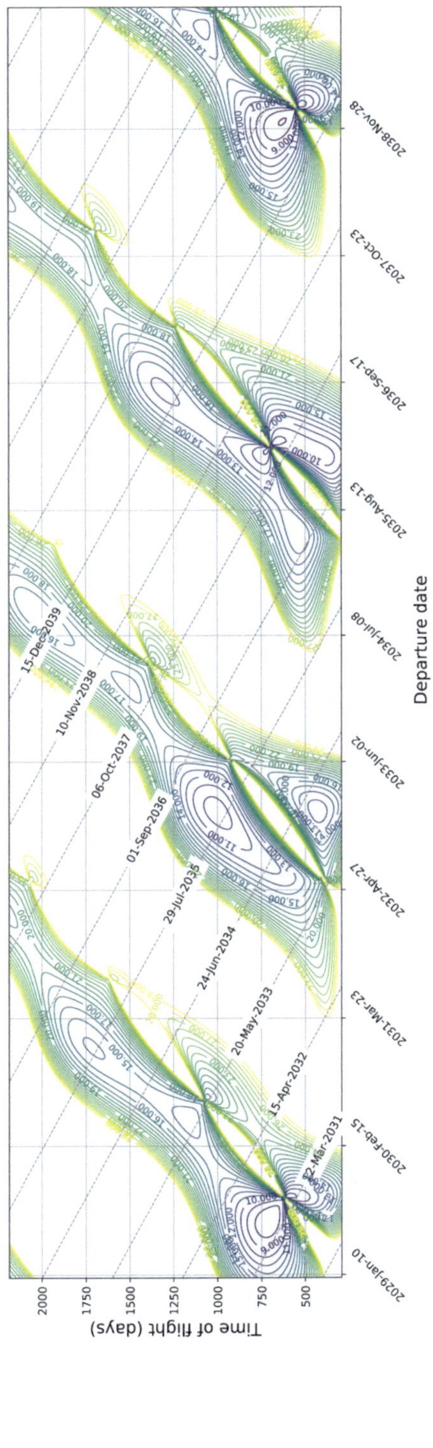


**Fig. B.9** Outbound porkchop plot for all mission opportunities between 2030 and 2035

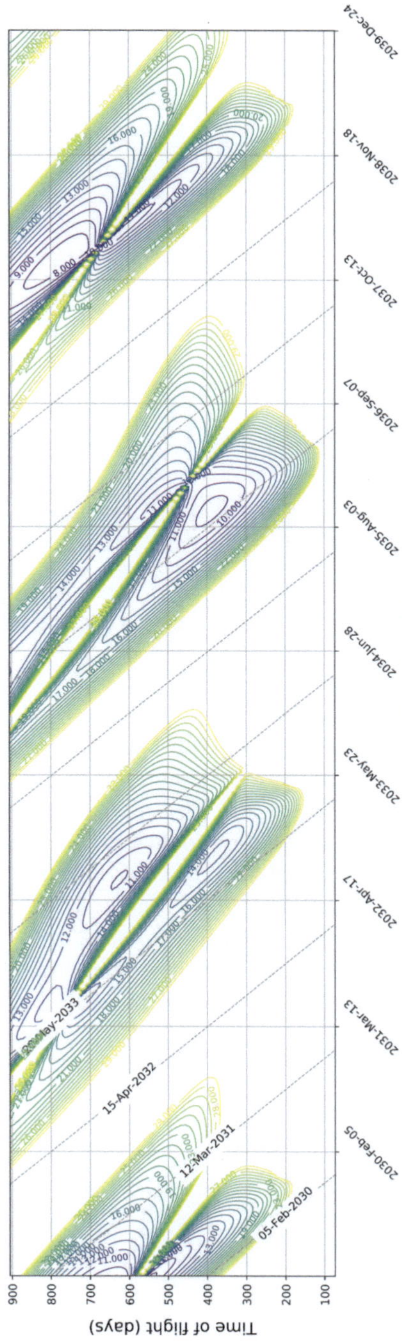


**Fig. B.10** Return porkchop plot for all mission opportunities between 2030 and 2035

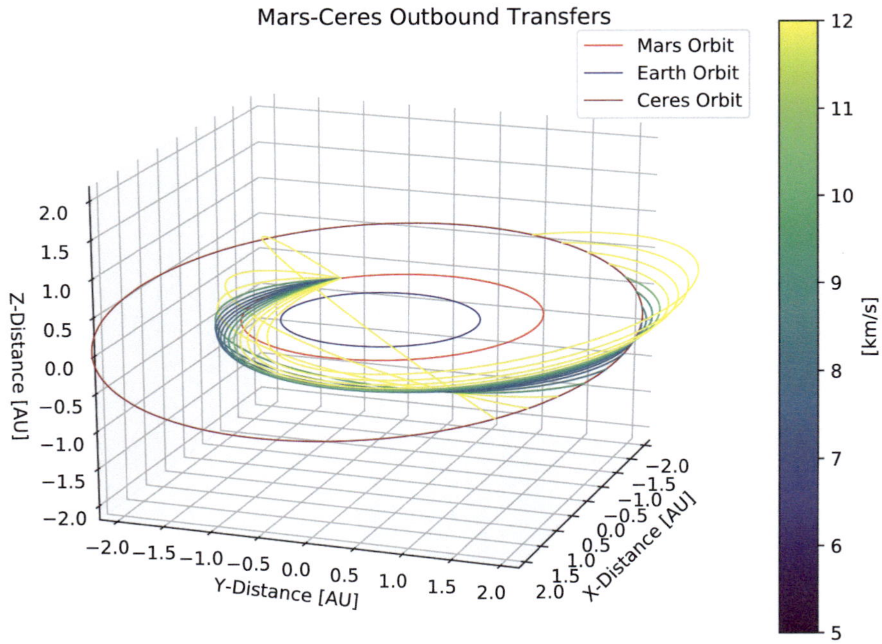


**Fig. B.11** The 3D Keplerian transfer orbits for the outbound Mars-Ceres mission with a start date in July 2029

plots for the outbound and return missions for the entire window starting in 2029 and ending in 2040. The earliest return opportunity is in May 2032, where the $\Delta v$ ranges between 10 km $s^{-1}$ and 14 km $s^{-1}$. These transfers have a rather large time of flight, of about 2 years, leading to a return date in the middle of 2034. Another opportunity with such $\Delta v$'s starts at the end of 2035 and has a time of flight around 1 year, with a return date in 2036.

Figures B.11 and B.12 show the 3D Keplerian orbits of the outbound and return transfers at the earliest considered opportunities in the mission window (2029 and 2032). In Fig. B.11 transfers of $\Delta\theta = 180°$ are visibly out-of-plane, and as expected, require a large $\Delta v$. The return transfers show a similar behaviour.

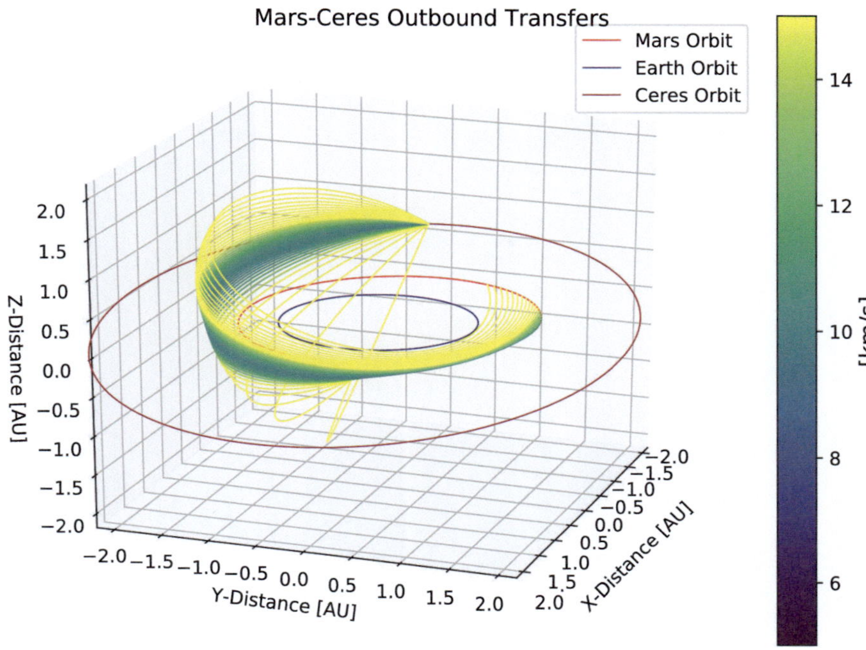


**Fig. B.12** The 3D Keplerian transfer orbits for the return Mars-Ceres mission with a start date in November 2031

## References


1. JPL Horizons On-Line Ephemeris System. (2024). https://ssd.jpl.nasa.gov/horizons/app.html#/
2. Battin, R.H. (1999) *An introduction to the mathematics and methods of astrodynamics*. Revised Edition, AIAA, Education Series, Reston, Virginia.
3. Prussing, J. E., & Conway, B. A. (1993). *Orbital mechanics*. Oxford University Press. https://books.google.de/books?id=96xCER34THAC
4. https://nssdc.gsfc.nasa.gov/planetary/factsheet/
5. Krause, H. G. L., Aeronautics, U. States. N., Administration, S., & Center, G. C. M. S. F. (1964). Astrorelativity. *National Aeronautics*. https://books.google.de/books?id=iUzNRhOriwoC

# Appendix C
# Newtonian and Relativistic Derivation of the Tsiolkovsky Equation

In the following we will derive the rocket equation for Newtonian as well as relativistic mechanics from first principles.

## C.1 Heuristic Derivation of Tsiolkovsky Equation in Newtonian Mechanics

We assume that the rocket is in free space (without external forces, such as gravity drag or radiation pressure). Observed from an inertial reference system the rocket shall be at rest at $t_0$ as it did not yet ignite its engine. Applying the linear momentum equation to our rocket leads to:

$$I_{\mathrm{t}} = m_{\mathrm{t}} v_{\mathrm{c}}, \tag{C.1}$$

$I_{\mathrm{t}}$ total momentum including propellant inside and outside the rocket at any time,
$m_{\mathrm{t}}$ total mass of the rocket including the propellant inside and outside the rocket,
$v_{\mathrm{c}}$ velocity of the centre of mass.

Since the rocket is at rest relative to our observer, the total momentum is zero and the law of momentum conservation demands that it remains zero as long as no external forces are acting and, even after the ignition of rocket's engine.

$$I_{\mathrm{t}} = 0. \tag{C.2}$$

Despite this null result it is possible to achieve meaningful conclusions. Assume that the pilot starts the engine which ejects in the timespan $\Delta t = t_1 - t_0$ the propellant mass $\Delta m_{\mathrm{e}}$ with the velocity $v_{\mathrm{e}}$ out of the rocket into space. The change in momentum can be written as the sum of two contributors:



F. Gamgami, *Space Propulsion and Spaceship Design*, Springer Praxis Books,
https://doi.org/10.1007/978-3-031-71336-1

$$\Delta I_{\mathrm{t}} = (m_{\mathrm{R}} - \Delta m_{\mathrm{e}})\Delta v_{\mathrm{R}} + \Delta m_{\mathrm{e}} v_{\mathrm{e}} = 0, \tag{C.3}$$

$m_{\mathrm{R}}$ total rocket mass at time t prior to ejection of $\Delta m_{\mathrm{e}}$,
$\Delta m_{\mathrm{e}}$ ejected mass increment over time span $\Delta t$,
$\Delta v_{\mathrm{R}}$ rocket's incremental velocity change over $\Delta t$ within inertial reference frame,
$v_{\mathrm{e}}$ velocity of the ejected mass $\Delta m_{\mathrm{e}}$ within inertial reference frame.

The first term gives the change in momentum of the rocket, while the second term gives the momentum of the ejected mass. The increased complexity of this equation compared to A-C 2 it does not alter the fact that the change in total momentum is also zero as is the absolute value. This is equivalent to the fact that the rocket's centre of mass still remains at the same spot and thus is not moving. It is conceptually important to distinguish between $m_{\mathrm{R}}$ and $\Delta m_{\mathrm{exh}}$. The relation can be understood and derived considering the mass conservation of the total system:

$$m_{\mathrm{t}} = m_{\mathrm{R}}(t) + m_{\mathrm{e}}(t) = const, \tag{C.4}$$

$$m_{\mathrm{t}} = m_{\mathrm{R}}(t + \Delta t) + m_{\mathrm{e}}(t + \Delta t) = const. \tag{C.5}$$

Subtraction of both equation leads to:

$$m_{\mathrm{R}}(t + \Delta t) + m_{\mathrm{e}}(t + \Delta t) - m_{\mathrm{R}}(t) - m_{\mathrm{e}}(t) = 0, \tag{C.6}$$

$$m_{\mathrm{R}}(t + \Delta t) - m_{\mathrm{R}}(t) + m_{\mathrm{e}}(t + \Delta t) - m_{\mathrm{e}}(t) = 0. \tag{C.7}$$

$$\Delta m_{\mathrm{R}} + \Delta m_{\mathrm{e}} = 0, \tag{C.8}$$

$$\Delta m_{\mathrm{R}} = -\Delta m_{\mathrm{e}}. \tag{C.9}$$

This is the mathematical formulation of the trivial fact that the increase of the mass of the exhaust gas is of the same magnitude by which the rocket mass decreases. This results helps us to simplify the momentum equation C.3 by replacing $\Delta m_{\mathrm{exh}}$:

$$\Delta I_{\mathrm{t}} = m_{\mathrm{R}}\Delta v_{\mathrm{R}} - \Delta m_{\mathrm{e}}\Delta v_{\mathrm{R}} + \Delta m_{\mathrm{e}} v_{\mathrm{e}},$$

$$\Delta I_{\mathrm{t}} = m_{\mathrm{R}}\Delta v_{\mathrm{R}} + \Delta m_{\mathrm{R}}\Delta v_{\mathrm{R}} - \Delta m_{\mathrm{R}} v_{\mathrm{e}}. \tag{C.10}$$

Remains the relation between $v_{\mathrm{e}}$ and $\Delta v_{\mathrm{R}}$ to be described. A co-moving reference frame that is at rest with respect to the rocket, would observe another velocity of the exhaust gas, namely $c_{\mathrm{e}}$. This is the same exhaust velocity you would observe, if the same engine is tested on ground. Since the two reference frames are moving relative to each other with the velocity $\Delta v_{\mathrm{R}}$, the relation is given by the well-known Galileo transformation (Fig. C.1).

$$\mathbf{v}_{\mathrm{e}} = \mathbf{c}_{\mathrm{e}} + \Delta \mathbf{v}_{\mathrm{R}}, \tag{C.11}$$

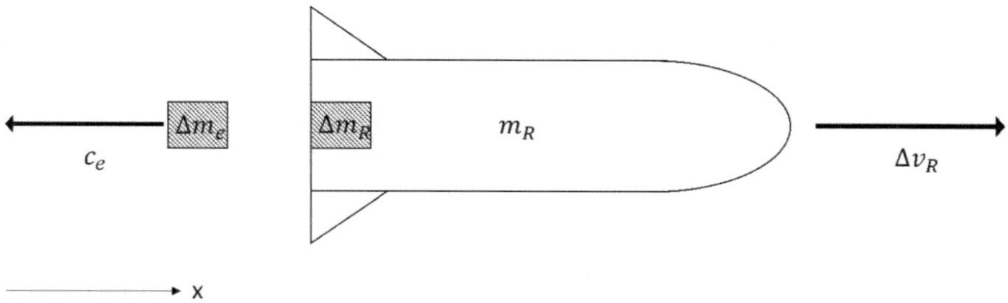


**Fig. C.1** Depiction of a spaceship ejecting mass to propel itself in a force-free environment

$\mathbf{c}_e$ velocity of the exhaust gas (ejected mass) as observed from the rocket's reference frame,

The vector form can be omitted considering the relative direction of $c_e$ to $\Delta v_R$ and the definition of positive x direction:

$$v_e = -c_e + \Delta v_R. \tag{C.12}$$

Inserting this into the equation of motion Eq. C.10 helps us to further simplify it:

$$\Delta I_t = m_R \Delta v_R + \Delta m_R \Delta v_R - \Delta m_R(-c_e + \Delta v_R) = 0.$$
$$\Delta I_t = m_R \Delta v_R + \Delta m_R \Delta v_R - \Delta m_R \Delta v_R + \Delta m_R c_e = 0. \tag{C.13}$$

Leading to:

$$m_R \Delta v_R + \Delta m_R c_e = 0, \tag{C.14}$$
$$m_R \Delta v_R = -\Delta m_R c_e. \tag{C.15}$$
$$\Delta v_R = -c_e(\Delta m_R)/m_R. \tag{C.16}$$

Decreasing the timespan $\Delta t$ towards zero ($\Delta t \to 0$) allows us to replace the difference by a differential $\Delta \to d$:

$$dv_R = -c_e \frac{dm_R}{m_R}. \tag{C.17}$$

Integration yields the famous Tsiolkovsky equation:

$$\int_{0}^{EOM} dv_R = -c_e \int_{0}^{EOM} \frac{1}{m_R} dm_R,$$

$$\Delta v_{R,EOM} = c_e \ln\left(\frac{m_0}{m_{EOM}}\right), \tag{C.18}$$

| | |
|---|---|
| $m_{EOM}$ | mass of the rocket after at end of manoeuvre (EOM), |
| $\Delta v_{R,EOM}$ | achieved velocity change of the rocket after at (EOM), |

In Chap. 6, we have omitted the index $R$ for the velocity, this is justified since $v_e$ has been eliminated from the equation and any mix-up of velocities is ruled out. Further, do not confuse the $\Delta$ symbol in the final Tsiolkovsky equation and the $\Delta$ in the derivation process which became a differential d. The $\Delta$ in the final Tsiolkovsky equation C.18 is a macroscopic difference between two arbitrary properties in time and space.

## *C.1.1 A Formal Approach Towards the Tsiolkovsky Equation*

We will show a second method which is mathematically more formal and which will help us to derive the relativistic equation later on.[3]

The total linear momentum of the system that is composed of the rocket (index $R$) and the exhaust gas (index $e$) is constant :

$$I_t = m_R v_R + m_e v_e = const, \tag{C.19}$$

The total momentum is constant and its differential $dI_t$ is 0:

$$dI_t = d(m_R v_R) + d(m_e v_e), \tag{C.20}$$

$$dI_t = dm_R v_R + m_R dv_R + dm_e v_e + m_e dv_e. \tag{C.21}$$

The first two terms describe the change of the rocket's momentum due to (a) internal mass change and (b) velocity change of the rocket. The last two terms are related to the change in momentum of the exhaust gas; the first of the two says that each ejected mass element $dm_e$ adds to the momentum of the exhaust gas. However, the last term says that the entire exhaust gas changes momentum, if the exhaust velocity changes. This is not wrong but it is not the case. Every mass that is ejected has the instantaneous velocity $v_e$, as in the third term. The last term vanishes since the

[3] Above we assumed that the rocket was at rest prior to engine ignition. This assumption can be generalised and is not mandatory as demonstrated here.

velocity of ejected mass does not further change in free space.[4] The equation reduces to:

$$dI_t = dm_R v_R + m_R dv_R + dm_e v_e. \tag{C.22}$$

Considering the two relations derived above:

$$\textit{Mass conservation (A-C 6)}: \quad dm_e = -dm_R, \tag{C.23}$$
$$\textit{Galileo velocity transformation (A-C 8)}: \quad v_e = v_R - c_e. \tag{C.24}$$

Inserting into A-C 14 and rearranging:

$$dI_t = dm_R v_R + m_R dv_R - dm_R(v_R - c_e),$$
$$dI_t = m_R dv_R + dm_R c_e. \tag{C.25}$$

Since $dI_t = 0$ it is possible separate the variables:

$$dv_R = -c_e \frac{(dm_R)}{m_R}. \tag{C.26}$$

Integration leads then to the Tsiolkovsky equation:

$$\Delta v_{R,EOM} = c_e \ln\left(\frac{m_0}{m_{EOM}}\right). \tag{C.27}$$

## C.2 Derivation of Tsiolkovsky Equation in Relativistic Mechanics

The classical Tsiolkovsky equation is only valid for spaceship velocities much smaller than the speed of light. It loses its validity for relativistic velocities as does the entire Newtonian mechanics from which it is derived. Even if mankind will not be any time soon in a lucky position to deal with relativistic spaceflight, the relativistic form of the rocket equation shall be discussed for completeness.

The derivation of the relativistic rocket equation presented here is follows the same logic as the formal approach presented above and differs only in two distinct points from the classical derivation. The reader is encourage to follow the derivation.[5]

---

[4] Thanks to our heuristic derivation of the change in momentum before, we unintentionally avoided the unflattering situation to explain why the last term actually vanishes. Note that other derivations of the Tsiolkovsky equation start directly with Eq. C.14 and thus omit its explanation.

[5] A more rigorous derivation of relativistic kinematics and dynamics of rockets has been provided by Helmut G. Krause [1].

The special relativity of Einstein makes two postulates that have been proven correct in numerous experiments.[6]

**The postulate of *relativity***: It is impossible to measure or detect unaccelerated translatory motion of a system through free space.
**The postulate of the *constancy of the velocity of light***: The velocity of light in free space is the same for all observers, independent of the relative velocity of the source of light and the observer.

The first postulates means that all inertial reference frames are equivalent. Thus, the observer in one of each reference frames cannot tell who is moving and who is at rest. The second postulates states that all reference frames will measure the same velocity of light, irrespective of the movement of the light source. While the first is within scope of human experience, it is the second postulate that leads to mind boggling consequences.

The starting point of the derivation is again the conservation of momentum formulated in an inertial reference system (see eq. A-C 14):

$$dI_t = dm_R v_R + m_R dv_R + dm_e v_e. \tag{C.28}$$

If the reader recalls both derivations of the Newtonian Tsiolkovsky equation, it becomes obvious that the guiding principle was actually twofold.

Simplify the equation be reducing the equation such that it contains only independent variables.

Transform the variables from the inertial reference frame to the rocket's reference frame.

As above, the relation between $dm_e$ and $dm_R$ can be obtained from the law of mass conservation:

$$\text{mass conservation}: \qquad dm_e = -dm_R. \tag{C.29}$$

Regarding the kinematics of the velocity transformation it is not any more possible to apply the Galileo transformation, since the involved velocities are a significant fraction of the speed of light. Instead, the famous *Lorentz transformation* needs to be applied.[7]

[6] Albert Einstein (1955$^\dagger$) revolutionised physics with his theories of special and general relativity, fundamentally changing our understanding of space, time and gravity. He explained the photoelectric effect and provided crucial evidence for quantum theory, for which he was awarded the Nobel Prize in Physics in 1921. His famous equation $E = mc^2$ established the equivalence of mass and energy, influencing the development of nuclear energy and modern physics. Interestingly, he never accepted the Copenhagen interpretation of quantum mechanics. His doubts are coined in the phrase: 'God does not play dice with the universe'.

[7] The Dutch physicist Hendrik Lorentz (1928$^\dagger$) made remarkable contributions to the electromagnetic theory of light and the theory of electrons. He formulated the Lorentz transformations, which describe how the measurements of time and space change for observers moving relative to one another, forming a cornerstone for Einstein's theory of special relativity. Lorentz was awarded the

$$\text{Einstein's velocity addition theorem: :} \qquad v_e = \frac{v_R - c_e}{1 - \dfrac{c_e v_R}{c^2}}, \tag{C.30}$$

$c$ speed of light in vacuum, 299,792,458 metres per second.

For clarity we will not yet insert the relativistic velocity transformation into the momentum equation, but replace $dm_e$ and rearrange the result first:

$$dm_R v_R + m_R dv_R - dm_R v_e = 0,$$
$$dv_R = (v_e - v_R)\frac{dm_R}{m_R}. \tag{C.31}$$

A second peculiarity of special relativity besides velocity transformation is the *mass transformation*, which is required in order to switch to the rocket's reference frame (index $f$):

$$\frac{dm_R}{m_R} = \left(1 - \frac{c_e v_R}{c^2}\right)\frac{dm_{R,f}}{m_{R,f}}. \tag{C.32}$$

Inserting this relation and the relativistic velocity addition theorem into Eq. C.31 leads to:

$$dv_R = \left(\frac{v_R - c_e}{1 - \frac{c_e v_R}{c^2}} - v_R\right)\left(1 - \frac{c_e v_R}{c^2}\right)\frac{dm_{R,f}}{m_{R,f}},$$
$$dv_R = \left(v_R - c_e - v_R \cdot \left(1 - \frac{c_e v_R}{c^2}\right)\right)\frac{dm_{R,f}}{m_{R,f}},$$
$$dv_R = \left(v_R - c_e - v_R + \frac{c_e v_R^2}{c^2}\right)\frac{dm_{R,f}}{m_{R,f}},$$
$$dv_R = \left(-c_e + \frac{c_e v_R^2}{c^2}\right)\frac{dm_{R,f}}{m_{R,f}},$$
$$dv_R = -c_e\left(1 - \frac{v_R^2}{c^2}\right)\frac{dm_{R,f}}{m_{R,f}}. \tag{C.33}$$

Compare Eqs. C.33 to C.26. The main mathematical difference is that $v_R$ appears now on both sides which complicates slightly the integration.

Separation of variables yields:

$$\frac{-1}{c_e\left(1 - \frac{v_R^2}{c^2}\right)} dv_R = \frac{dm_{R,f}}{m_{R,f}}. \tag{C.34}$$

---

Nobel Prize in Physics in 1902, which he shared with Pieter Zeeman, for their research on the influence of magnetism on radiation phenomena.

Integration from 0 to EOM:

$$\int_{0}^{\text{EOM}} \frac{1}{m_{\text{R,f}}} dm_{\text{R,f}} = \frac{-1}{c_{\text{e}}} \int_{0}^{\text{EOM}} \frac{1}{\left(1 - \frac{v_{\text{R}}^2}{c^2}\right)} dv_{\text{R}}, \tag{C.35}$$

$$\ln \frac{m_{R,EOM,f}}{m_{R,f0}} = \frac{-c}{2c_{e,0}} \ln \left( \frac{1 + \frac{\Delta v_{\text{R,EOM}}}{c}}{1 - \frac{\Delta v_{\text{R,EOM}}}{c}} \right). \tag{C.36}$$

We will omit the index f for clarity. Removing the logarithmic by the inverse function:

$$\frac{m_{\text{R,EOM}}}{m_{\text{R}}} = \left( \frac{1 + \frac{\Delta v_{\text{R,EOM}}}{c}}{1 - \frac{\Delta v_{\text{R,EOM}}}{c}} \right)^{\frac{-c}{2c_{e,0}}}. \tag{C.37}$$

The reciprocal leads us to the relativist version of the classical Tsiolkovsky equation:

$$\frac{m_{\text{R}}}{m_{\text{R,EOM}}} = \left( \frac{1 + \frac{\Delta v_{\text{R,EOM}}}{c}}{1 - \frac{\Delta v_{\text{R,EOM}}}{c}} \right)^{\frac{c}{2c_{\text{e}}}}. \tag{C.38}$$

And with the mass ration $\mu = \frac{m_{\text{R}}}{m_{\text{R,EOM}}}$ it is possible to formulate the relativistic velocity change $\Delta v$ in concise form:

$$\frac{\Delta v_{\text{R,EOM}}}{c} = \frac{\mu^{\frac{2c_{\text{e}}}{c}} - 1}{\mu^{\frac{2c_{\text{e}}}{c}} + 1}. \tag{C.39}$$

Figure C.2 depicts the velocity increase relative to the speed of light as a function of the mass fraction $\mu$. To see a significant change we have assumed a hypothetical propulsion system with a specific impulse of 25 Million seconds. This corresponds to an ejection velocity of 245,250,000 m/s, that is 82% of the speed of light.

The classical formulation predicts a velocity of the spacecraft above the speed of light (horizontal line) for mass fractions above 3. This prediction is obviously false. The correct behaviour is given by the relativistic equation which does not exceed but approaches the speed of light asymptotically. In order to observe a noticeable relativistic effect, we had to assume a very high specific impulses and thus exhaust velocity. The issue of generating the huge amount of energy not to speak of the mechanism needed to achieve these enormous exhaust velocities is beyond current human comprehension.

Beside the technological challenges related to propulsion systems of the IME category, relativistic space travel is very peculiar and completely beyond humanities

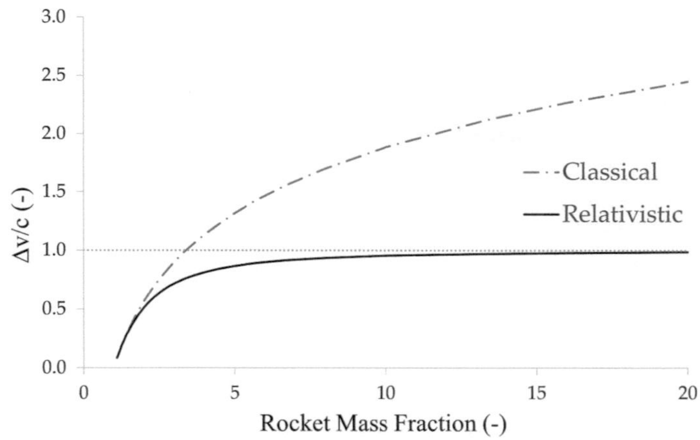


**Fig. C.2** Comparison of achievable $\Delta v$ between classical and relativistic rocket equation. A rocket mass fraction of 3 corresponds to a propellant mass fraction $\mu_{pl} = m_{pl}/m_{\mathrm{R}}$ of 70%

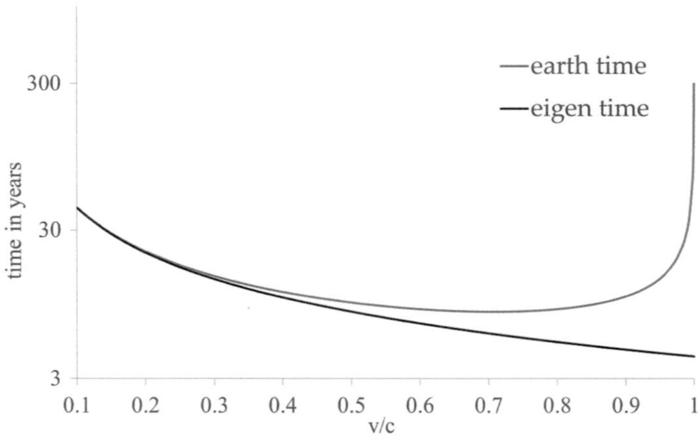


**Fig. C.3** Comparison of elapsed times for a mission to the Alpha Centauri star system, with a distance of 4.3 light years (ly)

experience. Figure C.3 shows two travel times for a mission to the Alpha Centauri system[8]: the elapsed time in the spacecraft (proper time) and the elapsed time on Earth (Earth time) as function of travel speed. Travelling with a speed of 10 % the speed of light translates to a travel time of 42.5 years as measured by the crew (travellers proper time). Relativistic time dilatation is low, merely 2.6 months but becomes significant at higher velocities:

[8] Alpha Centauri a hierarchical three body system consisting of Alpha Centauri A and B as well as Proxima Centauri.

$$\Delta t_{\text{Earth}} = \frac{\Delta t_{\text{prop}}}{\sqrt{1 - \frac{v^2}{c^2}}}. \tag{C.40}$$

A spaceship capable to reach 50 % speed of light – currently in the realm of science fiction – would still require 17 years for a round trip to the Alpha Centauri system. It is doubtful that *suitable* candidates could be found to engage in a decades long mission. They would leave family members and friends knowing that even when returning these people will be not the same if at all alive. Therefore, in addition to advancements in propulsion technology, humanity needs to solve the biological, sociological and maybe most importantly psychological challenges before such endeavours could take place.

## Reference

1. Krause, H. G. L., Aeronautics, U. States. N., Administration, S., & Center, G. C. M. S. F. (1964). Astrorelativity. *National Aeronautics*.

# Appendix D
# Sizing of Solar Power Generator for C-One

We will apply the algorithm established in Sect. 8.3.4 to the reference mission C-One.

*Step 1:* Define BOL and EOL power demand.
According to Table 10.16 a BOL power demand of 200 kW and EOL demand of 30 kW is required. The EOL demand is in general the design driver, which leads to overcapacity for BOL.

*Step 2:* Select solar cell type and define operating environment.
We adopt the IMM-$\alpha$ solar cell technology as baseline. Performance figures and the assumed form factor are provided in Table 10.17. The temperature was computed by applying the thermal model introduced in Sect. 8.3.2. The degradation of cell efficiency is considered via Table 8.11.

*Step 3*: Define the BOL output voltage of the array, consider a provision of 10 % for voltage losses due to the PVA diodes.
Since the X3 thruster requires at least 300 V input voltage, we will this figure as BOL output voltage of the PVA in order to reduce voltage conversion losses in the power processing unit. This yields a required BOL voltage of 330 V.

*Step 4*: Compute the number of cells in row, $\mathrm{N_{cr}}$ with EOL cell performance, $V_{\mathrm{mp}}$:

$$\mathrm{N_{cr}} = \frac{330\,\mathrm{V}}{5.34\,V} = 95. \tag{D.1}$$

*Step 5*: Compute the required total PVA current, $I_{\mathrm{pva}}$ with the EOL power demand:

$$I_{\mathrm{pva}} = \frac{30\,kW}{330\,\mathrm{V}} = 58\,A. \tag{D.2}$$



F. Gamgami, *Space Propulsion and Spaceship Design*, Springer Praxis Books,
https://doi.org/10.1007/978-3-031-71336-1

*Step 6*: Define number of wings, $N_w$, number of booms per wing, $N_b$ and number of blankets per boom, $N_{bk}$.
These numbers are closely related to the overall layout of the solar power generator and subject to an excessive trade-off. They however, do not change the total dimension of the PVA. A reasonable baseline for the STEV has 2 wings, 4 booms per wing and 2 blankets per boom.

*Step 7*: Compute the current per blanket, $I_{bk}$:

$$I_{bk} = \frac{I_{pva}}{N_w \cdot N_b \cdot N_{bk}} = \frac{58\,A}{16} = 3.6\,\text{A}. \tag{D.3}$$

*Step 8*: Compute number of parallel strings per blanket, $N_{s,bk}$:

$$N_{s,bk} = \frac{I_{bk}}{I_{mpp}} = \frac{3.6\,\text{A}}{0.024\,\text{A}} = 155. \tag{D.4}$$

*Step 9*: Compute the total number of parallel strings and add 2 strings , $N_{s,tot}$:

$$N_{s,tot} = N_{s,bk} \cdot N_w \cdot N_b \cdot N_{bk} + 2 = 2482. \tag{D.5}$$

# Glossary

**Adapted Nozzle:** A nozzle with an exit pressure that matches the ambient pressure

**Adiabatic:** A thermodynamic process in which no heat transfer of the respective system with its surround occurs.

**Aerobraking:** An orbital manoeuvre in which the spacecraft dips into the denser part of the planet's upper atmosphere to reduce its velocity, thereby achieving or at least facilitating planetary capture.

**Airmass Zero:** A condition in which solar radiation is measured outside the Earth's atmosphere, representing the solar irradiance in space.

**Aneutronic:** A process in nuclear physics in which very little energy is carried away by neutrons.

**Apocentre:** Most distant point of an object in a bound orbit around its primary object. The suffix 'centre' can be specified according to the primary object: *Aphelion* for the Sun, *Apogee* for the Earth, *Apoastron* for a star, *Apojove* for Jupiter and so forth.

**Apsis, Pl. Apsides:** The term refers, without distinction, to both the most distant point, *Apocentre*, and the nearest point, *Pericentre*, in an orbit relative to a central body (i.e. Sun, planet, star, black hole, etc.).

**Asteroid:** A celestial body revolving around our Sun, composed of rocky material with varying concentrations of metals, with a diameter below 1,000 km.

**Astronomical Unit:** Defined as the mean distance between Earth and Sun, 149,597,870,700 m, and abbreviated as AU.

**Atomisation:** A process of breaking up a liquid propellant into fine droplets or a mist via specially designed injector elements. This enhances mixing and the quality of combustion.



F. Gamgami, *Space Propulsion and Spaceship Design*, Springer Praxis Books,
https://doi.org/10.1007/978-3-031-71336-1

**Autogenous Pressurisation:** The use of gaseous propellant to pressurise the tank. The gas is tapped-off from the feed-system of pump-fed engines . Cryogenic propellant is specifically suited for this.

**Back-Up:** Refers to a unit (e.g. thruster) that is nominally not needed and is only used in case the nominal unit fails or has a malfunction.

**Ballistic:** Refers to the unpowered flight in a gravitational field, where external forces on the body, other than gravity, can be neglected.

**Ballistic Factor:** Defined as the ratio of a body's mass to the product of its cross-sectional area and drag coefficient. Large values imply that gravity determines its trajectory rather than the atmosphere. Small values imply that the orbital lifetime is strongly limited and that the body will rather sooner than later enter into the atmosphere.

**Barycentre:** The barycentre is the centre of mass around which two or more bodies orbit due to their mutual gravitational attraction. It is important to note that the barycentre is not attached to a body; rather, it is a location. However, in the presence of a dominant mass, such as our Sun in our solar system, the location of the barycentre will be within the physical sphere of the dominant body but not in its geometric centre.

**Baseline:** The totality of information that delineates the actual state of a system or the entirety of a project, including specifications, designs, analyses, plans, schedules, and so forth.

**Binary:** Denotes a stellar system that is composed of two stars revolving around their barycentre.

**Bi-Prop:** Stand for bi-propellant and refers in space jargon to a storable propellant combination of fuel and oxidiser.

**Birthing:** The process by which a module in space is mated with a larger system (e.g. space station) by means of a special device (e.g. robot arm). The module stays passive during this procedure but is cooperative, meaning it stands still or takes an attitude that helps the external device to conduct the operation. Examples are JAXA's Kounotori Cargo module and the US Cygnus.

**Black Body:** A physical concept of a body that absorbs all energy it receives and emits energy as a function of its temperature according to the Stefan-Boltzmann law.

**Blow-Down:** A form of propellant supply in which the propellant is stored under pressure within the tank, which reduces during depletion causing the thrust of the engines to decrease as well.

**Boil-Off:** A process in which cryogenic propellant evaporation in the tank due to parasitic heat fluxes into the tank.

**Boil-Off Losses:** Propellant gas generated by the process of 'boil-off', which must be vented into space in order to prevent the tank from becoming over-pressurised. It is of paramount importance to limit boil-off losses in order to ensure the long-term storage of cryogenic propellant in space.

**Bookkeeping:** A method for propellant mass estimation and management.

**Bound Orbit:** An orbit with a specific energy lower than zero, i.e. circular or elliptical, causing the body to revolve around the barycentre of the system. Bound orbits are not closed, i.e. Keplerian, if external forces act and/or the gravitational field is not spherical symmetric.

**Buckling:** A structural failure caused by compression loads in which the body cannot maintain its shape and collapses. Thin walled structures, like launcher stages, are particularly vulnerable.

**Burst Pressure:** A pressure limit which, if exceeded, leads to physical damage to the pressurised structure, e.g. piping and valves. This causes leakage of the fluid, .e.g. propellant or pressurant. Engineering standards require a burst pressure that is 50% larger than MEOP.

**C3:** Specific excess energy of a mass on hyperbolic orbit. It is equivalent to its velocity squared at infinity. The parameter is important in spaceflight since it is used to characterise the escape trajectory.

**Caloric Perfect Gas:** An idealisation in which the specific heat capacities $c_p$ and $c_v$ are constant, specifically not a function of temperature.

**Chill down Losses:** A process whereby the feed system is flushed with cryogenic propellant in order to pre-condition it, by lowering its temperature, for the forthcoming operation. The used propellant is lost for the subsequent combustion process.

**Cislunar:** The term denotes a domain in space that comprises the Earth, the Moon, their respective Lagrange points.

**Clearance:** A distance or a spatial provision that needs to stay free for either safety reason or to allow late access after integration, for instance, for maintenance.

**Collimation:** The process by which a high energetic gas is aligned to form a directed jet (in case of a thermal engine) or beam (in case of an electric engine).

**Configuration Management:** A discipline within Systems Engineering. Its main task is maintaining a *configured baseline*. This is achieved by tracking changes and maintaining the system's *timely* consistency (not necessarily its technical consistency), thereby securing *traceability*. A baseline is the entire body of technical and programmatic data established over the course of a project, including user requirements, technical specifications, analyses, plans, and schedules. It evolves and reaches consistency asymptotically.

**Continuous Transfer:** Orbital Transfer whereby the vehicle continuously accelerates or decelerates.

**Corpuscular Radiation:** A form of radiation composed by high energetic particles that are harmful to humans and machines.

**Docking:** The process by which a module in space mates with a larger system (e.g. space station) by its own based on its sensor suit and actuators. The other object is passive. Examples are: Roscosmos Progress, ESA's ATV (Automated Transfer Vehicle) & Space X's Dragon Capsule.

**Escape Velocity:** The velocity required relative to the barycentre to escape its gravitational sphere of influence.

**Fail to Open:** A failure Mode of a flow control equipment that is supposed to open in order to enable the propellant or pressurant to flow. This failure usually prevents a manoeuvre from being carried out as planned.

**Fail to Close:** A failure mode of a flow control equipment that is supposed to close in order to prevent the propellant or pressurant to flow into a segment of the feed system. This failure can result in various operating errors, e.g. preventing a manoeuvre from being stopped as planned.

**Fault Tolerance:** A concept referring to a system's ability to absorb one (single fault tolerance) or two (dual fault tolerance) failures in the design. A fault-tolerant system is capable to detect, isolate, and recover from failures via sensors and redundant units throughout the functional chain.

**Fluence:** The total number of particles (e.g. electrons) crossing a surface per unit area. It is the time integral of the respective flux.

**Fluid:** The term fluid refers to substances in the liquid, gas, and plasma states of matter. Unlike solids, fluids do not have a fixed shape. Instead, they are characterised by their ability to flow and conform to the shape of their containers.

**Flux:** Amount of a quantity crossing a surface per unit time and unit area. This quantity could be the number of particles (e.g. photons or ions), heat or momentum. Flux is a vector quantity.

**Geoid:** The surface elevation of the ocean, assuming that only gravity and the Earth's rotation were acting, but not wind and tides. It can be interpreted as the average height of the actual time-dependent height.

**Gravity Assist:** A manoeuvre by which energy is exchanged between the planetary object and the spacecraft (or natural body) by which the orbital parameter of the spacecraft are changed. This is in general used to enhance the velocity of the spacecraft, thereby reducing flight time, or to change its inclination.

**Gravity Turn:** A flight strategy for ascending (or landing) from a planetary body in which the primary change in trajectory angle is due to the gravitational field, other than initialisation.

**Habitable Zone:** A zone in terms of distance from our Sun in which water can exist in liquid form, if the hosting celestial body fulfils certain conditions with respect to its atmosphere and global magnetic field.

**Hohmann Transfer:** The energetically most efficient transfer between two bound orbits, if the ratio of the final to the initial semi major axis is below 11.94.

**Hydrolox:** Refers to the cryogenic propellant combination of liquid hydrogen and liquid oxygen, LH2-LOX. It is the most energetic non-toxic propellant combination.

**Isentropic:** A thermodynamic process that is adiabatic and frictionless. The entropy of the system does not change.

**Laval-Nozzle:** A specific nozzle configuration comprising a convergent and divergent section, which enables the generation of supersonic flows when the pressure gradient to the ambient environment is sufficiently high.

**Leakage:** The unwanted flow of a fluid out of a pressurised compartment, e.g. pipe segment or tank, into another compartment (*internal leakage*) or out of the system (*external leakage*). Leakage could occur, for instance, if a seal in a valve does not perform nominally.

**Line of Apsides:** An imaginary line that connects the two apsides of an orbit and crosses the barycentre: pericentre and apocentre.

**Load Path:** The term refers to the path within a structural design taken by the load. For example, thrust through the engine frame (also called a thrust frame) into the structure, or inertial forces during ascent through the primary structure of payload via the payload launch adapter into the launcher.

**Luminosity (Sun):** The total amount of energy emitted by the Sun in a second, $L_{\odot} = 3.846 \times 10^{26}$ W.

**Methalox:** Refers to the cryogenic propellant combination of liquid hydrogen and liquid methane, LH2-LCH4.

**MEOP:** Stands for 'Maximum Expected Operating Pressure' and refers to the maximum pressure that a pressurised component experiences under nominal conditions.

**Moderator:** A material within a nuclear reactor required to slow down the velocity of the neutrons in order to maintain a critical chain reaction.

**Multi-Junction Solar Cells:** A type of solar cell that is constructed from a combination of different types of semiconductor materials, arranged in a layered structure, which enables it to absorb and convert the maximum possible amount of incident irradiation into electricity.

**Off-the-Shelf:** The term describes equipment that is readily available and can be purchased to be used in a system without additional development or qualification costs. In practice, this is rarely the case, and additional engineering effort is required to ensure compliance with the requisite specifications. If the equipment is available on the commercial market (for example, the automotive market), it is referred to as commercial off-the-shelf (COTS).

**Over Expansion:** A nozzle with an exit pressure that is larger than the ambient pressure. This is always the case for engines that operate in space, e.g. satellite propulsion, or the upper atmosphere, e.g. upper stage engines.

**Pericentre:** Nearest point of an object in a bound orbit around its primary object. The suffix 'centre' can be specified according to the primary object: *Perihelion* for the Sun, *Perigee* for the Earth, *Periastron* for a star, *Perijove* for Jupiter and so forth.

**Plume:** The cloud generated by the exhaust gas of an engine (thermal or electrical) in space.

**Pressure-Fed:** A form of propellant supply that enables constant inlet conditions for the engines in terms of pressure and mass flow without the use of active devices. This is achieved by pressurising the propellant tank with a dedicated gas, called *pressurant*, that is separately stored in high-pressure vessels.

**Proof Pressure:** A pressure limit which, if exceeded, leads to irreversible damage to the pressurised structure, e.g. piping and valves. This does not cause leakage of the fluid, but nominal operation cannot be guaranteed. Engineering standards require a burst pressure that is 25 % larger than MEOP. The range between MEOP and the proof pressure may be regarded as an additional margin for short-term non-nominal operational conditions.

**Pump-Fed:** A form of propellant supply that involves turbopumps or electric pumps to rise the pressure level from storage condition to engine inlet conditions.

**Reflector:** A material within a nuclear reactor required to reflect neutrons from the reactor back into it. It is a mean to maintain a critical chain reaction.

**Regenerative Cooling:** An active cooling mechanism by which part of the propellant (usually fuel) is tapped off and channelled through the thrust chamber.

**Regolith:** A geological term that denotes the upper most layer on top of the bedrock. On Earth, this layer includes soil that comprises biological elements. Regolith on other celestial objects differ strongly due to the different environmental conditions they experience.

**Residuals:** Remaining propellant in the tank that cannot be extracted anymore, typically 2%.

**Secular Perturbation:** A perturbation in a celestial two-body system whose effect is on average unidirectional, causing an increase or decrease in an orbital parameter, such as atmospheric drag on the semi-major axis or solar photon pressure on eccentricity. In contrast, periodic perturbations cancel out over one or more periods.

**Siderial Period:** The time a celestial body within the solar system needs to return to the same or a similar position relative to the Sun as seen by an observer at rest with respect to the barycentre.

**Space Probe:** A spacecraft send to other objects in our solar system or beyond.

**Specific Impulse:** A measure of how efficient propellant is used to achieve a $\Delta v$. It is, next to thrust, the most important parameter of propulsion systems based on mass expulsion, i.e. internal momentum exchange.

**Sphere of Influence:** The space in which the gravity of a celestial body is the dominating gravitational force. Other gravitating bodies can be considered as perturbations. The term is only meaningful, in multi body system of at least two objects. Also called *region of influence*.

**Synodic Period:** The time a celestial body within the solar system needs to return to the same or a similar position relative to the Sun as seen by an observer on Earth. It is important to note that the Earth has moved in the meantime.

**Systems Engineering:** It is not a branch of engineering in the technological sense but a methodology by which knowledge from various branches is orchestrated to form a cohesive system that meets customer requirements.

**Thrust:** The force that acts on the vehicle to accelerate it and is generated by the propulsion system. If thrust is achieved by mass expulsion, i.e. internal momentum exchange, it points *in opposite direction* to the ejected exhaust gas. If thrust is achieved by an external source, e.g. solar photonic pressure, via a solar sail, it acts *in direction* of the external force.

**Thrust Chamber:** The space in a chemical engine where the injected propellant mixes and reacts to generate hot gas. It comprises the combustion chamber, the convergent part of the nozzle and the initial segment of the divergent part, which extends beyond the nozzle throat.

**Thruster:** In space propulsion jargon engines with small thrust, i.e. below 20 N without clearly defined threshold.

**Unbound Orbit:** An elliptic or hyperbolic orbit in which the body passes the barycentre only once before leaving its gravitational sphere of influence.

**Under Expansion:** A nozzle with an exit pressure that is above the ambient pressure. This is typically the case for first stage engines of launchers.

**Vernier Engines:** Also 'Vernier Thruster', a set of engines surrounding the main engine(s) that are often, but not necessarily, steerable via a gimbal for pitch and yaw control.

**Vessel:** The term denotes a high-pressure tank, which is typically used to store gas.

**Vis-Viva Integral:** Latin for 'living force'. It denotes the specific orbital energy equation in a two-body problem.

**Weak Stability Transfer:** An orbital transfer method by which little $\Delta v$ is needed at the expense of a longer transfer time. The method uses perturbations induced by a third body in a two-body system, e.g. Sun perturbations in the Earth-Moon system.

**Wet Mass:** Mass of spacecraft that is fully loaded with propellant. In general it is equivalent to the launch mass but not, if fuelling (complete or partially) is planned in orbit)

**Working Hypothesis:** The hypothesis must be in principle verifiable but initially it is a provisional assumption that guides an initial analysis (or research in general) and provides direction. In the course of the design, the working hypothesis may be refined, or rejected at a later stage.

MIX
Papier aus verantwortungsvollen Quellen
Paper from responsible sources
FSC® C105338

If you have any concerns about our products,
you can contact us on
**ProductSafety@springernature.com**

In case Publisher is established outside the EU,
the EU authorized representative is:
**Springer Nature Customer Service Center GmbH**
**Europaplatz 3, 69115 Heidelberg, Germany**

Printed by Libri Plureos GmbH
in Hamburg, Germany